Student's Solutions Manual

to accompany

Prealgebra

Fourth Edition

Stefan Baratto
Clackamas Community College

Barry Bergman
Clackamas Community College

Don Hutchison
Clackamas Community College

Prepared by
Avrio Knowledge Group, Inc.

Mc Graw Hill Education

STUDENT'S SOLUTIONS MANUAL TO ACCOMPANY
PREALGEBRA, FOURTH EDITION

Published by McGraw-Hill Education, 2 Penn Plaza, New York, NY 10121. Copyright © 2014 by McGraw-Hill Education. All rights reserved.
Printed in the United States of America. Previous editions © 2010, 2007, 2003. No part of this publication may be reproduced or distributed in
any form or by any means, or stored in a database or retrieval system, without the prior written consent of McGraw-Hill Education, including, but
not limited to, in any network or other electronic storage or transmission, or broadcast for distance learning.

Some ancillaries, including electronic and print components, may not be available to customers outside the United States.

This book is printed on acid-free paper.

1 2 3 4 5 6 7 8 9 0 QVS/QVS 1 0 9 8 7 6 5 4 3

ISBN: 978-0-07-757450-5
MHID: 0-07-757450-8

All credits appearing on page or at the end of the book are considered to be an extension of the copyright page.

The Internet addresses listed in the text were accurate at the time of publication. The inclusion of a website does not indicate an endorsement by
the authors or McGraw-Hill Education, and McGraw-Hill Education does not guarantee the accuracy of the information presented at these sites.

www.mhhe.com

Contents

Preface

This *Student's Solutions Manual* contains comprehensive, worked-out solutions to all odd-numbered exercises in the book, including the Section Exercises, Prerequisite Checks, Summary Exercises, Chapter Tests, and Cumulative Reviews. All steps in the solutions follow the solving style of the Examples in the book.

Use this resource to help check your progress on homework assignments, identify any errors in your solving methodology, and to help yourself prepare for exams.

Chapter 1
Whole Numbers

Exercises 1.1

< Objective 1 >

1. 456
6 ones, 5 tens, 4 hundreds
$456 = 400 + 50 + 6$
or $(4 \times 100) + (5 \times 10) + (6 \times 1)$

3. 5,073
3 ones, 7 tens, 0 hundreds, 5 thousands
$5,073 = 5,000 + 70 + 3$
or $(5 \times 1,000) + (7 \times 10) + (3 \times 1)$

5. 1,500
0 ones, 0 tens, 5 hundreds, 1 thousand
$1,500 = 1,000 + 500$
or $(1 \times 1,000) + (5 \times 100)$

< Objective 2 >

7. 416
6 ones, 1 ten, 4 hundreds
The place value of 4 is hundreds.

9. 56,489
9 ones, 8 tens, 4 hundreds, 6 thousands,
5 ten thousands
The place value of 6 is thousands.

11. 3,052
2 ones, 5 tens, 0 hundreds, 3 thousands
The place value of 0 is hundreds.

13. 43,729
9 ones, 2 tens, 7 hundreds, 3 thousands,
4 ten-thousands
  (a) The digit 3 tells the number of thousands
  (b) The digit 2 tells the number of tens.

15. 1,403,602
2 ones, 0 tens, 6 hundreds, 3 thousands,
0 ten thousands, 4 hundred thousands,
1 million
  (a) The digit 4 tells the number of hundred thousands.
  (b) The digit 2 tells the number of ones.

< Objective 3 >

17. $\underset{\text{thousands}}{5}$, $\underset{\text{ones}}{618}$
Five thousand, six hundred eighteen

19. $\underset{\text{millions}}{1}$, $\underset{\text{thousands}}{532}$, $\underset{\text{ones}}{657}$
One million, five hundred thirty-two thousand, six hundred fifty-seven

21. $\underset{\text{thousands}}{200}$, $\underset{\text{ones}}{304}$
Two hundred thousand, three hundred four

< Objective 4 >

23. Two hundred fifty-three thousand, four hundred eighty-three in standard form is 253,483.

25. Two million, three hundred eight thousand, forty seven in standard form is 2,308,047.

27. Five hundred two million, seventy-eight thousand in standard form is 502,078,000.

29. One million, four hundred forty thousand dollars in standard form is $1,440,000.

31. Four hundred fifty-nine thousand, eight hundred in standard form is 459,800.

33. The population of San Diego, CA in 2010 is listed as 1,307. Because the population is given in thousands, the population of San Diego in 2010 is 1,307,000.

35. The population of Philadelphia in 2010 is listed as 1,526. Because the population is given in thousands, the population of Philadelphia in 2010 is 1,526,000.

37.

$2,565
5 ones
6 tens
5 hundreds
2 thousands

Inci should write the amount of the check as two thousand, five hundred sixty-five.

39. 480,000

41. (a) $795 \times 1,000 = 795,000$ units

   (b) $1,135 \times 1,000 = 1,135,000$ units

   (c) $910 \times 1,000 = 910,000$ units

43. Arranging the scrambled place values in decreasing order, we have:

3 ten thousands $= 3 \times 10,000$

4 thousands $= 4 \times 1000$

2 hundreds $= 2 \times 100$

1 ten $= 1 \times 10$

5 ones $= 5 \times 1$

The number represented is:
$(3 \times 10,000) + (4 \times 1,000) + (2 \times 100)$
$+ (1 \times 10) + (5 \times 1)$
$= 30,000 + 4,000 + 200 + 10 + 5 = 34,215$

45. For the word names of the numbers from one to one thousand, the first letter of each word name or group of word names is given below:

1	O
2	T
3	T
4	F
5	F
6	S
7	S
8	E
9	N
10	T
11	E
12	T
13	T
14	F
15	F
16	S
17	S
18	E
19	N
20-39	T
40-59	F
60-79	S
80-89	E
90-99	N
100-199	O
200-399	T
400-599	F
600-799	S
800-899	E
900-999	N
1,000	O

Alphabetizing this list would show that a word name beginning with E would appear first on the list. Of the word names beginning with E, the word "eight" is first alphabetically. Thus, "eight" would appear first in the alphabetized list.

47. The largest 5 digit number that can be made using the digits 6, 3, and 9 if each digit is to be used at least once is 99,963.

49. Above and Beyond

Exercises 1.2

< Objective 1 >

1. $5 + 1 = 6$

3. $8 + 4 = 12$

5. $8 + 8 = 16$

7. $7 + 3 = 10$

< Objective 2 >

9. The order of the addends, 5 and 8, was changed. The sum remains the same by the commutative property of addition.

11. The grouping of the addends was changed. The sum remains the same by the associative property of addition.

13. The order of the addends, 7 and 6, was changed. The sum remains the same by the commutative property of addition.

15. The order of the addends, 5 and $(2 + 3)$ was changed. The sum remains the same by the commutative property of addition.

< Objective 4 >

17.
$$\begin{array}{r} 2,792 \\ +\ \ 205 \\ \hline 2,997 \end{array}$$

19.
$$\begin{array}{r} 2,345 \\ +\ 6,053 \\ \hline 8,398 \end{array}$$

21.
$$\begin{array}{r} 2,531 \\ +\ 5,354 \\ \hline 7,885 \end{array}$$

23.
$$\begin{array}{r} 21,314 \\ +\ 43,042 \\ \hline 64,356 \end{array}$$

25.
$$\begin{array}{r} \overset{\text{1 11}}{3,490} \\ 548 \\ +\ \ \ \ 25 \\ \hline 4,063 \end{array}$$

27.
$$\begin{array}{r} \overset{\text{1 23}}{2,289} \\ 38 \\ 578 \\ +\ 3,498 \\ \hline 6,394 \end{array}$$

29.
$$\begin{array}{r} \overset{\text{1 1}}{23,458} \\ +\ 32,623 \\ \hline 56,081 \end{array}$$

31.
$$\begin{array}{r} 46 \\ +\ 32 \\ \hline 78 \end{array}$$

33.
$$\begin{array}{r} \overset{\text{1 1}}{4,032} \\ +\ 2,289 \\ \hline 6,321 \end{array}$$

35.
$$\begin{array}{r} \overset{\text{1 1}}{32} \\ 867 \\ +\ 42,085 \\ \hline 42,984 \end{array}$$

37. In $5 + 4 = 9$
5 is an *addend*, 4 is an *addend*, and 9 is a *sum*.

< Objective 5 >

39. $5\ \text{ft} + 7\ \text{ft} + 6\ \text{ft} + 4\ \text{ft} = 22\ \text{ft}$

41. $8 \text{ yd} + 7 \text{ yd} + 6 \text{ yd} = 21 \text{ yd}$

43. $10 \text{ in.} + 3 \text{ in.} + 10 \text{ in.} + 3 \text{ in.} = 26 \text{ in.}$

45. **(a)** $1,213 + 356 = 1,569$

    (b) $23 + 2,845 + 5 + 589 = 3,462$

    (c) $\left(2,195 + 348 + 640 + 59 + 23,785\right)$

        $= 27,027$

    (d) $125 + 34 = 159$

    (e) $457 + 96 = 553$

47. **(a)** The total cost of the purchase.
    (b) He bought 3 items. The laptop cost $2,120, the printer cost $379, and the software cost $589.

49. **(a)** The total amount of grapes shipped in the 3-month period.
    (b) The vineyard shipped 4,200 lb, 5,970 lb, and 4,850 lb grapes over the 3-month period.

< Objective 3 >

51.
$$\begin{array}{r} \overset{2\ 2\ 1}{26,895} \\ 54,200 \\ +\ 69,950 \\ \hline 151,045 \end{array}$$
Tral invested $151,045 in the three cars.

53.
$$\begin{array}{r} 42 \\ +\ 46 \\ \hline 88 \end{array}$$
The total score for the round is 88.

55.
$$\begin{array}{r} \overset{1}{4,200} \\ 5,970 \\ +\ 4,850 \\ \hline 15,020 \end{array}$$
They shipped 15,020 lb of grapes in the 3-month period.

57.
$$\begin{array}{r} 325 \\ +\ 273 \\ \hline 598 \end{array}$$
They drove 598 mi in the 2 days.

59.
$$\begin{array}{r} 18,250 \\ +\ \ \ \ 445 \\ \hline 18,695 \end{array}$$
The total price of Emma's car was $18,695.

61.

Department	Oct.	Nov.	Dec.	Dept. Totals
Office	$31,714	$32,512	$30,826	$95,052
Production	85,146 ·	87,479	81,234	253,859
Sales	34,568	37,612	33,455	105,635
Warehouse	16,588	11,368	13,567	41,523
Monthly Totals	**$168,016**	**$168,971**	**$159,082**	**$496,069**

63. **(a)** Number of organic farms in California $= 4,234$
Number of organic farms in Oregon $= 657$
Number of organic farms in Washington $= 886$
Since we want a total, use addition. Write $4,234 + 657 + 886 = 4,234$ farms.

    (b) Wisconsin devotes to organic farming $= 195,603$ acres
New York devotes to organic farming $= 168,428$ acres
Since we want a total, use addition. Write $195,603 + 168,428 = 364,031$ acres.

    (c) Since we want a total, use addition. Write $1,148,650 + 132,764 + 281,970$

        $+ 105,133 + 155,613$

      $= 1,824,130.$

The combined sales from these five states in 2008 were $1,824,130,000.

65. **(a)** Each number after the first is 7 more than the previous number. The next four numbers in the sequence are 33, 40, 47, 54.
(b) Each number after the first is 6 more than the previous number. The next four numbers in the sequence are 32, 38, 44, 50.
(c) Each number after the first is 6 more than the previous number. The next four numbers are 31, 37, 43, 49.
(d) Each number after the first is 8 more than the previous number. The next four numbers are 41, 49, 57, 65.

67. Enter: $[C]3295153[+]573128[+]21257$
$[+]2586241[+]5291[=]$
Display: 6,481,070

69.

Branch	Mon	Tues	Wed	Thurs	Fri.	Weekly Totals
Downtown	487	356	429	278	834	2,384
Suburban	236	255	254	198	423	1,366
Westside	345	278	323	257	563	1,766
Daily Totals	**1,068**	**889**	**1,006**	**733**	**1,820**	**5,516**

71. $248 \text{ ft} + 124 \text{ ft} + 428 \text{ ft} + 162 \text{ ft} = 962 \text{ ft}$

73. $48 \text{ in.} + 8 \text{ in.} = 56 \text{ in.}$

75. Answers may vary. One example is shown.

8	3	4
1	5	9
6	7	2

Exercises 1.3

< Objectives 1 and 3 >

1.
$$\begin{array}{r} 347 \\ -\,201 \\ \hline 146 \end{array}$$
Check:
$$\begin{array}{r} 146 \\ +\,201 \\ \hline 347 \end{array}$$

3.
$$\begin{array}{r} 689 \\ -\,245 \\ \hline 444 \end{array}$$
Check:
$$\begin{array}{r} 444 \\ +\,245 \\ \hline 689 \end{array}$$

5.
$$\begin{array}{r} 3{,}446 \\ -\,2{,}326 \\ \hline 1{,}120 \end{array}$$
Check:
$$\begin{array}{r} 1{,}120 \\ +\,2{,}326 \\ \hline 3{,}446 \end{array}$$

7.
$$\begin{array}{r} \overset{5}{\cancel{6}}\,{}^{1}4 \\ -\,2\ 7 \\ \hline 3\ 7 \end{array}$$
Check:
$$\begin{array}{r} 37 \\ +\,27 \\ \hline 64 \end{array}$$

9.
$$\begin{array}{r} 6\,\overset{1}{\cancel{2}}\,{}^{1}7 \\ -\,3\ 5\ 8 \\ \hline 9 \end{array}$$

$$\begin{array}{r} \overset{5}{\cancel{6}}\,\overset{11}{\cancel{2}}\,{}^{1}7 \\ -\,3\ 5\ 8 \\ \hline 2\ 6\ 9 \end{array}$$
Check:
$$\begin{array}{r} 269 \\ +\,358 \\ \hline 627 \end{array}$$

11.
$$\begin{array}{r} 6{,}4\,\overset{1}{\cancel{2}}\,{}^{1}3 \\ -\,3{,}6\ 7\ 8 \\ \hline 5 \end{array}$$

$$\begin{array}{r} 6{,}\overset{3}{\cancel{4}}\,\overset{11}{\cancel{2}}\,{}^{1}3 \\ -\,3{,}6\ 7\ 8 \\ \hline 4\ 5 \end{array}$$

$$\begin{array}{r} \overset{5}{\cancel{6}}{,}\overset{13}{\cancel{4}}\,\overset{11}{\cancel{2}}\,{}^{1}3 \\ -\,3{,}6\ 7\ 8 \\ \hline 2{,}7\ 4\ 5 \end{array}$$
Check:
$$\begin{array}{r} 2{,}745 \\ +\,3{,}678 \\ \hline 6{,}423 \end{array}$$

13.

$$\begin{array}{r} 6,0\overset{2}{\cancel{3}}{}^{1}4 \\ -\ 2,5\ 6\ 9 \\ \hline 5 \end{array}$$

$$\begin{array}{r} \overset{5}{\cancel{6}},{}^{1}0\overset{2}{\cancel{3}}{}^{1}4 \\ -\ 2,5\ 6\ 9 \\ \hline 5 \end{array}$$

$$\begin{array}{r} \overset{5}{\cancel{6}},\overset{10}{\cancel{0}}\overset{12}{\cancel{3}}{}^{1}4 \\ -\ 2,5\ 6\ 9 \\ \hline 3,\ 4\ 6\ 5 \end{array}$$

Check: 3,465
          + 2,569
          _____
            6,034

15.

$$\begin{array}{r} \overset{3}{\cancel{4}},{}^{1}000 \\ -\ 2,\ 345 \\ \hline \end{array}$$

$$\begin{array}{r} \overset{3}{\cancel{4}},\overset{\cancel{10}}{\cancel{0}}{}^{1}00 \\ -\ 2,3\ 45 \\ \hline \end{array}$$

$$\begin{array}{r} \overset{3}{\cancel{4}},\overset{9}{\cancel{0}}\overset{9}{\cancel{0}}{}^{1}0 \\ -\ 2,3\ 4\ 5 \\ \hline 1,\ 6\ 5\ 5 \end{array}$$

Check: 1,655
          + 2,345
          _____
            4,000

17.

$$\begin{array}{r} 33,4\overset{7}{\cancel{8}}{}^{1}6 \\ -\ 14,0\ 4\ 7 \\ \hline 4\ 3\ 9 \end{array}$$

$$\begin{array}{r} \overset{2}{\cancel{3}}{}^{1}3,4\overset{7}{\cancel{8}}{}^{1}6 \\ -\ 1\ 4,0\ 4\ 7 \\ \hline 1\ 9,4\ 3\ 9 \end{array}$$

Check: 19,439
          + 14,047
          _____
            33,486

19.

$$\begin{array}{r} 2\overset{8}{\cancel{9}},{}^{1}400 \\ -\ 1\ 7,\ 900 \\ \hline 1\ 1,\ 500 \end{array}$$

Check: 11,500
          + 17,900
          _____
            29,400

21.

$$\begin{array}{r} 58 \\ -\ 5 \\ \hline 53 \end{array}$$

Check: 55
          + 3
          _____
            58

23.

$$\begin{array}{r} 148 \\ -\ 23 \\ \hline 125 \end{array}$$

Check: 125
          + 23
          _____
            148

25.

$$\begin{array}{r} 1\overset{1}{\cancel{2}}{}^{1}7 \\ -\ 6\ 9 \\ \hline 5\ 8 \end{array}$$

Check: 69
          + 58
          _____
            127

27.

$$\begin{array}{r} 32,8\overset{6}{\cancel{7}}{}^{1}1 \\ -\ 9\ 7\ 6 \\ \hline 5 \end{array}$$

$$\begin{array}{r} 32,\overset{7}{\cancel{8}}\overset{16}{\overset{6}{\cancel{7}}}{}^{1}1 \\ -\ 9\ 7\ 6 \\ \hline 9\ 5 \end{array}$$

$$\begin{array}{r} 3\overset{1}{\cancel{2}},\overset{17}{\overset{7}{\cancel{8}}}\overset{16}{\overset{6}{\cancel{7}}}{}^{1}1 \\ -\ 9\ 7\ 6 \\ \hline 3\ 1,\ 8\ 9\ 5 \end{array}$$

Check: 31,895
          + 976
          _____
            32,871

29.

$$\begin{array}{r} 19 \\ -\ 7 \\ \hline 12 \end{array}$$

Check: 12
          + 7
          _____
            19

31.

$$\begin{array}{r} 4,0\overset{2}{\cancel{3}}{}^{1}2 \\ -\ 2,2\ 8\ 9 \\ \hline 3 \end{array}$$

$$\begin{array}{r} \overset{3}{\cancel{4}},0\overset{9}{\overset{10}{\cancel{3}}}\overset{12}{\overset{2}{\cancel{2}}} \\ -\ 2,2\ 8\ 9 \\ \hline 1,\ 7\ 4\ 3 \end{array}$$

Check: 1,743
          + 2,289
          _____
            4,032

33.

$$\begin{array}{r} 2,\overset{2}{\cancel{3}}\overset{9}{\overset{10}{\cancel{0}}}{}^{1}1 \\ -\ 9\ 8 \\ \hline 2,\ 2\ 0\ 3 \end{array}$$

Check: 2,203
          + 98
          _____
            2,301

35.

$$\begin{array}{r} \overset{7}{\cancel{8}},{}^{1}5\overset{0}{\cancel{1}}{}^{1}6 \\ -\ 6\ 0\ 9 \\ \hline 7,\ 9\ 0\ 7 \end{array}$$

Check: 7,907
          + 609
          _____
            8,516

37. In $9-6=3$, 9 is the *minuend*, 6 is the *subtrahend*, and 3 is the *difference*. $3+6=9$ is the related addition statement.

39. 25 less than 76 is written $76-25=51$.

41. The difference between 97 and 43 is written $97-43=54$.

43. 298 decreased by 47 is written $298-47=251$.

45. Yes, the difference $8\text{ mi}-4\text{ mi}$ produces a meaningful result since the units, miles, are identical.

47. No, the sum $7\text{ ft}+11\text{ in.}$ does not produce a meaningful result since the units, feet and inches, are not identical.

49. Yes, the difference $17\text{ yd}-10\text{ yd}$ produces a meaningful result since the units, yards, are identical.

< Objective 2 >

51.

Starting elevation	1,053
Increase in 123 ft	+ 123
	1,176
Decrease in 98 ft	− 98
	1,078
Increase in 63 ft	+ 63
	1,141

The final elevation of the hiker is 1,141 ft.

53.

Starting elevation	7,302
Decrease in 623 ft	− 623
	6,679
Decrease in 123 ft	− 123
	6,556
Increase in 307 ft	+ 307
	6,863

The final elevation of the hiker is 6,863 ft.

55. Tony's score was 23 points less than Shaka's score of 87. 23 less than 87 is written $87-23=64$. Tony's score on the test was 64.

57. We want the difference between the heights of the buildings. The difference between 1,454 and 1,250 is written $1,454-1,250=204$. The Wills Tower is 204 ft taller than the Empire State Building.

59. $655-\text{Smaller number}=134$

$\text{Smaller number}=655-134=\ 521$

The smaller number is 521

61. First, find the total amount of Margaret's earnings.
$\$480+\$108=\$588$
Next, find the total amount of her deductions.
$\$153+\$36=\$189$
Her take-home pay is the difference between these totals, written $\$588-\$189=\$399$.
Margaret's take-home pay was $399.

63. First, find the total number of miles accumulated.
$13,850\text{ mi}+2,800\text{ mi}+1,475\text{ mi}$
$\qquad +4,280\text{ mi}$
$=22,405\text{ mi}$
Carmen must fly the difference between 30,000 and the total number of miles accumulated, is written
$30,000\text{ mi}-22,405\text{ mi}=7,595\text{ mi}$. Carmen must fly an additional 7,595 mi for her free trip.

65.

Monthly income	$3,240
House payment	− $1,343
Balance	$1,897
Car payment	− $283
Balance	$1,614
Food	− $512
Balance	$1,102
Clothing	− $189
Balance	$913

67. **(a)** First, find the combined value of both types of lettuce.

$512 + $725 = $1,237$

The combined value of both types of lettuce = $1,237 (in millions).

Next, find the difference between the combined value of both types of lettuce and value of broccoli.

$1,237 − $297 = 940

The combined value of both types of lettuce is $940,000,000 greater than broccoli.

(b) First, find the combined value of lettuces and broccoli.

$512 + $725 + $297 = $1,534$

The combined value of lettuces and broccoli = $1,534 (in millions).

Next, find the difference between the value of lettuces and broccoli combined and the value of strawberries.

$1,534 − $751 = 783

The value of lettuces and broccoli combined is $783,000,000 greater than the value of strawberries.

69. Enter: 5830 − 3987 =

Display: 1843

71. Enter: 534 + 678 − 235 =

Display: 977

73. Enter: 5830 − 3987 =

Display: 1843

75. Enter: 534 + 678 − 235 =

Display: 977

77. Gallons used (Diesel):

Enter: 73255 − 28387 =

Display: 44868

Gallons used (Unleaded):

Enter: 82349 − 19653 =

Display 62696

Gallons used (Super unleaded):

Enter: 81258 − 8654 =

Display: 72604

Total gallons used:

Enter: 44868 + 62696 + 72604 =

Display: 180168

Completed table

	Diesel	Unleaded	Super Unleaded	Total
Beginning reading	73,255	82,349	81,258	
End reading	28,387	19,653	8,654	
Gallons used	**44,868**	**62,696**	**72,604**	**180,168**

79. Land area of California = 155,959 mi^2
Land area of Oregon = 95,997 mi^2

Enter: 155959 − 95997 =

Display: 59962

California is 59,962 mi^2 larger than Oregon.

81. Land area of Oregon = 95,997 mi^2
Land area of Washington = 66,544 mi^2

Enter: 95997 − 66544 =

Display: 29453

Oregon is 29,453 mi^2 larger than Washington.

83. Land area of Alabama = 52,419 mi^2
Land area of Louisiana = 51,843 mi^2

Enter: 52419 − 51843 =

Display: 576

Alabama is 576 mi^2 larger than Louisiana.

85. First, find the sum of the concentrations of chloride and bicarbonate.

93 mEq/L + 24 mEq/L = 117 mEq/L

The difference between this sum and the concentration of sodium is written

140 mEq/L − 117 mEq/L = 23 mEq/L.

87. $\$500 - \$150 = \$350$

89. 14 lb − 10 lb = 4 lb

91. Above and Beyond

93. Adding the diagonal from bottom left to top right gives $8 + 5 + 2 = 15$. Each row, each column, and each diagonal must add up to 15.

a	7	2
b	5	c
8	d	e

a must be 6 since $6 + 7 + 2 = 15$
b must be 1 since $6 + 1 + 8 = 15$
c must be 9 since $1 + 5 + 9 = 15$
d must be 3 since $7 + 5 + 3 = 15$
e must be 4 since $8 + 3 + 4 = 15$
As a check, note that all rows, columns, and diagonals add to 15.

6	7	2
1	5	9
8	3	4

95. Adding along the diagonal from top left to bottom right gives $16 + 10 + 7 + 1 = 34$. Each row, each column, and each diagonal must add up to 34.

16	3	a	13
b	10	11	c
9	6	7	d
4	e	f	1

a must be 2 since $16 + 3 + 2 + 13 = 34$
b must be 5 since $16 + 5 + 9 + 4 = 34$
c must be 8 since $5 + 10 + 11 + 8 = 34$
d must be 12 since $9 + 6 + 7 + 12 = 34$
e must be 15 since $3 + 10 + 6 + 15 = 34$
f must be 14 since $2 + 11 + 7 + 14 = 34$
As a check, note that all rows, columns, and diagonals add to 34.

16	3	**2**	13
5	10	11	**8**
9	6	7	**12**
4	**15**	**14**	1

Exercises 1.4

< Objective 1 >

1. The digit to the right of the tens place, 8, is 5 or more. So round up. 38 is rounded to 40.

3. The digit to the right of the tens place, 3, is less than 5. Thus, the tens digit remains the same and the digit to its right becomes zero. 253 is rounded to 250.

5. The digit to the right of the tens place, 6, is 5 or more. So round up. The tens digit is 9, it is replaced with zero and the next digit to the left is increased by 1. 696 is rounded to 700.

7. The digit to the right of the tens place, 3, is less than 5. Thus, the tens digit remains the same and the digit to its right becomes zero. 2,493 is rounded to 2,490.

9. The digit to the right of the hundreds place, 8, is 5 or more. So round up. 683 is rounded to 700.

11. The digit to the right of the hundreds place, 4, is less than 5. Thus, the hundreds digit remains the same and the digits to its right become zero. 6,741 is rounded to 6,700.

13. The digit to the right of the hundreds place, 6, is 5 or more. So round up. The hundred digit is 9, it is replaced with zero and the next digit to the left is increased by 1. 5,962 is rounded to 6,000.

15. The digit to the right of the hundreds place, 0, is less than 5. Thus, the hundreds digit remains the same and the digits to its right become zero. 12,908 is rounded to 12,900.

17. The digit to the right of the thousands place, 3, is less than 5. Thus, the thousands digit remains the same and the digits to its right become zero. 4,352 is rounded to 4,000.

19. The digit to the right of the thousands place, 9, is 5 or more. So round up. 4,927 is rounded to 5,000.

21. The digit to the right of the thousands place, 4, is less than 5. Thus, the thousands digit remains the same and the digits to its right become zero. 23,429 is rounded to 23,400.

23. The digit to the right of the thousands place, 2, is less than 5. Thus, the thousands digit remains the same and the digits to its right become zero. 9,206 is rounded to 9,000.

25. The digit to the right of the thousands place, 8, is 5 or more. So round up. The thousands digit is 9, it is replaced with zero and the next digit to the left is increased by 1. 129,816 is rounded to 130,000.

27. The digit to the right of the ten thousands place, 7, is 5 or more. So round up. 787,000 is rounded to 790,000.

29. The digit to the right of the millions place, 8, is 5 or more. So round up. 21,800,000 is rounded to 22,000,000.

31. The digit in the tens place, 1, is less than 5. 12 is rounded down to 0.

33. The digit in the hundreds place, 7, is more than 5. 741 is rounded up to 1,000.

< Objective 2 >

35. Estimate: 60   Actual sum: 58
                    30   27
                  + 30   + 33
                   120   118

By rounding to the nearest ten and adding quickly, the estimated sum is 120. Because this is close to the sum calculated, 118, the sum seems reasonable.

37. Estimate: 80   Actual difference: 83
                  − 30   − 27
                    50   56

By rounding to the nearest ten and subtracting quickly, the estimated difference is 50. Because this is close to the difference calculated, 56, the difference seems reasonable.

39. Estimate: 400   Actual sum: 379
                  1,200   1,215
                  + 500   + 528
                  2,100   2,122

By rounding to the nearest hundred and adding quickly, the estimated sum is 2,100. Because this is close to the sum calculated, 2,122, the sum seems reasonable.

41. Estimate: 900   Actual difference: 915

$$\begin{array}{r} 900 \\ -\ 400 \\ \hline 500 \end{array}$$

$$\begin{array}{r} 915 \\ -\ 411 \\ \hline 504 \end{array}$$

By rounding to the nearest hundred and subtracting quickly, the estimated difference is 500. Because this is close to the difference calculated, 504, the difference seems reasonable.

43. Estimate: 2,000   Actual sum: 2,238

$$\begin{array}{r} 2,000 \\ 4,000 \\ +\ 5,000 \\ \hline 11,000 \end{array}$$

$$\begin{array}{r} 2,238 \\ 3,925 \\ +\ 5,217 \\ \hline 11,380 \end{array}$$

By rounding to the nearest thousand and adding quickly, the estimated sum is 11,000. Because this is close to the sum calculated, 11,380, the sum seems reasonable.

45. Estimate: 5,000   Actual difference: 4,822

$$\begin{array}{r} 5,000 \\ -2,000 \\ \hline 3,000 \end{array}$$

$$\begin{array}{r} 4,822 \\ -2,134 \\ \hline 2,688 \end{array}$$

By rounding to the nearest thousand and subtracting quickly, the estimated difference is 3,000. Because this is close to the difference calculated, 2,688, the difference seems reasonable.

47. Rounding each cost to the nearest whole dollar, estimate the total by finding the sum.
$3+2+11+5+3+2+10+3=39$
The total amount of the lunch check is approximately $39.

49. Rounding each score to the nearest ten, estimate the total by finding the sum
$80+90+80+70+100=420$.
Oscar's total score is approximately 420.

51. Rounding the cost of each item to the nearest whole dollar, estimate the total by finding the sum $33+10+68+126+18=255$.
Mrs. Gonzalez's total cost was approximately $255.

< Objective 3 >

53. Since 500 lies to the right of 400 on the number line, $500 > 400$.

55. Since 100 lies to the left of 1,000 on the number line, $100 < 1,000$.

57.

Appliance	Power Required (W/hr)	Rounded Watts
Clock radio	10	0
Electric Blanket	100	100
Clothes Washer	500	500
Toaster Oven	1,225	1,200
Laptop	50	100
Hair Dryer	1,875	1,900
DVD Player	25	0
Estimated Total		**3,800**

59. $400+200+300=900$ screws

61.

(a) 55 is the smallest possible whole number that, when rounded to the nearest ten, rounds to 60.
(b) Since 65, when rounded to the nearest ten, rounds up to 70, 64 is the largest possible whole number that rounds, to the nearest ten, to 60.

63. The whole numbers that, when rounded to the nearest ten, round to 40 are: 35, 36, 37, 38, 39, 40, 41, 42, 43, 44. If each of these numbers represents a number of blue marbles, then the corresponding numbers of green marbles would be 25, 24, 23, 22, 21, 20, 19, 18, 17, 16, respectively. With the exception of 25, all of these numbers, when rounded to the nearest ten, round to 20 as specified. Since there cannot be 25 green marbles, there cannot be 35 blue marbles. Thus, there are 36, 37, 38, 39, 40, 41, 42, 43, or 44 blue marbles.

65. The digit to the right of the ten millions place, 2, is less than 5. Thus, the ten millions digit remains the same and the digits to its right become zero. 312,160,918 people is rounded down to 310,000,000 people.

Exercises 1.5

< Objective 1 >

1.
$$5 \\ \underline{\times\,3} \\ 15$$

3.
$$6 \\ \underline{\times\,0} \\ 0$$

5.
$$\overset{3}{4}8 \\ \underline{\times\ 4} \\ 192$$

7.
$$5\overset{4}{0}8 \\ \underline{\times\ \ 6} \\ 3{,}048$$

9.
$$\overset{3}{\overset{4}{7}}5 \\ \underline{\times\ 68} \\ 600 \\ \underline{4500} \\ 5{,}100$$

11.
$$\overset{1\ 3}{\overset{2\ 6}{3}}27 \\ \underline{\times\ \ 59} \\ 2943 \\ \underline{16350} \\ 19{,}293$$

13.
$$\overset{6\ 4}{\overset{3\ 2}{4}}{,}075 \\ \underline{\times\ \ \ 84} \\ 16300 \\ \underline{326000} \\ 342{,}300$$

15.
$$\overset{1\ 2}{1}24 \\ \underline{\times\ 225} \\ 620 \\ 2480 \\ \underline{24800} \\ 27{,}900$$

17.
$$\overset{1\ 2}{\overset{1\ 4}{\overset{3\ 7}{6}}}39 \\ \underline{\times\,358} \\ 5112 \\ 31950 \\ \underline{191700} \\ 228{,}762$$

19.
$$\overset{2\ 2}{\overset{3\ 4}{6}}68 \\ \underline{\times\,305} \\ 3340 \\ \underline{200400} \\ 203{,}740$$

21.
$$\overset{\ \ \ 1}{\overset{1\ 1\ 6}{3}}{,}219 \\ \underline{\times\ \ 207} \\ 22533 \\ \underline{643800} \\ 666{,}333$$

23.
$$\begin{array}{r} {}^{1\ 1}_{\ 2} \\ {}_{2\ 3} \\ 3,158 \\ \times\,2,034 \\ \hline 12632 \\ 94740 \\ 6316000 \\ \hline 6,423,372 \end{array}$$

25.
$$\begin{array}{r} {}^{3} \\ 58 \\ \times\ \ 40 \\ \hline 2,320 \end{array}$$

27.
$$\begin{array}{r} {}^{6} \\ 907 \\ \times\ \ 900 \\ \hline 816,300 \end{array}$$

29.
$$\begin{array}{r} {}^{1} \\ 362 \\ \times\,310 \\ \hline 000 \\ 3620 \\ 108600 \\ \hline 112,220 \end{array}$$

31.
$$\begin{array}{r} 18 \\ \times\ \ 1 \\ \hline 18 \end{array}$$

33.
$$\begin{array}{r} 64 \\ \times\ \ 0 \\ \hline 0 \end{array}$$

35.
$$\begin{array}{r} {}^{2} \\ 304 \\ \times\ \ 7 \\ \hline 2,128 \end{array}$$

37.
$$\begin{array}{r} {}^{1} \\ 551 \\ \times\ \ 21 \\ \hline 551 \\ 11020 \\ \hline 11,571 \end{array}$$

< Objective 2 >

39. The order of the factors, 5 and 8, was changed. The product remains the same by the commutative property of multiplication.

41. The grouping of the factors was changed. The product remains the same by the associative property of multiplication.

43. The multiplicative identity property states that the product of 1 and any number is just that number. The product of 5 and 1 illustrates the multiplicative identity property.

45. The multiplicative property of 0 states that the product of 0 and any number is zero. The product of 8 and 0 illustrates the multiplicative property of 0.

47. The commutative property of multiplication states that we can multiply two numbers in either order and get the same result. Changing the order of the factors of 3×8 gives 8×3. So the commutative property of multiplication tells us $7+(3\times8)=7+(8\times3)$.

< Objective 3 >

49. There were 34 truck shipments carrying 8 cars each. Write $34\times8=272$. Thus, 272 cars were shipped.

51. There are 14 rows with 24 spaces in each row. Write $14\times24=336$. Thus, 336 cars can be parked in the lot.

53. There are 28 days with 15 stoves made each day. Write $28 \times 15 = 420$. In 28 days, 420 stoves can be made.

55. In 1 second, sound travels 1,088 feet. The thunder is heard in 15 seconds after seeing a lightning flash. Therefore, the lightning flash is $(1,088 \times 15)$ ft $= 16,320$ ft far away.

57. There are 500 sheets in 1 ream of paper. An office machine uses 29 reams in 1 week. Write 29×500. In one week, 14,500 sheets are used.

59. Estimate:
$$\begin{array}{r} 40 \\ \times\ 20 \\ \hline 800 \end{array}$$

61. Estimate:
$$\begin{array}{r} 400 \\ \times\ \ 500 \\ \hline 200,000 \end{array}$$

63. There are about 50 rows with about 40 seats per row. Write $50 \times 40 = 2,000$. There are about 2,000 seats in the theater.

65. There are about 130 days with about 50 sleds manufactured each day. Write $130 \times 50 = 6,500$. This company can make about 6,500 sleds.

< Objective 5 >

67. $6 \text{ yd} \times 6 \text{ yd} = 36 \text{ yd}^2$

69. $3 \text{ in.} \times 6 \text{ in.} = 18 \text{ in.}^2$

71. First, find area of the larger rectangle. The length is $2 \text{ in.} + 2 \text{ in.} + 3 \text{ in.} = 7 \text{ in.}$ and the width is 5 in.
The area is $7 \text{ in.} \times 5 \text{ in.} = 35 \text{ in.}^2$
Next, find the area of the missing rectangle.
$2 \text{ in.} \times 2 \text{ in.} = 4 \text{ in.}^2$
The difference between the two areas is:
$35 \text{ in.}^2 - 4 \text{ in.}^2 = 31 \text{ in.}^2$

73. $V = LWH$
$6 \text{ ft} \times 6 \text{ ft} \times 6 \text{ ft} = 216 \text{ ft}^3$

75. With a scientific calculator:
$82 \boxed{\times} 46 \boxed{=}$
With a graphing calculator:
$82 \boxed{\times} 46 \boxed{\text{ENTER}}$
Result: 3,772

77. With a scientific calculator:
$148 \boxed{\times} 593 \boxed{=}$
With a graphing calculator:
$148 \boxed{\times} 593 \boxed{\text{ENTER}}$
Result: 87,764

79. With a scientific calculator:
$66 \boxed{\times} 16088 \boxed{=}$
With a graphing calculator:
$66 \boxed{\times} 16088 \boxed{\text{ENTER}}$
Result: 1,061,808

81. With a scientific calculator:
$5 \boxed{\times} 19 \boxed{\times} 72 \boxed{=}$
With a graphing calculator:
$5 \boxed{\times} 19 \boxed{\times} 72 \boxed{\text{ENTER}}$
Result: 6,840

83. With a scientific calculator:
$52 \boxed{\times} \boxed{(} 14 \boxed{+} 41 \boxed{)} \boxed{=}$
With a graphing calculator:
$52 \boxed{\times} \boxed{(} 14 \boxed{+} 41 \boxed{)} \boxed{\text{ENTER}}$
Result: 2,860

85. With a scientific calculator:
$847 \boxed{\times} \boxed{(} 12 \boxed{+} 459 \boxed{)} \boxed{=}$
With a graphing calculator:
$847 \boxed{\times} \boxed{(} 12 \boxed{+} 459 \boxed{)} \boxed{\text{ENTER}}$
Result: 398,937

87. $125 \times 6 = 750$ mL

89. $10 \times 1,000 = 10,000$ cents
$10,000 \div 100 = \$100$

91. $50 \times 25 = 1,250$ resistors

93. The total earning of 6 machinists in an hour
$= 6 \times 21 = \$126$.
The total earning of 3 assembly workers in an hour $= 3 \times 12 = \$36$.
The total earning of 1 supervisor maintenance person = \$28.
The shop's payroll for a hour
$= \$126 + \$36 + \$28 = \$190$.
The shop's payroll for a 40-hour week
$= 40 \times 190 = \$7,600$.

95. Above and Beyond

97. Above and Beyond

99. Above and Beyond

Exercises 1.6

1. In $48 \div 8 = 6$, 8 is the *divisor*, 48 is the *dividend*, and 6 is the *quotient*.

< Objective 1 >

3. $\begin{array}{cccc} 36 & 27 & 18 & 9 \\ -9 & -9 & -9 & -9 \\ \hline 27 & 18 & 9 & 0 \end{array}$

because 9 can be subtracted from 36 four times, $36 \div 9 = 4$.

5. Stephanie wants to plant 63 plants with 9 plants per row, therefore $63 \div 9 = 7$.
Stephanie will have 7 rows.

< Objective 2 >

7. Since $36 \div 4 = 9$, 36 pages $\div 4 = 9$ pages.

9. Since $4,900 \div 7 = 700$,
$4,900$ km $\div 7 = 700$ km.

11. Since $160 \div 4 = 40$, 160 mi $\div 4$ hr $= 40$ mi/hr.

13. Since $3,720 \div 5 = 744$,
$3,720$ hr $\div 5$ mo $= 744$ hr/mo.

15. $9\overline{)54}$ with quotient 6 Check: $9 \times 6 = 54$

17. $6\overline{)42}$ with quotient 7 Check: $6 \times 7 = 42$

19. $4\overline{)32}$ with quotient 8 Check: $4 \times 8 = 32$

21. $\begin{array}{r} 8 \\ 5\overline{)43} \\ \underline{40} \\ 3 \end{array}$

We have $43 \div 5 = 8$ r3.
Check: $43 = 5 \times 8 + 3$

23. $\begin{array}{r} 7 \\ 9\overline{)65} \\ \underline{63} \\ 2 \end{array}$

We have $65 \div 9 = 7$ r2.
Check: $65 = 9 \times 7 + 2$

25. $\begin{array}{r} 7 \\ 8\overline{)57} \\ \underline{56} \\ 1 \end{array}$

We have $57 \div 8 = 7$ r1.
Check: $57 = 8 \times 7 + 1$

27. $\begin{array}{r} 0 \\ 5\overline{)0} \\ \underline{0} \\ 0 \end{array}$ $0 \div 5 = 0$ because $0 = 5 \times 0$.

29. $0\overline{)4}$ $4 \div 0$ is undefined since there is no number that can be multiplied by 0 to give a product of 4.

31. $6\overline{)0}$ $0 \div 6 = 0$ because $0 = 6 \times 0$.

$$\frac{0}{0}$$

$$0$$

33.
$$5\overline{)83}$$
$$\underline{5}$$
$$33$$
$$\underline{30}$$
$$3$$

We have $83 \div 5 = 16\ \text{r}3$.
Check: $83 = 5 \times 16 + 3$

35.
$$3\overline{)162}$$
$$\underline{15}$$
$$12$$
$$\underline{12}$$
$$0$$

We have $162 \div 3 = 54$.
Check: $162 = 3 \times 54$

37.
$$8\overline{)293}$$
$$\underline{24}$$
$$53$$
$$\underline{48}$$
$$5$$

We have $93 \div 8 = 36\ \text{r}5$.
Check: $293 = 8 \times 36 + 5$

39.
$$8\overline{)3,136}$$
$$\underline{24}$$
$$73$$
$$\underline{72}$$
$$16$$
$$\underline{16}$$
$$0$$

We have $3,136 \div 8 = 392$.
Check: $3,136 = 8 \times 392$

41.
$$8\overline{)5,438}$$
$$\underline{48}$$
$$63$$
$$\underline{56}$$
$$78$$
$$\underline{72}$$
$$6$$

We have $5,438 \div 8 = 679\ \text{r}6$.
Check: $5,438 = 8 \times 679 + 6$

43.
$$8\overline{)22,153}$$
$$\underline{16}$$
$$61$$
$$\underline{56}$$
$$55$$
$$\underline{48}$$
$$73$$
$$\underline{72}$$
$$1$$

We have $22,153 \div 8 = 2769\ \text{r}5$.
Check: $22,153 = 8 \times 2769 + 1$

45.
$$45\overline{)2,367}$$
$$\underline{225}$$
$$117$$
$$\underline{90}$$
$$27$$

We have $2,367 \div 45 = 52\ \text{r}27$.
Check: $2,367 = 45 \times 52 + 27$

47.

$$34\overline{)8{,}748}$$ quotient 257

$$\underline{68}$$
$$194$$
$$\underline{170}$$
$$248$$
$$\underline{238}$$
$$10$$

We have $8{,}748 \div 34 = 257 \text{ r}10$.

Check: $8{,}748 = 34 \times 257 + 10$

49.

$$42\overline{)7{,}902}$$ quotient 188

$$\underline{42}$$
$$370$$
$$\underline{336}$$
$$342$$
$$\underline{336}$$
$$6$$

We have $7{,}902 \div 42 = 188 \text{ r}6$.

Check: $7{,}902 = 42 \times 188 + 6$

51.

$$8\overline{)1{,}672}$$ quotient 209

$$\underline{16}$$
$$072$$
$$\underline{72}$$
$$0$$

We have $1{,}672 \div 8 = 29$.

Check: $1{,}672 = 29 \times 8$

53.

$$53\overline{)46{,}653}$$ quotient 880

$$\underline{424}$$
$$425$$
$$\underline{424}$$
$$13$$

We have $46{,}653 \div 53 = 880 \text{ r}13$.

Check: $46{,}653 = 880 \times 53 + 13$

55.

$$280\overline{)6{,}720}$$ quotient 24

$$\underline{560}$$
$$1120$$
$$\underline{1120}$$
$$0$$

We have $6{,}720 \div 280 = 24$.

Check: $6{,}720 = 280 \times 24$

57.

$$156\overline{)125{,}580}$$ quotient 805

$$\underline{1248}$$
$$780$$
$$\underline{780}$$
$$0$$

We have $125{,}580 \div 156 = 805$.

Check: $125{,}580 = 805 \times 156$

59.

$$245\overline{)857{,}990}$$ quotient $3{,}502$

$$\underline{735}$$
$$1229$$
$$\underline{1225}$$
$$490$$
$$\underline{190}$$
$$0$$

We have $857{,}990 \div 245 = 3{,}502$.

Check: $857{,}990 = 3{,}502 \times 245$

< Objective 3 >

61. Estimate: 800 divided by 40

$$40\overline{)800}$$ quotient 20

63. Estimate: 5,000 divided by 100

$$100\overline{)5{,}000}$$ quotient 50

65. Estimate: 9,000 divided by 90

$$90\overline{)9{,}000}$$ quotient 100

67. Estimate: 3,900 divided by 130

$$130\overline{)3,900}^{30}$$

69. Estimate: 3,800 divided by 190

$$190\overline{)3,800}^{20}$$

< Objective 4 >

71. There are 63 candy bars in 7 boxes. Write $63 \div 7 = 9$. Thus, there are 9 candy bars per box.

73. There are 77 pictures with 8 pictures per page. Write $77 \div 8 = 9$ r5. Joaquin will fill 9 full pages with 5 pictures left over.

75. There are 1,702 calls from 37 phones. Write $1,702 \div 37 = 46$. There were 46 calls placed per phone.

77. There are 10,880 lines with 340 lines per minute. Write $10,880 \div 340 = 32$. It will take 32 min.

79. With a scientific calculator:

187452 $\div$ 36 $=$

With a graphing calculator:

187452 $\div$ 36 $\boxed{\text{ENTER}}$
Result: 5,207

81. With a scientific calculator:

583467 $\div$ 129 $=$

With a graphing calculator:

583467 $\div$ 129 $\boxed{\text{ENTER}}$
Result: 4,523

83. With a scientific calculator:

11349 $\div$ 52 $=$

Display: 218.25
The whole-number part of the quotient is 218.
To calculate remainder, enter:

218.25 $-$ 218 $=$ $\times$ 52 $=$

The remainder is 13.
With a scientific calculator:

11349 $\div$ 52 $\boxed{\text{ENTER}}$

The whole-number part of the quotient is 218.
To calculate remainder, enter:

218.25 $-$ 218 $\boxed{\text{ENTER}}$

$\times$ 52 $\boxed{\text{ENTER}}$
The remainder is 13.
Result: 218 r13

85. With a scientific calculator:

2786986 $\div$ 478 $=$

Display: 5830.5146443
The whole-number part of the quotient is 5,830. To calculate remainder, enter:

5830.5146443 $-$ 5830 $=$ $\times$ 478 $=$

The remainder is 26.
With a scientific calculator:

2786986 $\div$ 478 $\boxed{\text{ENTER}}$

Display: 5830.5146443
The whole-number part of the quotient is 5,830. To calculate remainder, enter:

5830.5146443 $-$ 5830 $\boxed{\text{ENTER}}$

$\times$ 478 $\boxed{\text{ENTER}}$
The remainder is 26.
Result: 5,830 r26

87. With a scientific calculator:

$2657463 \boxed{\div} \boxed{8102} \boxed{=}$

Display: 328.000864
The whole –number part of the quotient is 328. To calculate remainder, enter:

$328.000864 \boxed{-} \boxed{328} \boxed{=} \boxed{\times} \boxed{8102} \boxed{=}$

The remainder is 7.
With a scientific calculator:

$2657463 \boxed{\div} \boxed{8102} \boxed{\text{ENTER}}$

Display: 328.000864
The whole-number part of the quotient is 328. To calculate remainder, enter:

$328.000864 \boxed{-} \boxed{328} \boxed{\text{ENTER}}$

$\boxed{\times} \boxed{8102} \boxed{\text{ENTER}}$

The remainder is 7.
Result: 328 r7

89. $525 \div 75 = 7$ pills

91. First, find the number of computers to switch connections possible.
200 ft $\div$ 4 ft = 50
She can make 50 4-ft cables.
Next, find the total number of computers.
5 rows $\times$ 9 computers per row
= 45 computers.
Therefore, there is enough cable for the job.

93. If you buy all small packages, the number of packages needed is $10,000 \div 500 = 20$ small packages.
If you buy all large packages, the number of packages needed is $10,000 \div 1,250 = 8$ large packages.

95. $1,752 \text{ lb} \div 24 = 73 \text{ lb}$

97. Above and Beyond

99. Above and Beyond

Exercises 1.7

< Objectives 1–3 >

1. $3^2 = 3 \times 3 = 9$

3. $5^1 = 5$

5. $10^3 = 10 \times 10 \times 10 = 1,000$

7. $2 \times 4^3 = 2 \times 64 = 128$

9. $5 + 2^2 = 5 + 4 = 9$

11. $42 - 7 + 9 = 35 + 9 = 44$

13. $20 \div 5 \times 2 = 4 \times 2 = 8$

15. $(3+2)^3 - 20 = 5^3 - 20 = 125 - 20 = 105$

17. $(7-4)^4 - 30 = 3^4 - 30 = 81 - 30 = 51$

19. $8^2 \div 4^2 + 2 = 64 \div 16 + 2 = 4 + 2 = 6$

21. $24 - 6 \div 3 = 24 - 2 = 22$

23. $(24-6) \div 3 = 18 \div 3 = 6$

25. $12 + 3 \div (3^2 - 2.3) = 12 + 3 \div 3 = 12 + 1 = 13$

27. $8^2 - 2^4 \div 2 = 64 - 16 \div 2 = 64 - 8 = 56$

29. $30 \div 6 - 12 \div 3 = 5 - 4 = 1$

31. $16 - 12 \div 3 \bullet 2 + (16-12)^2 \bullet 3$
$= 16 - 12 \div 3 \bullet 2 + (4)^2 \bullet 3$
$= 16 - 12 \div 3 \bullet 2 + 16 \bullet 3 = 16 - 4 \bullet 2 + 16 \bullet 3$
$= 16 - 8 + 48 = 16 - 8 + 48 = 8 + 48 = 56$

33. $6 + 3 \times 2^4 - (12-7)(10-7)$
$= 6 + 3 \times 2^4 - (5)(3) = 6 + 3 \times 16 - (5)(3)$
$= 6 + 48 - 15 = 54 - 15 = 39$

35. $3 \times \left[(7-5)^3 - 8 \right] + 5 \times 2$

$= 3 \times \left[(2)^3 - 8 \right] + 5 \times 2 = 3 \times [8-8] + 5 \times 2$

$= 3 \times [0] + 5 \times 2 = 0 + 10 = 10$

37. With a scientific calculator:

12 $\boxed{x^y}$ 4 $\boxed{=}$

With a graphing calculator:

12 $\boxed{\wedge}$ 4 $\boxed{\text{ENTER}}$

Result: 20,736

39. With a scientific calculator:

6 $\boxed{\times}$ 15 $\boxed{x^y}$ 3 $\boxed{=}$

With a graphing calculator:

6 $\boxed{\times}$ 15 $\boxed{x^y}$ 3 $\boxed{\text{ENTER}}$

Result: 20,250

41. With a scientific calculator:

6 $\boxed{+}$ 14 $\boxed{\times}$ 37 $\boxed{=}$

With a graphing calculator:

6 $\boxed{+}$ 14 $\boxed{\times}$ 37 $\boxed{\text{ENTER}}$

Result: 524

43. With a scientific calculator:

8 $\boxed{x^y}$ 3 $\boxed{\times}$ $\boxed{(}$ 2 $\boxed{+}$ 4 $\boxed{\times}$ 6 $\boxed{x^y}$ 2 $\boxed{)}$

$\boxed{x^y}$ 3 $\boxed{\text{ENTER}}$

With a graphing calculator:

8 $\boxed{\wedge}$ 3 $\boxed{\times}$ $\boxed{(}$ 2 $\boxed{+}$ 4 $\boxed{\times}$ 6 $\boxed{\wedge}$ 2 $\boxed{)}$

$\boxed{\wedge}$ 3 $\boxed{\text{ENTER}}$

Result: 1,593,413,632

45. With a scientific calculator:

5 $\boxed{\times}$ $\boxed{(}$ 4 $\boxed{+}$ 7 $\boxed{)}$ $\boxed{=}$

With a graphing calculator:

5 $\boxed{\times}$ $\boxed{(}$ 4 $\boxed{+}$ 7 $\boxed{)}$ $\boxed{\text{ENTER}}$

Result: 55

47. Enter: 38 $\boxed{\times}$ 528 $\boxed{=}$

Display: 20064

Enter: 33 $\boxed{\times}$ 647 $\boxed{=}$

Display: 21351

Enter: 19 $\boxed{\times}$ 912 $\boxed{=}$

Display: 17328

Enter: 20064 $\boxed{+}$ 21351 $\boxed{+}$ 17328 $\boxed{=}$

Display: 58743

Model	Number Sold	Profit per Sale	Monthly Profit
Subcompact	38	$528	$20,064
Compact	33	647	21,351
Standard	19	912	17,328
Total Monthly Profit			**$58,743**

49. $43 \times 100,000 = 4,300,000$ ohms

The third band is the most important. Misreading it would lead to errors of powers of ten.

51. (a) $P = IV = (13)(110) = 1,430$ watts

(b) $P = \dfrac{V^2}{R} = \dfrac{(220)^2}{22} = \dfrac{48,400}{22}$

$= 2,200$ watts

(c) $P = I^2 R = (25)^2 (12) = (625)(12)$

$= 5,625$ watts

53. $6^2 + 8^2 \overset{?}{=} 10^2$

$36 + 64 \overset{?}{=} 100$

$100 = 100$

Yes, 6, 8, 10 is a Pythagorean triple.

55. $5^2 + 12^2 \overset{?}{=} 13^2$

$25 + 144 \overset{?}{=} 169$

$169 = 169$

Yes, 5, 12, 13 is a Pythagorean triple.

57. $8^2 + 16^2 \overset{?}{=} 18^2$

$64 + 256 \overset{?}{=} 324$

$320 \neq 324$

No, 8, 16, 18 is not a Pythagorean triple.

59. Above and Beyond

Summary Exercises

1. 5,674
4 ones, 7 tens, 6 hundreds, 5 thousands
The place value of 6 is hundreds.

3. $\underset{\text{thousands}}{27}$, $\underset{\text{ones}}{428}$

Twenty-seven thousand, four hundred twenty-eight

5. Thirty-seven thousand, five hundred eighty-three in standard form is 37,583.

7. The order of the addends was changed. The sum remains the same by the commutative property of addition.

9.
$$\begin{array}{r} \overset{2\,1}{784} \\ 385 \\ + 247 \\ \hline 1,416 \end{array}$$

11.
$$\begin{array}{r} \overset{1\,3\,2}{367} \\ 289 \\ 1,463 \\ + 2,682 \\ \hline 4,801 \end{array}$$

13. (a) The total number of passengers.
(b) There were five flights. There were 173, 212, 185, 197, and 202 passengers on the flights.

15.
$$\begin{array}{r} \overset{2\,1}{173} \\ 212 \\ 185 \\ 197 \\ + 202 \\ \hline 969 \end{array}$$
The total number of passengers was 969.

17. 34 decreased by 7 is written as $34 - 7 = 27$.

19. The product of 9 and 5, divided by 3, is written as $(9 \times 5) \div 3 = 45 \div 3 = 15$.

21.
$$\begin{array}{r} 3\,\overset{7}{\cancel{8}},\overset{1}{4}00 \\ - 1\,9,600 \\ \hline 800 \end{array}$$

$$\begin{array}{r} \overset{2}{\cancel{3}}\,\overset{17}{\cancel{8}},\overset{1}{4}00 \\ - 1\,9,\,600 \\ \hline 1\,8,\,800 \end{array}$$

23.
$$\begin{array}{r} 2,6\,\overset{7}{\cancel{8}}\,\overset{1}{2} \\ - \quad 1\,0\,8 \\ \hline 2,5\,7\,4 \end{array}$$

25.

Amount owed	$795
Minus 1st payment	− $75
	$720
Minus 2nd payment	− $125
	$595
Minus 3rd payment	− $90
	$505
Plus interest charged	− $31
	$536

$536 remains to be paid on the account.

27. The hundreds digit is 9. The digit to the right, 7, is 5 or more. So round up. Since this digit is 9, replace the 9 with 0 and increase the next digit to the left by 1. 6975 is rounded up to 7000.

29. The ten thousands digit is 4. The digit to the right, 8, is 5 or more. So round up. 548,239 is rounded up to 550,000.

31. Since 60 lies to the left of 70 on the number line, $60 < 70$.

33. $P = 5 \text{ ft} + 2 \text{ ft} + 2 \text{ ft} + 5 \text{ ft} + 2 \text{ ft} + 2 \text{ ft}$
$= 18 \text{ ft}$

35. The order of the factors, 7 and 8, was changed. The product remains the same by the commutative property of multiplication.

37. The grouping of the factors was changed. The product remains the same by the associative property of multiplication.

39.
$$\begin{array}{r} \overset{2}{\overset{1}{5}}8 \\ \times\, 32 \\ \hline 116 \\ 1740 \\ \hline 1,856 \end{array}$$

41.
$$\begin{array}{r} \overset{3\,3}{\overset{7\,7}{3}}78 \\ \times\, 409 \\ \hline 3402 \\ 151200 \\ \hline 154,602 \end{array}$$

43.

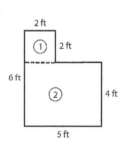

Area of region 1 $= 2 \text{ ft} \times 2 \text{ ft} = 4 \text{ ft}^2$
Area of region 2 $= 5 \text{ ft} \times 4 \text{ ft} = 20 \text{ ft}^2$
Total Area = Area of region 1
    + Area of region 2
$= 4 \text{ ft}^2 + 20 \text{ ft}^2 = 24 \text{ ft}^2$

45. First, find the area of the room.
$5 \text{ yd} \times 7 \text{ yd} = 35 \text{ yd}^2$
Now, multiply it by the cost per square yard.
$35 \times \$18 = \$630$
The total cost of the materials is $630.

47. $0 \div 8 = 0$ because $0 = 8 \times 0$.

49.
$$\begin{array}{r} 308 \\ 8\overline{)2,469} \\ \underline{24} \\ 06 \\ \underline{0} \\ 69 \\ \underline{64} \\ 5 \end{array}$$
We have $2,469 \div 8 = 308 \text{ r}5$.

51.
$$\begin{array}{r} 497 \\ 64\overline{)31,809} \\ \underline{256} \\ 620 \\ \underline{576} \\ 449 \\ \underline{448} \\ 1 \end{array}$$
We have $31,809 \div 64 = 497 \text{ r}1$.

53. First, determine the number of miles traveled.
$26,215 - 25,235 = 980$
Now, divide by the number of gallons used.
$980 \div 35 = 28$
Hasina's mileage for the trip was 28 mi/gal.

55. Estimate: 400 divided by 40
$$\begin{array}{r} 10 \\ 40\overline{)400} \end{array}$$

57. $5 \times 2^3 = 5 \times 8 = 40$

59. $4 + 8 \times 3 = 4 + 24 = 28$

61. $(4+8)\times3=12\times3=36$

63. $8\div4\times2-2+1=2\times2-2+1=4-2+1$
$$=2+1=3$$

65. $(3\times4)^2-100\div5\times6=12^2-100\div5\times6$
$$=144-100\div5\times6$$
$$=144-20\times6$$
$$=144-120=24$$

Chapter Test 1

1. $\underset{\text{thousands}}{\underline{302}}$, $\underset{\text{ones}}{\underline{525}}$

Three hundred two thousand, five hundred twenty-five

3. Two million, four hundred thirty thousand in standard form is 2,430,000.

5.
$$\overset{2\;1}{489}$$
$$562$$
$$613$$
$$+\;254$$
$$\overline{1,918}$$

7.
$$\overset{4}{\underset{}{\overset{5}{89}}}$$
$$\times\;56$$
$$\overline{534}$$
$$\underline{4450}$$
$$4,984$$

9.
$$289$$
$$-\;54$$
$$\overline{235}$$

11.
$$32,\overset{2}{\cancel{3}}{}^{1}45$$
$$-\;\;1,575$$
$$\overline{0}$$

$$3\overset{1}{\cancel{2}},\overset{12}{\cancel{3}}{}^{1}45$$
$$-\;\;1,575$$
$$\overline{70}$$

$$3\overset{1}{\cancel{2}},\overset{12}{\cancel{3}}{}^{1}45$$
$$-\;\;1,575$$
$$\overline{30,770}$$

13.
$$\begin{array}{r}266\\8\overline{)2,135}\\\underline{16}\\53\\\underline{48}\\55\\\underline{48}\\7\end{array}$$

We have $2,132\div8=266\text{ r}7$.

15. $(3+4)^2-(2+3^2-1)=(7)^2-(2+9-1)$
$$=(7)^2-(11-1)$$
$$=(7)^2-(10)=49-10$$
$$=39$$

17. The factor 4 was multiplied by each number inside the parentheses; then these products were added. The result is the same by the distributive property of multiplication over addition.

19. The grouping of the factors was changed. The product remains the same by the associative property of multiplication.

21. Since 49 lies to the right of 47 on the number line, $49>47$.

23.
$$
\begin{array}{r}
943 \\
3,281 \\
778 \\
2,112 \\
+\ \ 570 \\
\hline
7,700
\end{array}
$$

25.

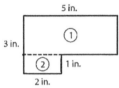

Area of region 1: $5 \text{ in.} \times 2 \text{ in.} = 10 \text{ in.}^2$
Area of region 2: $1 \text{ in.} \times 2 \text{ in.} = 2 \text{ in.}^2$
Total area = Area of region 1
    + Area of region 2
$= 10 \text{ in.}^2 + 2 \text{ in.}^2 = 12 \text{ in.}^2$

27. There are 25 new vans at \$22,350 per van.
Write $25 \times 22,350 = 558,750$.
The total cost of the order will be \$558,750.

29.
$$
\begin{array}{r}
^{1\ 1\ \ 1\ 1} \\
12,438 \\
14,325 \\
14,581 \\
+\ 14,634 \\
\hline
55,978
\end{array}
$$

Chapter 2
Introductions to Integers and Algebra

Prerequisite Check

1.
$$\begin{array}{r} \overset{1}{1}2 \\ +\ 9 \\ \hline 21 \end{array}$$

3.
$$\begin{array}{r} 23 \\ \times 16 \\ \hline 138 \\ 23\ \ \\ \hline 368 \end{array}$$

5. $3 \times 8^2 - 5 = 3 \times 64 - 5 = 192 - 5 = 187$

7. Commutative property of addition

9. Associative property of multiplication

11. $10 - 8 = 2$

13. Perimeter $= 2l + 2w = 2 \cdot 16 + 2 \cdot 9 = 32 + 18$
$$= 50 \text{ in.}$$

Exercises 2.1

< Objective 1 >

1. The numbers $5, -15, 18, -8, 3$ are represented on the number line as

3. The numbers $-3, 6, -10, 1, -16$ are represented on the number line as,

< Objective 2 >

5. From smallest to largest, the numbers are $-7, -5, -1, 0, 2, 3, 8$.

7. From smallest to largest, the numbers are $-11, -6, -2, 1, 4, 5, 9$.

9. From smallest to largest, the numbers are $-7, -6, -3, 3, 6, 7$.

11. From smallest to largest, the numbers are $-62, -40, -18, 0, 12, 36$.

< Objective 3 >

13. -6 is the minimum and 15 is the maximum.

15. -15 is the minimum and 21 is the maximum.

17. -2 is the minimum and 5 is the maximum.

19. -1.366 is the minimum and 4.088 is the maximum.

21. The opposite of 15 is -15.

23. The opposite of -19 is 19.

25. The opposite of -7 is 7.

27. The opposite of 11 is -11.

< Objective 4 >

29. $|17| = 17$

31. $|-10| = 10$

33. $-|3| = -3$

35. $-|-8| = -8$

37. $|-2| + |3| = 2 + 3 = 5$

39. $|-9| + |9| = 9 + 9 = 18$

41. $|4| - |-4| = 4 - 4 = 0$

43. $|15| - |8| = 15 - 8 = 7$

45. $|15 - 8| = |7| = 7$

47. $|-9| + |2| = 9 + 2 = 11$

49. $|-8| - |-7| = 8 - 7 = 1$

51. $-9 < -6$

53. $|-9| = 9$ and $9 > 6$. Therefore, $|-9| > 6$.

55. $-5 > -9$

57. $-20 < -10$

59. $|3| = 3$

61. $-4 < |-4|$

63. 400 ft above sea level: +400 ft

65. A loss of $200: –$200

67. A decrease in population of 25,000: –25,000 people

69. The withdrawal of $50: –$50

71. A temperature decrease of 10°: –10°F

73. An increase of 75 points: +75 points

75. Positive trade balance of $90,000,000: +$90,000,000

77. The absolute value of –10 is 10.

79. The absolute value of –7 is 7.

81. The opposite of 30 is –30.

83. The absolute value of the opposite of 3 is 3.

85. The opposite of the absolute value of –7 is –7.

87. True.
The whole numbers are 0, 1, 2, 3, 4, 5, …
The integers are … –3, –2, –1, 0, 1, 2, 3, …
Therefore, all the whole numbers are integers and the statement is true.

89. False

91. never

93. sometimes

95. –5 cm

97. The order of elevations from the smallest to the largest is,
–84, –45, –18, –13, 4, 27, 37, 49, 59, 66, 92.

99.

Battery	Variance from 10,000 (in Ω)	Measured Resistance	Ascending Order
Resistor 1	175	10,175	9,698 Ω
Resistor 2	−60	9,940	9,812 Ω
Resistor 3	−188	9,812	9,935 Ω
Resistor 4	10	10,010	9,940 Ω
Resistor 5	218	10,218	10,010 Ω
Resistor 6	−65	9,935	10,175 Ω
Resistor 7	−302	9,698	10,218 Ω

101. $|6 + (-2)| = 4$

or

$|6| + (-2) = 4$

or

$6 + (-|2|) = 4$

103. $\left|6\right|+\left|(-2)\right|=8$

or

$6+\left|(-2)\right|=8$

105. $\left|-6\right|+2=8$

or

$\left|-6\right|+\left|2\right|=8$

107. **(a)** The numbers -3 and 4 are plotted on the number line as,

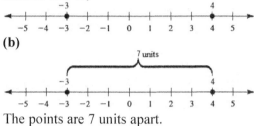

(b)

The points are 7 units apart.

109. Above and Beyond

Exercises 2.2

< Objectives 1 and 2 >

1. Add the absolute values $(3+6=9)$ and give the sum the sign $(+)$ of the original numbers: $3+6=9$.

3. Add the absolute values $(11+5=16)$ and give the sum the sign $(+)$ of the original numbers: $11+5=16$.

5. Add the absolute values $(2+3=5)$ and give the sum the sign $(-)$ of the original numbers: $-2+(-3)=-5$.

7. Subtract the absolute values $(9-3=6)$. The sum has the sign $(+)$ of the number with the larger absolute value, 9: $9+(-3)=6$.

9. Subtract the absolute values $(14-8=6)$. The sum has the sign $(-)$ of the number with the larger absolute value, -14: $8+(-14)=-6$.

11. Subtract the absolute values $(17-4=13)$. The sum has the sign $(+)$ of the number with the larger absolute value, 17: $-4+17=13$.

13. Subtract the absolute values $(15-8=7)$. The sum has the sign $(-)$ of the number with the larger absolute value, 15: $-15+6=-17$.

15. Add the absolute values $(13+24=37)$ and give the sum the sign $(-)$ of the original numbers.: $-10+(-24)=-37$.

17. Subtract the absolute values $(24-13=11)$. The sum has the sign $(+)$ of the number with the larger absolute value, 24: $-13+24=11$.

19. Subtract the absolute values $(45-36=9)$. The sum has the sign $(-)$ of the number with the larger absolute value, 45: $36+(-45)=-9$.

21. Add the absolute values $(458+179=637)$ and give the sum the sign $(-)$ of the original numbers: $-458+(-179)=-637$

23. Subtract the absolute values $(452-243=189)$. The sum has the sign $(+)$ of the number with the larger absolute value, 432: $432+(-243)=189$.

25. Subtract the absolute values $(689-471=218)$. The sum has the sign $(-)$ of the number with the larger absolute value, 689: $-689+471=-218$.

27. Subtract the absolute values $(1,104 - 732 = 372)$. The sum has the sign (+) of the number with the larger absolute value, $1,104$: $-732 + (1,104) = 372$.

29. Subtract the absolute values $(7,332 - 2,417 = 4,915)$. The sum has the sign (–) of the number with the larger absolute value, $7,332$: $2,417 + (-7,332) = -4,915$.

31. Add the absolute values $(1,056 + 4,879 = 5,935)$ and give the sum the sign (–) of the original numbers: $-1,056 + (-4,879) = -5,935$.

33. $-9 + 0 = -9$

35. $-14 + 14 = 0$

37. $-9 + (-17) + 9 = -9 + 9 + (-17)$
$$= (-9 + 9) + (-17)$$
$$= 0 + (-17) = -17$$

39. $8 + (-12) + (-4) = [8 + (-12)] + (-4)$
$$= -4 + (-4) = -8$$

41. $(-8) + (-17) + 5 = [(-8) + (-17)] + 5$
$$= -25 + 5 = -20$$

43. $-234 + 76 + (-48) = [-234 + (-48)] + 76$
$$= -282 + 76 = -206$$

45. $2 + 5 + (-11) + 4 = 2 + 5 + 4 + (-11)$
$$= (2 + 5 + 4) + (-11)$$
$$= 11 + (-11) = 0$$

47. $1 + (-2) + 3 + (-4) = -2 + 1 + 3 + (-4)$
$$= -2 + (1 + 3) + (-4)$$
$$= -2 + 4 + (-4)$$
$$= -2 + [4 + (-4)]$$
$$= -2 + 0 = -2$$

49. $-4 + 6 + (-3) + 0 = [-4 + 6] + (-3) + 0$
$$= (2) + (-3) + 0$$
$$= [(2) + (-3)] + 0$$
$$= (-1) + 0 = -1$$

51. $14 + (-7) + (-11) + 5$
$$= [14 + (-7)] + (-11) + 5 = (7) + (-11) + 5$$
$$= [7 + (-11)] + 5 = -4 + 5 = 1$$

53. $-272 + 951 + (-333) + (-129)$
$$= [-272 + 951] + (-333) + (-129)$$
$$= 679 + (-333) + (-129)$$
$$= [679 + (-333)] + (-129) = 346 + (-129)$$
$$= 217$$

55. $|3 + (-4)| = |-1| = 1$

57. $|-17 + 8| = |-9| = 9$

59. $|-5 + (-6)| = |-11| = 11$

61. $|-3 + 2 + (-4)| = |(-3 + 2) + (-4)|$
$$= |-1 + (-4)| = |-5| = 5$$

63. $|2 + (-3)| + |(-3) + 2| = |-1| + |-1| = 1 + 1 = 2$

< Objective 3 >

65. The points earned and lost by a football team are represented as,

$$(+3)+(-7)+(+3)+(-2)+(-3)$$
$$=\left[(+3)+(-7)\right]+(+3)+(-2)+(-3)$$
$$=\left[(-4)+(+3)\right]+(-2)+(-3)$$
$$=\left[(-1)+(-2)\right]+(-3)=(-3)+(-3)=-6$$

Therefore, the team lost by 6 points.

67. $8+(-23)=-15$

The low temperature was $-15°F$.

69. True

71. False

73. always

75. always

77. The integer expression that represents the change in the budget is,

$$50,000+(-1,000)+(-9,550)+(-542)$$
$$+(443)+(-123)+(-150)$$
$$=\left[50,000+(-1,000)\right]+(-9,550)+(-542)$$
$$+(443)+(-123)+(-150)$$
$$=\left[49,000+(-9,550)\right]+(-542)+(443)$$
$$+(-123)+(-150)$$
$$=\left[39,450+(-542)\right]+(443)+(-123)+(-150)$$
$$=\left[38,908+(443)\right]+(-123)+(-150)$$
$$=\left[39,351+(-123)\right]+(-150)$$
$$=39,228+(-150)=\$39,078$$

The amount of money left with Amir is $39,078.

79. Integer expression representing the change in the pressure of the reservoir is,

$$126+(-12)+(-7)+32+(-17)+(-15)+31$$
$$+(-4)+(-14)$$
$$=\left[126+(-12)\right]+(-7)+32+(-17)+(-15)$$
$$+31+(-4)+(-14)$$
$$=\left[114+(-7)\right]+32+(-17)+(-15)+31+(-4)$$
$$+(-14)$$
$$=\left[107+32\right]+(-17)+(-15)+31+(-4)+(-14)$$
$$=\left[139+(-17)\right]+(-15)+31+(-4)+(-14)$$
$$=\left[122+(-15)\right]+31+(-4)+(-14)$$
$$=\left[107+31\right]+(-4)+(-14)$$
$$=\left[138+(-4)\right]+(-14)=134+(-14)=120$$

The pressure in the tank at the end of the shift is 120 lb/in.^{2}

81. $24+(-12)=12$

The resulting voltage = 12 V.

83. $|-3|+|7|=10$

or

$|-3|+7=10$

85. $\left|-6+7+(-4)\right|=\left|(-6+7)+(-4)\right|=\left|1+(-4)\right|$
$$=\left|(-3)\right|=3$$

87. **(a)** Subtract the absolute values, $(14-8=6)$. The sum has the sign (+) of the number with the larger absolute value, 14: $14+(-8)=6$.

(b) Subtract the absolute values $(14-8=6)$ and give the sign (+) of the number with the larger absolute value, 14: $14-8=6$.

(c) They are the same.

(d) Above and Beyond

(e) Above and Beyond

89. Above and Beyond

Exercises 2.3

< Objectives 1 and 2 >

1. $21 - 13 = 21 + (-13) = 8$

3. $82 - 45 = 82 + (-45) = 37$

5. $8 - 10 = 8 + (-10) = -2$

7. $24 - 45 = 24 + (-45) = -21$

9. $-5 - 3 = -5 + (-3) = -8$

11. $-19 - 14 = -9 + (-14) = -23$

13. $3 - (-4) = 3 + 4 = 7$

15. $5 - (-11) = 5 + 11 = 16$

17. $7 - (-12) = 7 + 12 = 19$

19. $-36 - (-24) = -36 + 24 = -12$

21. $-19 - (-27) = -19 + 27 = 8$

23. $-11 - (-11) = -11 + 11 = 0$

25. $0 - (-8) = 0 + 8 = 8$

27. $-23 - (-18) = -23 + 18 = -5$

29. $-48 - (-61) = -48 + 61 = 13$

31. $-15 - (-34) + 8 = -15 + 34 + 8 = -15 + 42$
$\qquad = 27$

33. $54 + (-36) - 18 = \left[54 + (-36)\right] - 18$
$\qquad = 18 - 18 = 0$

35. $-64 + (-22) - (-18)$
$\qquad = \left[-64 + (-22)\right] - (-18) = -86 + 18 = -68$

37. $36 - 91 + (-34) - (-12)$
$\qquad = \left[36 - 91\right] + \left[(-34) - (-12)\right] = -55 - 22$
$\qquad = -77$

39. $448 + (-622) - 320 - (-216)$
$\qquad = \left[448 + (-622)\right] + \left[-320 - (-216)\right]$
$\qquad = -174 - 104 = -278$

< Objective 2 >

41. $\$853 + \$70 + \$70 = \$993$
His correct balance was \$993.

43. To find the temperature range in the United States on April 7, 2012, we take the high temperature, 91°F, and subtract the low temperature, −18°F, on that day. That gives us: $91 - (-18) = 91 + 18 = 109$. The temperature range in the United States on April 7, 2012 was 109°F.

45. To find the historical temperature range in the United States, we take the high temperature, 134°, and subtract the low temperature, −80°. That gives us: $134 - (-80) = 134 + 80 = 214$. The historical temperature range for the United States is 214°F.

47. To find the historical temperature range in the state of Minnesota, we take the high temperature, 114°, and subtract the low temperature, −60°. That gives us: $114 - (-60) = 119 + 60 = 174$. The historical temperature range for the state of Minnesota is 174°F.

49. To find the historical temperature range in the state of Ohio, we take the high temperature, 45°, and subtract the low temperature, −39°. That gives us: $45 - (-39) = 45 + 39 = 84$. The historical temperature range for the state of Ohio is 84°F.

51. $63 - (-21) = 63 + 21 = 84$

The total temperature drop was 84°.

53. $10 - (-5) = 10 + 5 = 15$

The difference between the high tide and the low tide is 15 ft.

55. $101 - (-8) = 101 + 8 = 109$

The difference between the high tide and the low tide (the exchange) is 109 in.

57. $(10 \cdot 50) + (12 \cdot 50) = 1,100$

$(100 \cdot 2) + (125 \cdot 2) + 150 + 175 + 225 + 250$

$= 1,250$

$1,250 - 1,100 = 150$

No. They received 1,100 lb of flour, but expect to need 1,125 lb.

59. 577 ft below sea level

61. sometimes

63. sometimes

65. With a scientific calculator:

56 $\boxed{+}$ 123 $\boxed{=}$

Display: 179

With a graphing calculator:

56 $\boxed{+}$ 123 $\boxed{\text{ENTER}}$

Display: 179

67. With a scientific calculator:

720 $\boxed{+}$ 458 $\boxed{+/-}$ $\boxed{=}$

Display: 262

With a graphing calculator:

729 $\boxed{+}$ $\boxed{(-)}$ 458 $\boxed{\text{ENTER}}$

Display: 262

69. With a scientific calculator:

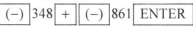
104 $\boxed{+/-}$ $\boxed{+}$ 783 $\boxed{=}$

Display: 679

With a graphing calculator:

$\boxed{(-)}$ 104 $\boxed{+}$ 783 $\boxed{\text{ENTER}}$

Display: 679

71. With a scientific calculator:

348 $\boxed{+/-}$ $\boxed{+}$ 861 $\boxed{+/-}$ $\boxed{=}$

Display: −1209

With a graphing calculator:

$\boxed{(-)}$ 348 $\boxed{+}$ $\boxed{(-)}$ 861 $\boxed{\text{ENTER}}$

Display: −1209

73. With a scientific calculator:

56 $\boxed{-}$ 123 $\boxed{=}$

Display: −67

With a graphing calculator:

56 $\boxed{-}$ 123 $\boxed{\text{ENTER}}$

Display: −67

75. With a scientific calculator:

720 $\boxed{-}$ 458 $\boxed{+/-}$ $\boxed{=}$

Display: 1178

With a graphing calculator:

720 $\boxed{-}$ $\boxed{(-)}$ 458 $\boxed{\text{ENTER}}$

Display: 1178

77. With a scientific calculator:

104 $\boxed{+/-}$ $\boxed{-}$ 783 $\boxed{=}$

Display: −887

With a graphing calculator:

$\boxed{(-)}$ 104 $\boxed{-}$ 783 $\boxed{\text{ENTER}}$

Display: −887

79. With a scientific calculator:

648 $\boxed{+/-}$ $\boxed{-}$ 861 $\boxed{+/-}$ $\boxed{=}$

Display: 513

With a graphing calculator:

$\boxed{(-)}$ 348 $\boxed{-}$ $\boxed{(-)}$ 861 $\boxed{\text{ENTER}}$

Display: 513

81. With a scientific calculator:

8 $\boxed{+}$ 4 $\boxed{-}$ 3 $\boxed{+/-}$ $\boxed{-}$ 2 $\boxed{=}$

Display: 13

With a graphing calculator:

8 $\boxed{+}$ 4 $\boxed{(-)}$ 3 $\boxed{-}$ 2 $\boxed{\text{ENTER}}$

Display: 13

83. With a scientific calculator:

145 $\boxed{-}$ 547 $\boxed{+/-}$ $\boxed{+}$ 92 $\boxed{+/-}$ $\boxed{-}$ 234 $\boxed{=}$

Display: 366

With a graphing calculator:

145 $\boxed{-}$ $\boxed{(-)}$ 547 $\boxed{+}$ $\boxed{(-)}$ 92 $\boxed{-}$ 234

$\boxed{\text{ENTER}}$

Display: 366

85. The difference in the elevation
$= 311 - 362 = -51 \text{ inches}$.

87. Change in steel inventory
$= 2,581 - 2,489 = +92 \text{ pounds}$.

89. Change in the rpm after loading
$= 4,250 - 5,400 = -1,150 \text{ rpm}$.

91. Above and Beyond

93. Above and Beyond

95. Above and Beyond

Exercises 2.4

< Objective 1 >

1. $4 \cdot 10 = 40$

3. $(5)(-12) = -60$

5. $(-8)(10) = -80$

7. $(-8)(9) = -72$

9. $(-11)(12) = -132$

11. $(-8)(-7) = 56$

13. $(-5)(-12) = 60$

15. $(1)(-18) = -18$

17. $(-5)(0) = 0$

< Objective 2 >

19. $\dfrac{70}{14} = 5$

21. $\dfrac{-20}{-4} = 5$

23. $-24 \div 8 = -3$

25. $\dfrac{50}{-5} = -10$

27. $\dfrac{0}{-8} = 0$

29. $-17 \div 1 = -17$

31. $\dfrac{-27}{-1} = 27$

33. $\dfrac{-10}{0}$ is undefined.

< Objective 3 >

35. $\left[(-5)(3)\right](-2) = (-15)(-2) = 30$

37. $\left[(8)(-3)\right](7) = (-24)(7) = -168$

39. $\left[(-3)(-5)\right](-2) = (15)(-2) = -30$

41. $\left[(-5)(-4)\right](2) = (20)(2) = 40$

43. $\left[(-9)(-12)\right](0) = (108)(0) = 0$

45. $-(-3) = 3$

47. $-(-(-1)) = -1$

49. $-(-(-(-(-123)))) = -123$

51. $-6^3 = -(6)^3 = -216$

53. $-6^2 = -(6)^2 = -36$

55. $(-8)^2 = (-8)(-8) = 64$

57. $-8^3 = -(8^3) = -(512) = -512$

59. $-10^2 = -(10^2) = -(100) = -100$

61. $-\dfrac{-52}{-13} = -4$

63. $5(7-2) = 5(5) = 25$

65. $2(5-8) = 2(-3) = -6$

67. $-3(9-7) = -3(2) = -6$

69. $-3(-2-5) = -3(-7) = 21$

71. $\dfrac{(-6)(-3)}{2} = \dfrac{18}{2} = 9$

73. $\dfrac{24}{-4-8} = \dfrac{24}{-12} = -2$

75. $\dfrac{55-19}{-12-6} = \dfrac{36}{-18} = -2$

77. $\dfrac{7-5}{2-2} = \dfrac{2}{0}$ is undefined.

79. $\dfrac{(-9)(-6)-10}{18-(-4)} = \dfrac{54-10}{18+4} = \dfrac{44}{22} = 2$

81. $(-2)(-7) + (2)(-3) = 14 + (-6) = 8$

83. $(-7)(3) - (-2)(-8) = -21 - 16 = -37$

85. $\dfrac{(3)(6)-(-4)(8)}{6-(-4)} = \dfrac{18-(-32)}{6+4} = \dfrac{50}{10} = 5$

87. $\dfrac{2(-5)+4(6-8)}{3(-4+2)} = \dfrac{2(-5)+4(-2)}{3(-2)}$

$\qquad = \dfrac{-10+(-8)}{-6} = \dfrac{-18}{-6} = 3$

89. $(-7)^2 - 17 = (-7)(-7) - 17 = 49 - 17 = 32$

91. $-7^2 - 17 = -(7)(7) - 17 = -49 - 17 = -66$

93. $(-4)^2 - (-2)(-5) = (-4)(-4) - (-2)(-5)$

$\qquad = 16 - (10) = 6$

95. $(-6)^2 - (-3)^2 = 36 - 9 = 27$

97. $(-8)^2 - 8^2 = 64 - 64 = 0$

99. $5 + 3 \cdot (4-6)^2 = 5 + 3 \cdot (-2)^2 = 5 + 3 \cdot 4$

$\qquad = 5 + 12 = 17$

101. $-20 \div 2 + 10 \cdot 2 = -10 + 20 = 10$

< Objective 4 >

103. For 14 correct answers and 4 incorrect answers, we have:
$5(14) - 2(4) = 70 - 8 = 62$. The 2 questions left blank are neither added nor subtracted, so the student scored a 62 on the exam.

105. $45 lost per hour over a 4-hour period would be: $(-45)(4) = -180$. The gambler lost $180.

107. Total amount of money left with Michelle
$= \$1,000 - \$100 - \$200 - \$56 = \$644$
Money left with her to use each week
$= \$644 \div 4 = \$161$

109. False

111. True

113. never

115. sometimes

117. The complete table is:

$4 \cdot 3$	12
$4 \cdot 2$	8
$4 \cdot 1$	4
$4 \cdot 0$	0
$4(-1)$	-4
$4(-2)$	-8
$4(-3)$	-12
$4(-4)$	-16

119. With a scientific calculator:
25 $\times$ 21 $+/-$ $=$
Result: -525
With a graphing calculator:
25 $\times$ $(-)$ 21 ENTER
Result: -525

121. With a scientific calculator:
392 $\div$ 14 $+/-$ $=$
Result: -28
With a graphing calculator:
392 $\div$ $(-)$ 14 ENTER
Result: -28

123. With a scientific calculator
4 $+/-$ y^x 5 $=$
Result: $-1,024$
With a graphing calculator
$($ $(-)$ 4 $)$ $\wedge$ 5 ENTER
Result: $-1,024$

125. With a scientific calculator
$-$ 5 y^x 4 $=$
Result: -625
With a graphing calculator
$-$ 5 $\wedge$ 4 ENTER
Result: -625

127. $3\left[8 + (-11)\right]^3$
With a scientific calculator:
3 $\times$ $($ 8 $+$ 11 $+/-$ $)$ x^y 3 $=$
Result: -81.
With a graphing calculator:
3 $\times$ $($ 8 $+$ $(-)$ 11 $)$ $\wedge$ 3 ENTER
Result: -81.

129. $\dfrac{15(-3) + (-5)(9)}{3 \cdot 2}$

With a scientific calculator:
$($ 15 $\times$ 3 $+/-$ $+$ 5 $+/-$ $\times$ 9 $)$
$\div$ $($ 3 $\times$ 2 $)$ $=$
Result: 15
With a graphing calculator:
$($ 15 $\times$ $(-)$ 3 $+$ $(-)$ 5 $\times$ 9 $)$
$\div$ $($ 3 $\times$ 2 $)$ ENTER
Result: 15

131.

Product	Profit (+) / Loss (-) (in $)	Units	Monthly Profit (+) / Loss (-) (in $)
A	18	127	2,286
B	-4	273	$-1,092$
C	11	201	2,211
D	38	377	14,326
E	-15	43	-645
Total Profit			17,086

The total profit for the month was $17,086.

133. Average amount of LP consumed each day
$= (1,789 - 676) \div 21 = 1,113 \div 21$
$= 53$ gal/day

135. Above and Beyond

137. Above and Beyond

Exercises 2.5

< Objective 1 >

1. The sum of c and d is written as $c + d$.

3. w plus z is written as $w + z$.

5. x increased by 2 is written as $x + 2$.

7. 10 more than y is written as $y + 10$.

9. a minus b is written as $a - b$.

11. b decreased by 7 is written as $b - 7$.

13. 6 less than r is written as $r - 6$.

15. w times z is written as wz.

17. The product of 5 and t is written as $5t$.

19. The product of 8, m, and n is written as $8mn$.

21. The product of 3 and the quantity p plus q is written as $3(p + q)$.

23. Twice the sum of x and y is written as $2(x + y)$.

25. The sum of twice x and y is written as $2x + y$.

27. Twice the difference of x and y is written as $2(x - y)$.

29. The quantity a plus b times the quantity a minus b is written as $(a + b)(a - b)$.

31. The product of m and 3 less than m is written as $m(m - 3)$.

33. x divided by 5 is written as $\dfrac{x}{5}$.

35. The quotient of a plus b, and 7 is written as $\dfrac{a + b}{7}$.

37. The difference of p and q, divided by 4 is written as $\dfrac{p - q}{4}$.

39. The sum of a and 3, divided by the difference of a and 3 is written as $\dfrac{a + 3}{a - 3}$.

41. 5 more than a number is written as $x + 5$.

43. 7 less than a number is written as $x - 7$.

45. 9 times a number is written as $9x$.

47. 6 more than 3 times a number is written as $3x + 6$.

49. Twice the sum of a number and 5 is written as $2(x + 5)$.

51. The product of 2 more than a number and 2 less than that same number is written as $(x + 2)(x - 2)$.

53. The quotient of a number and 7 is written as $\dfrac{x}{7}$.

55. The sum of a number and 5, divided by 8 is written as $\dfrac{x + 5}{8}$.

57. 6 more than a number divided by 6 less than that same number is written as $\dfrac{x + 6}{x - 6}$.

59. Four times the length of a side is written as $4s$.

61. Twice the sum of length L and the width W is written as $2(L+W)$.

63. One-half the product of the height h and the sum of two unequal sides b_1 and b_2 is written as $\dfrac{1}{2}h(b_1+b_2)$.

< Objective 2 >

65. $2(x+5)$ is an expression. Its meaning is clear.

67. $4+\div m$ is not an expression. The two operations in a row have no meaning.

69. $2b=6$ is not an expression. The equals sign is not an operation sign.

71. $2a+5b$ is an expression. Its meaning is clear.

< Objective 3 >

73. The other number expressed in terms of x, is $35-x$.

75. The expression for the interest earned is Prt.

77. The expression for the number of student tickets sold is $400-x$.

79. (b)

81. (b)

83. True

85. never

87. The standard dosage formula is $\dfrac{DQ}{H}$.

89. The formula that describes the number of carriage bolts sold last month is $H-284$.

91. Above and Beyond

Exercises 2.6

< Objectives 1 and 2 >

1. $3c-2b=3(-4)-2(5)=-12-10=-22$

3. $8b+2c=8(5)+2(-4)=40-8=32$

5. $-b^2+b=-(5)^2+5=-25+5=-20$

7. $3a^2=3(-2)^2=3(4)=12$

9. $c^2-2d=(-4)^2-2(6)=16-12=4$

11. $2a^2+3b^2=2(-2)^2+3(5)^2=2(4)+3(25)$
$=8+75=83$

13. $2(a+b)=2(-2+5)=2(3)=6$

15. $4(2a-d)=4(2(-2)-6)=4(-4-6)$
$=4(-10)=-40$

17. $a(b+3c)=-2(5+3(-4))=-2(5-12)$
$=-2(-7)=14$

19. $\dfrac{6d}{c}=\dfrac{6(6)}{-4}=\dfrac{36}{-4}=-9$

21. $\dfrac{3d+2c}{b}=\dfrac{3(6)+2(-4)}{5}=\dfrac{18-8}{5}=\dfrac{10}{5}=2$

23. $\dfrac{2b-3a}{c+2d}=\dfrac{2(5)-3(-2)}{-4+2(6)}=\dfrac{10+6}{-4+12}=\dfrac{16}{8}=2$

25. $d^2-b^2=(6)^2-(5)^2=36-25=11$

27. $(d-b)^2=(6-5)^2=1^2=1$

29. $(d-b)(d+b)=(6-5)(6+5)=(1)(11)$
$=11$

31. $d^3 - b^3 = (6)^3 - (5)^3 = 216 - 125 = 91$

33. $(d-b)^3 = (6-5)^3 = 1^3 = 1$

35. $(d-b)(d^2 + db + b^2)$
$= (6-5)(6^2 + (6)(5) + 5^2)$
$= (1)(36 + 30 + 25) = (1)(91) = 91$

37. $b^2 + a^2 = (5)^2 + (-2)^2 = 25 + 4 = 29$

39. $(b+a)^2 = (5 + (-2))^2 = 3^2 = 9$

41. $a^2 + 2ad + d^2 = (-2)^2 + 2(-2)(6) + (6)^2$
$= 4 - 24 + 36 = 16$

43. $x^2 - 2y^2 + z^2 = (-2)^2 - 2(-3)^2 + (4)^2$
$= 4 - 18 + 16 = 2$

45. $2xy - (x^2 - 2yz)$
$= 2(-2)(-3) - [(-2)^2 - 2(-3)(4)]$
$= 12 - (4 + 24) = 12 - 28 = -16$

47. $2y(z^2 - 2xy) + yz^2$
$= 2(-3)[(4)^2 - 2(-2)(-3)] + (-3)(4)^2$
$= -6(16 - 12) + (-3)(16) = -24 - 48 = -72$

< Objective 3 >

49. Area $(A) = \dfrac{1}{2}bh$

$A = \dfrac{1}{2}(4)(8) = 2(8) = 16 \text{ cm}^2$

51. $F = \dfrac{9}{5}C + 32 = \dfrac{9}{5}(-10) + 32 = -18 + 32$
$= 14°F$
The Fahrenheit equivalent is 14°F.

53. Height $= 110t - 16t^2 + 6$
$= 110 \cdot 2 - 16(2)^2 + 6 = 220 - 64 + 6$
$= 162$
The height of the ball is 162 ft.

55. Height $= 110t - 16t^2 + 6$
$= 110 \cdot 4 - 16(4)^2 + 6$
$= 440 - 256 + 6 = 190$
The height of the ball is 192 ft.

57. True
$$22 - 7 \overset{?}{=} 2(5) + 5$$
$$15 \overset{?}{=} 10 + 5$$
$$15 = 15$$
The given values for the variables make the statement true.

59. False
$$2[-4 + (-2)] \overset{?}{=} 2(-4) + (-2)$$
$$2(-6) \overset{?}{=} -8 - 2$$
$$-12 \neq -10$$
The given values for the variables make the statement false.

61. False; dividing the numerator by the denominator is the last step.

63. never

65. $3x(2y + z) = 3(-15)[2(-8) + 2]$
$= -45[-16 + 2] = -45(-14)$
$= 630$

67. $\dfrac{2z - 3y}{x + 1} = \dfrac{2(2) - 3(-8)}{-15 + 1} = \dfrac{4 + 24}{-14} = \dfrac{28}{-14}$
$= -2$

69. $x^2 - 4x^3 + 3x = (3)^2 - 4(3)^3 + 3(3)$
$= 9 - 108 + 9 = -90$

71. $x^2 - 4x^3 + 3x = (27)^2 - 4(27)^3 + 3(27)$

$\qquad = 729 - 78,732 + 81$

$\qquad = -77,922$

73. $-2t^2 + 13t + 1 = -2(1)^2 + 13(1) + 1$

$\qquad = -2(1) + 13(1) + 1$

$\qquad = (-2 + 13) + 1 = 11 + 1 = 12$

(For $t = 1$)

Therefore, the approximate concentration of the antihistamine after one hour is $12\mu g/mL$.

75. $\dfrac{rT}{5,252} = \dfrac{2,626 \times 6}{5,252} = \dfrac{15,756}{5,252} = 3$

77. Above and Beyond

Exercises 2.7

< Objective 1 >

1. $5a + 2$ has two terms: $5a$ and 2.

3. $4x^3$ has one term: $4x^3$.

5. $3x^2 + 3x + (-7)$ has three terms: $3x^2$, $3x$, and -7.

< Objective 2 >

7. In the group of terms $5ab$, $3b$, $3a$, $4ab$, the like terms are $5ab$ and $4ab$, since these terms contain exactly the same letters raised to the same powers.

9. In the group of terms $4xy^2$, $2x^2y$, $5x^2$, $-3x^2y$, $5y$, $6x^2y$, the like terms are $2x^2y$, $-3x^2y$, and $6x^2y$, since these terms contain exactly the same letters raised to the same powers.

< Objective 3 >

11. $3m + 7m = (3 + 7)m = 10m$

13. $7b^3 + 10b^3 = (7 + 10)b^3 = 17b^3$

15. $21xyz + 7xyz = (21 + 7)xyz = 28xyz$

17. $9z^2 + (-3z^2) = (9 - 3)z^2 = 6z^2$

19. $5a^3 + (-5a^3) = (5 - 5)a^3 = 0a^3 = 0$

21. $19n^2 + (-18n^2) = (19 - 18)n^2 = 1n^2 = n^2$

23. $21p^2q + (-6p^2q) = (21 - 6)p^2q = 15p^2q$

25. $10x^2 + (-7x^2) + 3x^2 = (10 + (-7) + 3)x^2$

$\qquad = 6x^2$

27. $9a + (-7a) + 4b = (9 - 7)a + 4b = 2a + 4b$

29. $7x + 5y + (-4x) + (-4y)$

$\qquad = (7x + (-4x)) + (5y + ((-4y))$

$\qquad = (7 - 4)x + (5 - 4)y = 3x + y$

31. $4a + 7b + 3 + (-2a) + 3b + (-2)$

$\qquad = (4a + (-2a)) + (7b + 3b) + (3 + -2)$

$\qquad = (4 - 2)a + (7 + 3)b + 1 = 2a + 10b + 1$

33. $5a^4 + 8a^4 = (5 + 8)a^4 = 13a^4$

35. $15a^3 - 12a^3 = (15 - 12)a^3 = 3a^3$

37. $(8x + 3x) - 4x = (8 + 3)x - 4x = (11 - 4)x$

$\qquad = 7x$

39. $(9mn^2 + 5mn^2) - 3mn^2$

$\qquad = (9 + 5)mn^2 - 3mn^2 = (14 - 3)mn^2$

$\qquad = 11mn^2$

41. $2(3x + 2) + 4 = 6x + (4 + 4) = 6x + 8$

43. $5(6a-2)+12a = 30a-10+12a$
$$= (30+12)a-10$$
$$= 42a-10$$

45. $4s+2(s+4)+4 = 4s+2s+8+4$
$$= (4+2)s+12 = 6s+12$$

47. $7a^2+3a = 7(2)^2+3(2) = 7(4)+6 = 28+6$
$$= 34$$

49. $3c^2+5c^2 = 8c^2 = 8(5)^2 = 8(25) = 200$

51. $5b+3a-2b = 3b+3a = 3(3)+3(2) = 9+6$
$$= 15$$

53. $5ac^2-2ac^2 = 3ac^2 = 3(2)(5)^2 = 3(2)(25)$
$$= 150$$

55. For a rectangle opposite sides are equal. To find the perimeter of the rectangle add all 4 sides:
$$2 \bullet (8x+9)+2 \bullet (6x-7)$$
$$= 16x+18+12x-14$$
$$= (16+12)x+(18-14) = 28x+4$$

57. $\left(3y^2+3y-2\right)-\left(y^2-8\right)$
$$= (3-1)y^2+3y-2-(-8)$$
$$= 2y^2+3y+(8-2) = 2y^2+3y+6$$
The remaining piece of beam measures $\left(2y^2+3y+6\right)$ m.

59. $x+(3x+3)+\left(2x^2-5x+1\right)$
$$= x+3x+3+2x^2-5x+1$$
$$= 2x^2+(1+3-5)x+3+1 = 2x^2-x+4$$
The expression for the perimeter of the triangle is $\left(2x^2-x+4\right)$ ft.

61. Adding lengths of each side,
$$6y+2+10+3y+10+5+(5y+2)+8y$$
$$= (6+3+5+8)y+2+10+10+5+2$$
$$= 22y+29$$
The expression for the perimeter of the figure is $(22y+29)$ cm.

63. $\left(90x-x^2\right)-(150+25x)$
$$= -x^2+(90-25)x-150 = -x^2+65x-150$$
Therefore, the expression that represents the profit is $-x^2+65x-150$.

65. False; the variables and their exponents must match.

67. always

69. $54p+32p = (54+32)p = 86p$
The total load that the beam can support is $86p$.

71. The store starts out with 4 Frisbees and 8 basketballs. Add the number of Frisbees and basketballs received in shipment and subtract any that are sold. That gives us:
4 Frisbees + 8 basketballs – 2 Frisbees
– 1 basketball + 6 Frisbees
– 3 Frisbees – 2 basketballs.
We will us f to represent the number of Frisbees and b to represent the number of basketballs and combine any like terms using the commutative property.
$$4f+8b-2f-1b+6f-3f-2b$$
$$= (4-2+6-3)f+(8-1-2)b = 5f+5b$$

73. Above and Beyond

75. Above and Beyond

77. Above and Beyond

Exercises 2.8

< Objective 1 >

1. $x + 4 = 9$

Right Side   Left side

$$x + 4 \overset{?}{=} 9$$
$$5 + 4 \overset{?}{=} 9$$
$$9 = 9$$

Because $9 = 9$ is a true statement, 5 is a solution of the equation.

3. $x - 15 = 6$

Right Side   Left side

$$x - 15 \overset{?}{=} 6$$
$$-21 - 15 \overset{?}{=} 6$$
$$-36 = 6$$

Because the two sides do not name the same number, we do not have a true statement, and −21 is not a solution.

5. $5 - x = 2$

Right Side   Left side

$$5 - x \overset{?}{=} 2$$
$$5 - 4 \overset{?}{=} 2$$
$$1 \neq 2$$

Because the two sides do not name the same number, we do not have a true statement, and 4 is not a solution.

7. $4 + (-x) = 6$

Right Side   Left side

$$4 - x \overset{?}{=} 6$$
$$4 - (-2) \overset{?}{=} 6$$
$$6 = 6$$

Because $6 = 6$ is a true statement, −2 is a solution of the equation.

9. $3x + 4 = 13$

Right Side   Left side

$$3x + 4 \overset{?}{=} 13$$
$$3(8) + 4 \overset{?}{=} 13$$
$$24 + 4 \overset{?}{=} 13$$
$$28 \neq 13$$

Because the two sides do not name the same number, we do not have a true statement, and 8 is not a solution.

11. $4x - 5 = 7$

Right Side   Left side

$$4x - 5 \overset{?}{=} 7$$
$$4(2) - 5 \overset{?}{=} 7$$
$$8 - 5 \overset{?}{=} 7$$
$$3 \neq 7$$

Because the two sides do not name the same number, we do not have a true statement, and 2 is not a solution.

13. $5 - 2x = 7$

Right Side   Left side

$$5 - 2x \overset{?}{=} 7$$
$$5 - 2(-1) \overset{?}{=} 7$$
$$5 + 2 \overset{?}{=} 7$$
$$7 = 7$$

Because $7 = 7$ is a true statement, −1 is a solution of the equation.

15. $4x - 5 = 2x + 3$

Right Side   Left side

$$4x - 5 \overset{?}{=} 2x + 3$$
$$4(4) - 5 \overset{?}{=} 2(4) + 3$$
$$16 - 5 \overset{?}{=} 8 + 3$$
$$11 = 11$$

Because $11 = 11$ is a true statement, 4 is a solution of the equation.

17. $x + 3 + 2x = 5 + x + 8$

Right Side   Left side

$x + 3 + 2x \overset{?}{=} 5 + x + 8$

$3x + 3 \overset{?}{=} x + 13$

$3(5) + 3 \overset{?}{=} 5 + 13$

$18 = 18$

Because $18 = 18$ is a true statement, 5 is a solution of the equation.

19. $2x + 1 = 9$ is a linear equation.

21. $2x - 8$ is an expression.

23. $7x + 2x + 8 - 3$ is an expression.

25. $2x - 8 = 3$ is a linear equation.

< Objective 2 >

27. $x + 9 = 11$ Check: $2 + 9 \overset{?}{=} 11$

$\underline{\quad -9 \quad -9}$ $11 = 11$

$x \quad = \quad 2$

The solution is 2.

29. $x + (-8) = 3$ Check: $11 + (-8) \overset{?}{=} 3$

$\underline{\quad +8 \quad +8}$ $3 = 3$

$x \qquad = 11$

The solution is 11.

31. $x + (-8) = -10$ Check: $-2 + (-8) \overset{?}{=} -10$

$\underline{\quad +8 \qquad +8}$ $-10 = -10$

$x \qquad = -2$

The solution is –2.

33. $x + 4 = -3$ Check: $-7 + 4 \overset{?}{=} -3$

$\underline{\quad -4 \quad -4}$ $-3 = -3$

$x \quad = -7$

The solution is –7.

35. $11 = x + 5$ Check: $11 \overset{?}{=} 6 + 5$

$\underline{\;-5 \qquad -5}$ $11 = 11$

$6 = x$

The solution is 6.

37. $4x = 3x + 4$ Check: $4(4) \overset{?}{=} 3(4) + 4$

$\underline{-3x \quad -3x}$ $16 = 16$

$x = \qquad 4$

The solution is 4.

39. $11x = 10x + (-10)$

$\underline{-10x \quad -10x}$

$x = \qquad\qquad -10$

Check: $11(-10) \overset{?}{=} 10(-10) + (-10)$

$-110 = -100 - 10$

$-110 = -110 - 10$

The solution is –10.

41. $6x + 3 = 5x$ Check: $6(-3) + 3 \overset{?}{=} 5(-3)$

$\underline{-5x \qquad -5x}$ $-18 + 3 = -15$

$x + 3 = \quad 0$ $-15 = -15$

$\underline{\qquad -3 \quad -3}$

$x \quad = \quad -3$

The solution is –3.

43. $8x + (-4) = 7x$

$\underline{-7x \qquad\qquad -7x}$

$x + (-4) = \quad 0$

$\underline{\qquad +4 \qquad +4}$

$x \qquad = \quad 4$

Check: $8(4) + (-4) \overset{?}{=} 7(4)$

$32 - 4 = 28$

$28 = 28$

The solution is 4.

45.

$$2x + 3 = x + 5$$

$$\underline{-x \qquad -x}$$

$$x + 3 = 5$$

$$\underline{-3 \qquad -3}$$

$$x = 2$$

The solution is 2.

Check: $2(2) + 3 \overset{?}{=} 2 + 5$

$$4 + 3 = 7$$

$$7 = 7$$

47.

$$5x + (-7) = 4x - 3$$

$$\underline{-4x \qquad -4x}$$

$$x + (-7) = -3$$

$$\underline{+7 \qquad +7}$$

$$x = 4$$

Check: $5(4) + (-7) \overset{?}{=} 4(4) + (-3)$

$$20 - 7 = 16 - 3$$

$$13 = 13$$

The solution is 4.

49.

$$7x + (-2) = 6x + 4$$

$$\underline{-6x \qquad -6x}$$

$$x + (-2) = 4$$

$$\underline{+2 \qquad +2}$$

$$x = 6$$

Check: $7(6) + (-2) \overset{?}{=} 6(6) + 4$

$$42 - 2 = 36 + 4$$

$$40 = 40$$

The solution is 6.

51.

$$3 + 6x + 2 = 3x + 11 + 2x$$

$$6x + 5 = 5x + 11$$

$$\underline{-5x \qquad -5x}$$

$$x + 5 = 11$$

$$\underline{-5 \qquad -5}$$

$$x = 6$$

Check: $3 + 6(6) + 2 \overset{?}{=} 3(6) + 11 + 2(6)$

$$3 + 36 + 2 \overset{?}{=} 18 + 11 + 12$$

$$41 = 41$$

The solution is 6.

53.

$$4x + 7 + 3x = 5x + 13 + x$$

$$7x + 7 = 6x + 13$$

$$\underline{-6x \qquad -6x}$$

$$x + 7 = 13$$

$$\underline{-7 \qquad -7}$$

$$x = 6$$

Check: $4(6) + 7 + 3(6) \overset{?}{=} 5(6) + 13 + 6$

$$24 + 7 + 18 \overset{?}{=} 30 + 13 + 6$$

$$49 = 49$$

The solution is 6.

55.

$$3x + (-5) + 2x + (-7) + x = 5x + 2$$

$$6x + (-12) = 5x + 2$$

$$\underline{-5x \qquad -5x}$$

$$x + (-12) = 2$$

$$\underline{+12 \qquad +12}$$

$$x = 14$$

Check:

$$3(14) + (-5) + 2(14) + (-7) + 14 \overset{?}{=} 5(14) + 2$$

$$42 + (-5) + 28 + (-7) + 14 \overset{?}{=} 70 + 2$$

$$72 = 72$$

The solution is 14.

57.

$$4(3x + 4) = 11x + (-2)$$

$$12x + 16 = 11x + (-2)$$

$$\underline{-11x \qquad -11x}$$

$$x + 16 = -2$$

$$\underline{-16 \qquad -16}$$

$$x = -18$$

Check: $4(3(-18) + 4) \overset{?}{=} 11(-18) + (-2)$

$$4(-54 + 4) \overset{?}{=} -198 + (-2)$$

$$-200 = -200$$

The solution is -18.

59.
$$3(7x+2) = 5(4x+1)+17$$

$$
\begin{array}{rcl}
21x + 6 &=& 20x + 5 + 17 \\
21x + 6 &=& 20x \quad + 22 \\
\underline{20x \qquad -20x} & & \\
x + 6 &=& \qquad 22 \\
\underline{\;-6 \qquad -6} & & \\
x &=& \qquad 16
\end{array}
$$

Check: $3\bigl(7(16)+2\bigr) \overset{?}{=} 5\bigl(4(16)+1\bigr)+17$

$$3(112+2) \overset{?}{=} 5(64+1)+17$$

$$342 = 342$$

The solution is 16.

61. $x+3 = 7$

63. $3x - 7 = 2x$

65. $2(x+5) = x+18$

< Objective 3 >

67.
$$
\begin{aligned}
x + 7 &= 33 \\
x + 7 - 7 &= 33 - 7 \\
x &= 26
\end{aligned}
$$
The number is 26.

69.
$$
\begin{aligned}
x - 15 &= 7 \\
x - 15 + 15 &= 7 + 15 \\
x &= 22
\end{aligned}
$$
The number is 22.

71. Let x be the number of votes received by the losing candidate. Then $x + 1{,}840$ is the total number of votes cast.
$$
\begin{aligned}
x + 1{,}840 &= 3{,}260 \\
x + 1{,}840 - 1{,}840 &= 3{,}260 - 1{,}840 \\
x &= 1{,}420
\end{aligned}
$$
The losing candidate received 1,420 votes.

73. Let \$$x$ be the cost of the dryer. Then $x + 360$ is the cost of the washer and dryer combined.
$$
\begin{aligned}
x + 360 &= 650 \\
x + 360 - 360 &= 650 - 360 \\
x &= 290
\end{aligned}
$$
The dryer costs \$290.

75. Let b be the number of bones he needs. Then $b + 50$ is the total number of bones he needs for the costume.
$$
\begin{array}{rcl}
b + 50 &=& 62 \\
\underline{\;-50 \quad -50} & & \\
b &=& 12
\end{array}
$$
Jeremiah needs 12 more bones.

77.
$$
\begin{array}{rcl}
8x + 5 &=& 9x + (-4) \\
\underline{\;+4 \qquad +4} & & \\
8x + 9 &=& 9x
\end{array}
$$
(c) is equivalent to the given equation.

79.
$$
\begin{array}{rcl}
12x + (-6) &=& 8x + 14 \\
\underline{-8x \qquad -8x} & & \\
4x + (-6) &=& \quad 14
\end{array}
$$
(a) is equivalent to the given equation.

81. True

83. True

85. sometimes

87. **(a)** Let c be the number of cases of carriage bolts sold and h be the number of cases of hex bolts sold. Then $c = h - 284$.
(b)
$$
\begin{array}{rcl}
c &=& h - 284 \\
2{,}680 &=& h - 284 \\
\underline{+284 \qquad +284} & & \\
2{,}964 &=& h
\end{array}
$$
They sold 2,964 cases.

89. Let p be the profit and s be the number of servers.

Then the equation for profit earned from the sale of servers is $p = 560s - 4,500$.

$p = 560s - 4,500 = 560(15) - 4,500$

$\quad = 8,400 - 4,500 = 3,900$

A profit of $3,900 is earned on the sale of 15 servers. Therefore, they did not earn a profit of $5,000.

91. Above and Beyond

93. Above and Beyond

95. Above and Beyond

97. Above and Beyond

Summary Exercises

1. 6, –18, –3, 2, 15, –9

3. 5 is the maximum and –6 is the minimum.

5. $|9| = 9$

7. $-|9| = -9$

9. $|12 - 8| = |4| = 4$

11. $-|8 - 12| = -|-4| = -4$

13. Add the absolute values $(3 + 8) = 11$ and give the sum the sign (–) of the original numbers: $-3 + (-8) = -11$.

15. $6 + (-6) = 0$

17. $-18 + 0 = -18$

19. $8 - 13 = 8 + (-13) = -5$

21. $10 - (-7) = 10 + 7 = 17$

23. $-9 - (-9) = -9 + 9 = 0$

25. $|4 - 8| = |4 + (-8)| = |-4| = 4$

27. $|-4 - 8| = |-4 + (-8)| = |-12| = 12$

29. $-6 - (-2) + 3 = [-6 - (-2)] + 3$
$\quad = (-6 + 2) + 3 = -4 + 3 = -1$

31. $-8 - (-7) = -8 + 7 = -1$

33. To find the temperature range in the United States on February 13, 2012, we take the high temperature, 84°F, and subtract the low temperature, −16°F on that day. That gives us: $84 - (-16) = 84 + 16 = 100°F$.

35. $(10)(-7) = -70$

37. $(-3)(-15) = 45$

39. $(0)(-8) = 0$

41. $\dfrac{80}{16} = 5$

43. $\dfrac{-81}{-9} = 9$

45. $\dfrac{32}{-8} = -4$

47. $\dfrac{-8 + 6}{-8 - (-10)} = \dfrac{-8 + 6}{-8 + 10} = \dfrac{-2}{2} = -1$

49. $\dfrac{(-5)^2 - (-2)^2}{-5 - (-2)} = \dfrac{25 - 4}{-5 + 2} = \dfrac{21}{-3} = -7$

51. $(2)(-3) - (-5)(-3) = -6 - 15 = -21$

53. 5 more than y is written as $y + 5$.

55. The product of 8 and a is written as $8a$.

57. 5 times the product of m and n is written as $5mn$.

59. 3 more than the product of 17 and x is written as $17x + 3$.

61. $3x + w = 3(-3) + 2 = -9 + 2 = -7$

63. $x + y + (-3z) = -3 + 6 + (-3(-4))$
$= -3 + 6 + 12 = 15$

65. $3x^2 + (-2w^2) = 3(-3)^2 + (-2(2)^2)$
$= 3(9) + (-2)(4) = 27 + (-8)$
$= 19$

67. $5(x^2 + (-w^2)) = 5((-3)^2 + (-2^2))$
$= 5(9 - 4) = 5(5) = 25$

69. $\dfrac{2x + (-4z)}{y + (-z)} = \dfrac{2(-3) + (-4(-4))}{6 + (-(-4))} = \dfrac{-6 + 16}{6 + 4}$
$= \dfrac{10}{10} = 1$

71. $\dfrac{x(y^2 + (-z^2))}{(y + z)(y + (-z))}$
$= \dfrac{-3(6^2 + (-(-4)^2))}{(6 + (-4))(6 + (-(-4)))}$
$= \dfrac{-3(36 + (-16))}{(2)(10)} = \dfrac{-3(20)}{20} = -3$

73. $4a^3 - 3a^2$ has two terms: $4a^3$ and $-3a^2$.

75. In the group of terms $5m^2$, $-3m$, $-4m^2$, $5m^3$, m^2 the like terms are $5m^2$, $-4m^2$, and m^2, since these terms contain exactly the same letters raised to the same powers.

77. $5c + 7c = 12c$

79. $4a + (-2a) = 2a$

81. $9xy + (-6xy) = 3xy$

83. $7a + 3b + 12a + (-2b)$
$= (7a + 12a) + (3b + (-2b)) = 19a + b$

85. $5x^3 + 17x^2 + (-2x^3) + (-8x^2)$
$= 5x^3 + (-2x^3) + 17x^2 + (-8x^2)$
$= 3x^3 + 9x^2$

87. $(2a^3 + 12a^3) - 4a^3 = 14a^3 - 4a^3 = 10a^3$

89. $7x + 2 = 16$ (2)

$7(2) + 2 \overset{?}{=} 16$

$14 + 2 \overset{?}{=} 16$

$16 = 16$
Because $16 = 16$ is a true statement, 2 is a solution of the equation.

91. $7x + (-2) = 2x + 8$ (2)

$7(2) + (-2) \overset{?}{=} 2(2) + 8$

$14 + (-2) \overset{?}{=} 4 + 8$

$12 = 12$
Because $12 = 12$ is a true statement, 2 is a solution of the equation.

93. $x + 5 + 3x = 2 + x + 23$ (6)
$4x + 5 = x + 25$

$4(6) + 5 \overset{?}{=} 6 + 25$

$29 \neq 31$
Because the two sides do not name the same number, we do not have a true statement, and 6 is not a solution of the equation.

95. $x + 5 = 7$ Check: $2 + 5 \overset{?}{=} 7$
$\underline{-5 \quad -5}$ $7 = 7$
$x = 2$
The solution is 2.

97.
$$5x = 4x + (-5)$$
$$\underline{-4x \quad -4x}$$
$$x = -5$$
Check: $5(-5) \overset{?}{=} 4(-5) + (-5)$

$$-25 \overset{?}{=} -20 + (-5)$$

$$-25 = -25$$
The solution is –5.

99.
$$5x + (-3) = 4x + 2$$
$$\underline{-4x \qquad -4x}$$
$$x + (-3) = 2$$
$$\underline{\quad +3 \qquad +3}$$
$$x = 5$$
Check: $5(5) + (-3) \overset{?}{=} 4(5) + 2$

$$25 + (-3) \overset{?}{=} 20 + 2$$

$$22 = 22$$
The solution is 5.

101.
$$7x + (-5) = 6x + (-4)$$
$$\underline{-6x \qquad -6x}$$
$$x + (-5) = -4$$
$$\underline{\quad +5 \qquad +5}$$
$$x = 1$$
Check: $7(1) + (-5) \overset{?}{=} 6(1) + (-4)$

$$7 + (-5) \overset{?}{=} 6 + (-4)$$

$$2 = 2$$
The solution is 1.

103.
$$4(2x + 3) = 7x + 5$$
$$8x + 12 = 7x + 5$$
$$\underline{-7x \qquad -7x}$$
$$x + 12 = 5$$
$$\underline{\quad -12 \qquad -12}$$
$$x = -7$$
Check: $4(2(-7) + 3) \overset{?}{=} 7(-7) + 5$

$$4(-14 + 3) \overset{?}{=} -49 + 5$$

$$4(-11) \overset{?}{=} -44$$

$$-44 = -44$$
The solution is –7.

Chapter Test 2

1. $5, -12, 4, -7, 18, -17$

3. $|7| = 7$

5. Add the absolute values $(8 + 5 = 13)$ and give the sum the sign (–) of the original numbers: $-8 + (-5) = -13$.

7. $9 - 15 = 9 + (-15) = -6$

9. $(-8)(5) = -40$

11. $\dfrac{75}{-3} = -25$

13. $|18 - 7| = |11| = 11$

15. Add the absolute values $(9 + 12 = 21)$ and give the sum the sign (–) of the original numbers: $(-9) + (-12) = -21$.

17. $-7 - (-7) = -7 + 7 = 0$

19. $\dfrac{-45}{9} = -5$

21. $-8 - (-3 + 7)^2 = -8 - (4)^2 = -8 - 16$
$= -8 + (-16) = -24$

23. 5 less than a is written as $a - 5$.

25. 4 times the sum of m and n is written as $4(m + n)$.

27. $7x + (-3) = 25$ (5)

$7(5) + (-3) \overset{?}{=} 25$

$35 + (-3) \overset{?}{=} 25$

$32 \neq 25$
Because the two sides do not name the same number, we do not have a true statement, and 5 is not a solution of the equation.

29. $4a + (-c) = 4(-2) + (-(-4)) = -8 + 4 = -4$

31. $6(2b + (-3c)) = 6(2(6) + (-3(-4)))$
$= 6(12 + 12) = 6(24) = 144$

33. $8a + 7a = (8 + 7)a = 15a$

35. $10x + 8y + 9x + (-3y)$
$= 10x + 9x + 8y + (-3y) = 19x + 5y$

37. $x + (-7) = 4$ Check: $11 + (-7) \overset{?}{=} 4$
$\quad\quad +7\quad +7$ $4 = 4$
$x \quad\quad\quad = 11$
The solution is 11.

39. $\begin{aligned} 9x + (-2) &= 8x + 5 \\ -8x \quad\quad\quad &\quad -8x \\ \hline x + (-2) &= \quad\quad 5 \\ +2 \quad\quad &\quad +2 \\ \hline x \quad\quad &= 7 \end{aligned}$

Check: $9(7) + (-2) \overset{?}{=} 8(7) + 5$

$63 + (-2) \overset{?}{=} 56 + 5$

$61 = 61$
The solution is 7.

41. The number of double lattes that shop must sell in order to break even $= \dfrac{210}{3} = 70$ double lattes.

Cumulative Review: Chapters 1–2

1. The place value of 7 is hundred thousands.

3. Two million, four hundred thirty thousand is written as 2,430,000.

5. $9 + 0 = 9$
The result of adding zero to a number is that number; this is the additive identity property.

7. $\begin{array}{r} \overset{2\;1}{593} \\ 275 \\ +\ \ 98 \\ \hline 966 \end{array}$

9. 5873
The number in the hundreds place is 8. The number to the right of the 8, 7, is greater than 4, so add 1 to the 8 and place all zeros to the right.
5873 is rounded to 5900.

11. ^{2}900
     3300
      800
     2100
   + 600
     7700

13. $\left|5\right|-\left|14\right|=5-14=5+(-14)=-9$

15. $-12+(-6)=-18$

17. $5+(-7)=-2$

19. $-8-(-8)=-8+8=0$

21. $(-8)(-12)=96$

23. $14\div(-7)=-2$

25. If $a=5$ and $d=-2$, then
   $6ad=6(5)(-2)=30(-2)=-60$.

27. If $c=4$ and $d=-2$, then
   $3(c-2d)=3\left[4-2(-2)\right]=3(4+4)=3(8)$
   $\qquad\qquad =24$.

29. $6x+14-3x+5=(6x-3x)+(14+5)$
   $\qquad\qquad\qquad\quad =(6-3)x+(14+5)$
   $\qquad\qquad\qquad\quad =3x+19$

31. $x-5=17$ $\qquad\qquad$ Check: $22-5\overset{?}{=}17$
   $\quad\underline{+5\ \ +5}$ $\qquad\qquad\qquad\quad 17=17$
   $\ \ x\quad =22$

Chapter 3
An Introduction to Fractions

Prerequisite Check

1. $30 \div 4 = 7 \text{ r}2$, therefore 4 does not divide exactly into 30.

3. $72 \div 6 = 12$ with no remainder, therefore 6 does divide exactly into 72.

5. $238 \div 2 = 119$ with no remainder, therefore 2 does divide exactly into 238.

7. The factors of 10 are: 1, 2, 5, and 10. These are the only whole numbers that divide exactly into 10.

9. The factors of 48 are: 1, 2, 3, 4, 6, 8, 12, 16, 24, and 48. These are the only whole numbers that divide exactly into 48.

11. $8 + (-7) = 1$

13. $12 - (-9) = 21$

15. $(-28) \div 4 = -7$

17. $\quad x - 6 = 4 \qquad$ Check: $10 - 6 \overset{?}{=} 4$
$x - 6 + 6 = 4 + 6 \qquad\qquad 4 = 4$
$\quad\quad x = 10$

19. $\quad\quad 4x - 5 = 3x - 8$
$4x - 3x - 5 = 3x - 3x - 8$
$\quad\quad x - 5 = -8$
$\quad x - 5 + 5 = -8 + 3$
$\quad\quad\quad x = -3$
Check: $4(-3) - 5 \overset{?}{=} 3(-3) - 8$
$\quad\quad -12 - 5 \overset{?}{=} -9 - 8$
$\quad\quad\quad -17 = -17$

Exercises 3.1

< Objective 1 >

1. $2 \times 2 = 4$
$1 \times 4 = 4$
1, 2, and 4 are all factors of 4.

3. $2 \times 5 = 10$
$1 \times 10 = 10$
1, 2, 5, and 10 are all the factors of 10.

5. $3 \times 5 = 15$
$1 \times 15 = 15$
1, 3, 5, and 15 are all the factors of 15.

7. $4 \times 6 = 24$
$3 \times 8 = 24$
$2 \times 12 = 24$
$1 \times 24 = 24$
1, 2, 3, 4, 6, 8, 12, and 24 are all the factors of 24.

9. $8 \times 8 = 64$
$4 \times 16 = 64$
$2 \times 32 = 64$
$1 \times 64 = 64$
1, 2, 4, 8, 16, 32, and 64 are all the factors of 64.

11. $1 \times 11 = 11$
1 and 11 are all the factors of 11.

13. $1 \times 135 = 135$

$3 \times 45 = 135$

$5 \times 27 = 135$

$9 \times 15 = 135$

1, 3, 5, 9, 15, 27, 45, and 135 are all the factors of 135.

15. $1 \times 256 = 256$

$2 \times 128 = 256$

$4 \times 64 = 256$

$8 \times 32 = 256$

$16 \times 16 = 256$

1, 2, 4, 8, 16, 32, 64, 128, and 256 are all the factors of 256

< Objective 2 >

17. Since each has exactly two factors, 1 and itself, the numbers 19, 23, 31, 59, 97, and 103 are prime numbers.

19. Since each has exactly two factors, 1 and itself, the numbers 31, 37, 41, 43, and 47 are all the prime numbers between 30 and 50.

< Objective 3 >

21. Since each number has a last digit of 0, 2, 4, 6, or 8, the numbers 72, 158, 260, 378, 570, 4,530, and 8,300 are all divisible by 2.

23. Since the final two digits are divisible by 4, the numbers 72, 260, and 8,300 are all divisible by 4.

25. Since each number is an even number and the sum of the digits of each is divisible by 3, the numbers 72, 378, 570, and 4,530 are all divisible by 6.

27. Since the last digit is zero, the numbers 260, 570, 4,530, and 8,300 are all divisible by 10.

29. Since the sum of the digits of each is divisible by 3, the numbers 36, 282, 741, and 6,285 are all divisible by 3.

31. Since the last digit of each is either 0 or 5, the numbers 1,840, 6,285, and 14,320 are divisible by 5.

33. Since the sum of the digits is divisible by 9, the number 36 is divisible by 9.

35. Yes. The ones digit is even, so -84 is divisible by 2.

37. Yes. The two-digit number formed by its final two digits is divisible by 4, so -84 is divisible by 4.

39. Yes. The number is even and the sum of its digits is divisible by 3, so -84 is divisible by 6.

41. No. The last digit is not zero, so -84 is not divisible by 10.

43. Yes. The sum of its digits is divisible by 3, so -135 is divisible by 3.

45. Yes. The ones digit is 5, so -135 is not divisible by 5.

47. Yes. The sum of its digits is divisible by 9, so -135 is divisible by 9.

49. Since $10 = 2 \times 5$, 350 is divisible by 10 if 350 is divisible by 2 and 5. Since its last digit is 0, 350 is divisible by 2 and 5. Yes, it is possible to have rows of 10 seats each. Since $15 = 3 \times 5$, 350 is divisible by 15 if 350 is divisible by 3 and 5. Since the sum of the digits, $3 + 5 + 0$, is 8, and 8 is not divisible by 3, 350 is not divisible by 3. No, it is not possible to have 15 rows of seats.

51. Above and Beyond.

53. Above and Beyond.

55. Above and Beyond.

57. Above and Beyond.

Exercises 3.2

< Objectives 1 and 2 >

1.

$$18$$
$$= 2 \times 9$$
$$= 2 \times 3 \times 3$$
$$18 = 2 \times 3 \times 3$$

3.

$$30$$
$$= 6 \times 5$$
$$= 2 \times 3 \times 5$$
$$30 = 2 \times 3 \times 5$$

5.

$$51$$
$$= 3 \times 17$$
$$51 = 3 \times 17$$

7. $2\overline{)66}^{\,33} \searrow 3\overline{)33}^{\,11}$

$$66 = 2 \times 3 \times 11$$

9. $2\overline{)130}^{\,65} \searrow 5\overline{)65}^{\,13}$

$$130 = 2 \times 5 \times 13$$

11. $3\overline{)315}^{\,105} \searrow 3\overline{)105}^{\,35} \searrow 5\overline{)35}^{\,7}$

$$315 = 3 \times 3 \times 5 \times 7$$

13. $5\overline{)225}$

$5\overline{)45}$

$3\overline{)9}$

3

$$225 = 3 \times 3 \times 5 \times 5$$

15. $3\overline{)189}$

$3\overline{)63}$

$3\overline{)21}$

7

$$189 = 3 \times 3 \times 3 \times 7$$

17. 48 has factors of 1, 2, 3, 4, 6, 8, 12, 16, 24, and 48. Two factors with a sum of 14 are 6 and 8.

19. 48 has factors of 1, 2, 3, 4, 6, 8, 12, 16, 24, and 48. Two factors with a difference of 8 are 4 and 12.

21. 24 has factors of 1, 2, 3, 4, 6, 8, 12, and 24. Two factors with a sum of 10 are 4 and 6.

23. 30 has factors of 1, 2, 3, 5, 6, 10, 15, and 30. Two factors with a difference of 1 are 5 and 6.

25. $-12 = -1 \times 12$

$$12$$
$$= 6 \times 2$$
$$= 2 \times 2 \times 3$$
$$-12 = -1 \times 2 \times 2 \times 3$$

27. $-40 = -1 \times 40$

$$40$$
$$= 5 \times 8$$
$$= 5 \times 2 \times 4$$
$$= 5 \times 2 \times 2 \times 2$$
$$-40 = -1 \times 2 \times 2 \times 2 \times 5$$

29. $-144 = -1 \times 144$

$2\overline{)144}^{\,72} \searrow 2\overline{)72}^{\,36} \searrow 2\overline{)36}^{\,18} \searrow 2\overline{)18}^{\,9} \searrow 3\overline{)9}^{\,3}$

$$-144 = -1 \times 2 \times 2 \times 2 \times 2 \times 3 \times 3$$

< Objective 3 >

31. The factors of each of the two numbers are
4: 1, 2, 4
6: 1, 2, 3, 6
The GCF is 2.

33. The factors of each of the two numbers are
10: 1, 2, 5, 10
15: 1, 3, 5, 15
The GCF is 5.

35. The factors of each of the two numbers are
21: 1, 3, 7, 21
24: 1, 2, 3, 4, 6, 8, 12, 24
The GCF is 3.

37. The prime factorizations of the numbers are
$20 = 1 \times 2 \times 2 \times 5$
$21 = 1 \times 3 \times 7$
The GCF is 1.

39. The prime factorizations of the numbers are
$18 = 2 \times 3 \times 3$
$24 = 2 \times 2 \times 2 \times 3$
The GCF is $2 \times 3 = 6$.

41. The prime factorizations of the numbers are
$18 = 2 \times 3 \times 3$
$54 = 2 \times 3 \times 3 \times 3$
The GCF is $2 \times 3 \times 3 = 18$.

< Objective 4 >

43. The factors of each of the numbers are
12: 1, 2, 3, 4, 6, 12
36: 1, 2, 3, 4, 6, 9, 12, 18, 36
60: 1, 2, 3, 4, 5, 6, 10, 12, 15, 20, 30, 60
The GCF is 12.

45. The factors of each of the numbers are
105: 1, 3, 5, 7, 15, 21, 35, 105
140: 1, 2, 4, 5, 7, 10, 14, 20, 28, 35, 70, 140
175: 1, 5, 7, 25, 35, 175
The GCF is 35.

47. The prime factorizations of each of the
numbers are
$25 = 5 \times 5$
$75 = 3 \times 5 \times 5$
$150 = 2 \times 3 \times 5 \times 5$
The GCF is $5 \times 5 = 25$.

49. always

51. never

53. (a) The factors of 28 are 1, 2, 4, 7, 14.
The sum of the factors is
$1 + 2 + 4 + 7 + 14 = 28$.
Therefore, 28 is a perfect number.
(b) 6

55. $6 = 2 \times 3$
$8 = 2 \times 2 \times 2$
The GCF of 6 and 8 is 2. Note that
$(6 \times 8) \div 2 = 24$, and the factors of 24 are
1, 2, 3, 4, 6, 8, 12, 24. Since no smaller
number in this list is divisible by both 6 and
8, 24 is the smallest number divisible by 6
and 8. Tom and Dick will both be off 24 days
after August 1, or August 25.

Exercises 3.3

< Objective 1 >

1. $\dfrac{6}{11}$ ← numerator
← denominator

3. $\dfrac{3}{11}$ ← numerator
← denominator

< Objective 2 >

5. We have shaded three of the four identical
parts. The fraction $\dfrac{3}{4}$ names the shaded part.

7. We have shaded five of the six identical parts.
The fraction $\dfrac{5}{6}$ names the shaded part.

9. We have shaded five of the five identical
parts. The fraction $\dfrac{5}{5}$ names the shaded part.

11. We have shaded eleven of the twelve identical parts. The fraction $\frac{11}{12}$ names the shaded part.

13. We have shaded five of the eight identical objects. The fraction $\frac{5}{8}$ names the shaded part.

15. If 7 questions were missed, then $20 - 7 = 13$ were correct. Thirteen of twenty correct means the part you got correct was $\frac{13}{20}$. Seven of twenty missed means the part you got wrong was $\frac{7}{20}$.

17. Of the 17 cars, 11 were sold. The part of the cars sold was $\frac{11}{17}$. If 11 cars were sold, then $17 - 11 = 6$ were not sold. The part of the cars not sold was $\frac{6}{17}$.

19. $\frac{2}{5} = 2 \div 5$

< Objective 3 >

21. Because the numerator is less than the denominator, $\frac{3}{5}$ is a proper fraction.

23. Because it is a sum of a whole number and a proper fraction, $2\frac{3}{5}$ is a mixed number.

25. Because the numerator is equal to the denominator, $\frac{6}{6}$ is an improper fraction.

27. Because the numerator is less than the denominator, $\frac{13}{17}$ is a proper fraction.

29. Because the numerator is larger than the denominator, $-\frac{11}{6}$ is an improper fraction.

31. Because the numerator is less than the denominator, $-\frac{12}{21}$ is a proper fraction.

33. The fully shaded square represents the whole number 1. For the remaining square, three of the four identical parts are shaded. The fraction $\frac{3}{4}$ names this shaded part. So the mixed number $1\frac{3}{4}$ or the improper fraction $\frac{7}{4}$ names the shaded portion of the diagram.

35. The fully shaded circles represent the whole number 3. For the remaining circle, five of the eight identical parts are shaded. The fraction $\frac{5}{8}$ names this shaded part. So the mixed number $3\frac{5}{8}$ or the improper fraction $\frac{29}{8}$ names the shaded portion of the diagram.

37. The fully shaded rectangles represent the whole number 5. For the remaining rectangle, one of the two identical parts is shaded. The fraction $\frac{1}{2}$ names this shaded part. So the mixed number $5\frac{1}{2}$ or the improper fraction $\frac{11}{2}$ names the shaded portion of the diagram.

39. The fully shaded rectangles represent the whole number 3. For the remaining rectangle, four of the five identical parts are shaded. The fraction $\frac{4}{5}$ names this shaded part. So the mixed number $3\frac{4}{5}$ or the improper fraction $\frac{19}{5}$ names the shaded portion of the diagram.

41.

The shaded rectangle in the square represents the fraction $\frac{1}{3}$.

43.

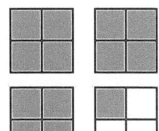

The 3 fully shaded squares and the one part shaded of the four identical parts in the remaining square represents the fraction $3\frac{1}{4}$.

45.

The 3 fully shaded circles represent the fraction $\frac{3}{1}$.

47. Because there are 4 quarters in each dollar, 64 quarters can be written $\frac{64}{4} = 16$. Clayton has $16.

49. Because there are 2 half gallons in each gallon, 35 half gallons can be written $\frac{35}{2} = 17\frac{1}{2}$. Manuel has $17\frac{1}{2}$ gal.

< Objective 4 >

51. $5\overline{)22}$ $\quad \frac{22}{5} = 4\frac{2}{5}$ $\leftarrow$ Remainder
$\quad \underline{20}$ $\qquad\qquad \downarrow 5 \leftarrow$ Original denominator
$\quad\quad 2$ $\qquad\qquad$ Quotient

53. $5\overline{)34}$ $\quad \frac{34}{5} = 6\frac{4}{5}$ $\leftarrow$ Remainder
$\quad \underline{30}$ $\qquad\qquad \downarrow 5 \leftarrow$ Original denominator
$\quad\quad 4$ $\qquad\qquad$ Quotient

55. $8\overline{)73}$ $\quad \frac{73}{8} = 9\frac{1}{8}$ $\leftarrow$ Remainder
$\quad \underline{72}$ $\qquad\qquad \downarrow 8 \leftarrow$ Original denominator
$\quad\quad 1$ $\qquad\qquad$ Quotient

57. $6\overline{)24}$ $\qquad \frac{24}{6} = 4$
$\quad \underline{24}$
$\quad\quad 0$

59. $1\overline{)9}$ $\qquad \frac{9}{1} = 9$
$\quad \underline{9}$
$\quad 0$

61. $-\frac{11}{11} = -\left(\frac{11}{11}\right)$

$\quad 11\overline{)11}$ $\qquad \frac{11}{11} = 1$
$\quad \underline{11}$
$\quad\quad 0$

$\quad -\frac{11}{11} = -\left(\frac{11}{11}\right) = -1$

63. $-\dfrac{12}{5} = -\left(\dfrac{12}{5}\right)$

$$5\overline{)12} \qquad \dfrac{12}{5} = 2\dfrac{2}{5} \qquad \leftarrow \text{Remainder}$$

2 ← Quotient

$\underline{10}$

$\ \ 2$

$\dfrac{12}{5} = 2\dfrac{2}{5}$ ← Original denominator

$-\dfrac{12}{5} = -\left(\dfrac{12}{5}\right) = -2\dfrac{2}{5}$

< Objective 5 >

65. $4\dfrac{2}{3} = \dfrac{(3\times 4)+2}{3} = \dfrac{14}{3}$

67. $8 = \dfrac{8}{1}$ since any fraction with a denominator of 1 is equal to the numerator alone.

69. $7\dfrac{6}{13} = \dfrac{(13\times 7)+6}{13} = \dfrac{97}{13}$

71. $10\dfrac{2}{5} = \dfrac{(5\times 10)+2}{5} = \dfrac{52}{5}$

73. $118\dfrac{3}{4} = \dfrac{(4\times 118)+3}{4} = \dfrac{475}{4}$

75. $4 = \dfrac{4}{1}$ since any fraction with a denominator of 1 is equal to the numerator alone.

77. $35 = \dfrac{35}{1}$ since any fraction with a denominator of 1 is equal to the numerator alone.

79. $-5 = -\dfrac{5}{1}$ since any fraction with a denominator of 1 is equal to the numerator alone

81. $\dfrac{3}{10}$

83. Number of machines that are operating, $36 - 5 = 31$. Therefore, the number of machines that are operating at one time are $\dfrac{31}{36}$.

85. Above and Beyond

Exercises 3.4

< Objective 1 >

1. Cross products $1\times 5 = 5$ and $3\times 3 = 9$ are not equal. The fractions are not equivalent.

3. Cross products $1\times 28 = 28$ and $7\times 4 = 28$ are equal. The fractions are equivalent.

5. Cross products $5\times 18 = 90$ and $6\times 15 = 90$ are equal. The fractions are equivalent.

7. Cross products $2\times 25 = 50$ and $21\times 4 = 84$ are not equal. The fractions are not equivalent.

9. Cross products $2\times 11 = 22$ and $7\times 3 = 21$ are not equal. The fractions are not equivalent.

11. Cross products $16\times 60 = 960$ and $24\times 40 = 960$ are equal. The fractions are equivalent.

< Objective 2 >

13. $\dfrac{15}{30} = \dfrac{3\times \cancel{5}}{2\times 3\times \cancel{5}} = \dfrac{1}{2}$

15. $\dfrac{8}{12} = \dfrac{\cancel{2}\times \cancel{2}\times 2}{\cancel{2}\times \cancel{2}\times 3} = \dfrac{2}{3}$

17. $\dfrac{10}{14} = \dfrac{\cancel{2}\times 5}{\cancel{2}\times 7} = \dfrac{5}{7}$

19. $\dfrac{12}{18} = \dfrac{\cancel{2} \times 2 \times \cancel{3}}{\cancel{2} \times \cancel{3} \times 3} = \dfrac{2}{3}$

21. $\dfrac{35}{40} = \dfrac{\cancel{5} \times 7}{2 \times 2 \times 2 \times \cancel{5}} = \dfrac{7}{8}$

23. $\dfrac{11}{44} = \dfrac{\cancel{11}}{4 \times \cancel{11}} = \dfrac{1}{4}$

25. $\dfrac{12}{36} = \dfrac{\cancel{2} \times \cancel{2} \times \cancel{3}}{\cancel{2} \times \cancel{2} \times \cancel{3} \times 3} = \dfrac{1}{3}$

27. $\dfrac{24}{27} = \dfrac{2 \times 2 \times 2 \times \cancel{3}}{\cancel{3} \times 3 \times 3} = \dfrac{8}{9}$

29. $\dfrac{32}{40} = \dfrac{\cancel{2} \times \cancel{2} \times \cancel{2} \times 2 \times 2}{\cancel{2} \times \cancel{2} \times \cancel{2} \times 5} = \dfrac{4}{5}$

31. $\dfrac{75}{105} = \dfrac{\cancel{3} \times \cancel{5} \times 5}{\cancel{3} \times \cancel{5} \times 7} = \dfrac{5}{7}$

33. $\dfrac{48}{60} = \dfrac{\cancel{2} \times \cancel{2} \times 2 \times 2 \times \cancel{3}}{\cancel{2} \times \cancel{2} \times \cancel{3} \times 5} = \dfrac{4}{5}$

35. $\dfrac{105}{135} = \dfrac{\cancel{3} \times \cancel{5} \times 7}{\cancel{3} \times 3 \times 3 \times \cancel{5}} = \dfrac{7}{9}$

37. $\dfrac{66}{110} = \dfrac{\cancel{2} \times 3 \times \cancel{11}}{\cancel{2} \times 5 \times \cancel{11}} = \dfrac{3}{5}$

39. $\dfrac{16}{21} = \dfrac{2 \times 2 \times 2 \times 2}{3 \times 7} = \dfrac{16}{21}$
There are no common factors, therefore, already in simplest form.

41. $\dfrac{31}{52} = \dfrac{31}{2 \times 2 \times 13} = \dfrac{31}{52}$
There are no common factors, therefore, already in simplest form.

43. $\dfrac{96}{132} = \dfrac{\cancel{2} \times \cancel{2} \times 2 \times 2 \times 2 \times \cancel{3}}{\cancel{2} \times \cancel{2} \times \cancel{2} \times 11} = \dfrac{8}{11}$

45. $\dfrac{85}{102} = \dfrac{5 \times \cancel{17}}{2 \times 3 \times \cancel{17}} = \dfrac{5}{6}$

47. $\dfrac{-7}{-8} = \dfrac{7}{8}$

49. $-\dfrac{-7}{-10} = \dfrac{7}{10}$

51. $\dfrac{-50}{-150} = \dfrac{50}{150} = \dfrac{\cancel{2} \times \cancel{5} \times \cancel{5}}{\cancel{2} \times 3 \times \cancel{5} \times \cancel{5}} = \dfrac{1}{3}$

53. $-\dfrac{72}{108} = -\dfrac{2 \times \cancel{2} \times \cancel{2} \times \cancel{3} \times \cancel{3}}{\cancel{2} \times \cancel{2} \times 3 \times \cancel{3} \times \cancel{3}} = -\dfrac{2}{3}$

55. $\dfrac{54}{72} \overset{?}{=} \dfrac{66}{88}$
$54 \times 88 = 4,752$
$72 \times 66 = 4,752$
Yes, Sam did get the same portion correct.

57. A quarter is 25¢ and a dollar is 100¢, so a quarter is $\dfrac{25}{100}$ of a dollar;
$\dfrac{25}{100} = \dfrac{\cancel{5} \times \cancel{5}}{2 \times 2 \times \cancel{5} \times \cancel{5}} = \dfrac{1}{4}$ of a dollar.

59. An hour is 60 min, so 15 min is $\dfrac{15}{60}$ of an hour; $\dfrac{15}{60} = \dfrac{\cancel{3} \times \cancel{5}}{2 \times 2 \times \cancel{3} \times \cancel{5}} = \dfrac{1}{4}$ of an hour.

61. $\dfrac{70}{100} = \dfrac{\cancel{2} \times \cancel{5} \times 7}{\cancel{2} \times 2 \times \cancel{5} \times 5} = \dfrac{7}{10}$ of a meter

63. $\dfrac{2}{8} = \dfrac{\cancel{2}}{\cancel{2} \times 2 \times 2} = \dfrac{1}{4}$ of the plugs were fouled.

65. With a scientific calculator:

28 $\boxed{\textbf{a b / c}}$ 40 $\boxed{=}$

With a graphing calculator:

28 $\boxed{\div}$ 40 $\boxed{\textbf{MATH}}$ $\boxed{\textbf{1: ▶ Frac}}$ $\boxed{\textbf{ENTER}}$

Display: $\dfrac{7}{10}$

67. With a scientific calculator:

96 $\boxed{\textbf{a b / c}}$ 144 $\boxed{=}$

With a graphing calculator:

96 $\boxed{\div}$ 144 $\boxed{\textbf{MATH}}$ $\boxed{\textbf{1: ▶ Frac}}$ $\boxed{\textbf{ENTER}}$

Display: $\dfrac{2}{3}$

69. With a scientific calculator:

299 $\boxed{\textbf{a b / c}}$ 391 $\boxed{=}$

With a graphing calculator:

299 $\boxed{\div}$ 391 $\boxed{\textbf{MATH}}$ $\boxed{\textbf{1: ▶ Frac}}$ $\boxed{\textbf{ENTER}}$

Display: $\dfrac{13}{17}$

71. $\dfrac{-175}{-280} = \dfrac{175}{280}$

With a scientific calculator:

175 $\boxed{\textbf{a b / c}}$ 280 $\boxed{=}$

With a graphing calculator:

175 $\boxed{\div}$ 280 $\boxed{\textbf{MATH}}$ $\boxed{\textbf{1: ▶ Frac}}$ $\boxed{\textbf{ENTER}}$

Display $\dfrac{5}{8}$

73. $-\dfrac{-13}{-468} = -\dfrac{13}{468}$

With a scientific calculator:

$\boxed{-}$ 13 $\boxed{\textbf{a b / c}}$ 468 $\boxed{=}$

With a graphing calculator:

$\boxed{-}$ 13 $\boxed{\div}$ 468 $\boxed{\textbf{MATH}}$ $\boxed{\textbf{1: ▶ Frac}}$

$\boxed{\textbf{ENTER}}$

Display: $-\dfrac{1}{36}$

75. $\dfrac{200 \text{ mg}}{300 \text{ mg}} = \dfrac{2 \times \cancel{100}}{3 \times \cancel{100}} = \dfrac{2}{3}$

77. $\dfrac{84 \text{ mm}}{156 \text{ mm}} = \dfrac{\cancel{2} \times \cancel{2} \times \cancel{3} \times 7}{\cancel{2} \times \cancel{2} \times \cancel{3} \times 13} = \dfrac{7}{13}$

69. $\dfrac{2}{12} = \dfrac{\cancel{2}}{2 \times 2 \times 3} = \dfrac{1}{6}$

$\dfrac{1}{6}$ part of the total staff are new.

71. Above and Beyond

73. **(a)** Five of the nine identical parts are shaded. The fraction $\dfrac{5}{9}$ names the shaded region.
(b) Ten of the eighteen identical parts are shaded. The fraction $\dfrac{10}{18}$ names the shaded region.

Exercises 3.5

< Objectives 1 and 2 >

1. $\dfrac{3}{4} \times \dfrac{5}{11} = \dfrac{3 \times 5}{4 \times 11} = \dfrac{15}{44}$

3. $\dfrac{3}{4} \times \dfrac{7}{11} = \dfrac{3 \times 7}{4 \times 11} = \dfrac{21}{44}$

5. $\dfrac{3}{5} \times \dfrac{5}{7} = \dfrac{3 \times \overset{1}{\cancel{5}}}{\underset{1}{\cancel{5}} \times 7} = \dfrac{3 \times 1}{1 \times 7} = \dfrac{3}{7}$

7. $\left(\dfrac{4}{9}\right)^2 = \dfrac{4}{9} \times \dfrac{4}{9} = \dfrac{16}{81}$

9. $\dfrac{3}{11} \times \dfrac{7}{9} = \dfrac{\overset{1}{\cancel{3}} \times 7}{11 \times \underset{3}{\cancel{9}}} = \dfrac{1 \times 7}{11 \times 3} = \dfrac{7}{33}$

11. $\dfrac{3}{10} \times \dfrac{5}{9} = \dfrac{\overset{1}{\cancel{3}} \times \overset{1}{\cancel{5}}}{\underset{2}{\cancel{10}} \times \underset{3}{\cancel{9}}} = \dfrac{1 \times 1}{2 \times 3} = \dfrac{1}{6}$

13. $\dfrac{7}{9} \times \dfrac{6}{5} = \dfrac{7 \times \overset{2}{\cancel{6}}}{\underset{3}{\cancel{9}} \times 5} = \dfrac{7 \times 2}{3 \times 5} = \dfrac{14}{15}$

15. $\dfrac{24}{33} \times \dfrac{55}{40} = \dfrac{\overset{3}{\cancel{24}} \times \overset{5}{\cancel{55}}}{\underset{3}{\cancel{33}} \times \underset{5}{\cancel{40}}} = \dfrac{\overset{1}{\cancel{3}} \times \overset{1}{\cancel{5}}}{\underset{1}{\cancel{3}} \times \underset{1}{\cancel{5}}} = \dfrac{1}{1} = 1$

17. $-\dfrac{6}{5} \bullet \dfrac{10}{21} = -\dfrac{\overset{2}{\cancel{6}} \times \overset{2}{\cancel{10}}}{\underset{1}{\cancel{5}} \times \underset{7}{\cancel{21}}} = -\dfrac{4}{7}$

19. $\dfrac{35}{10} \times \left(-\dfrac{45}{21}\right) = -\dfrac{\overset{5}{\cancel{35}} \times \overset{9}{\cancel{45}}}{\underset{2}{\cancel{10}} \times \underset{3}{\cancel{21}}} = -\dfrac{5 \times \overset{3}{\cancel{9}}}{2 \times \underset{1}{\cancel{3}}} = -\dfrac{15}{2}$

21. $-\dfrac{3}{10} \times \dfrac{70}{21} = -\dfrac{\overset{1}{\cancel{3}} \times \overset{7}{\cancel{70}}}{\underset{1}{\cancel{10}} \times \underset{7}{\cancel{21}}} = -\dfrac{7}{7} = -1$

23. $-\left(\dfrac{4}{6}\right)^2 = -\left(\dfrac{2}{3}\right)^2 = -\dfrac{2}{3} \times \dfrac{2}{3} = -\dfrac{4}{9}$

25. $\dfrac{2}{3} \times 2\dfrac{2}{5} = \dfrac{2}{3} \times \dfrac{12}{5} = \dfrac{2 \times \overset{4}{\cancel{12}}}{\underset{1}{\cancel{3}} \times 5} = \dfrac{2 \times 4}{1 \times 5} = \dfrac{8}{5} = 1\dfrac{3}{5}$

27. $\dfrac{2}{5} \times 3\dfrac{1}{4} = \dfrac{2}{5} \times \dfrac{13}{4} = \dfrac{\overset{1}{\cancel{2}} \times 13}{5 \times \underset{2}{\cancel{4}}} = \dfrac{1 \times 13}{5 \times 2} = \dfrac{13}{10} = 1\dfrac{3}{10}$

29. $2\dfrac{1}{3} \times 2\dfrac{1}{2} = \dfrac{7}{3} \times \dfrac{5}{2} = \dfrac{7 \times 5}{3 \times 2} = \dfrac{35}{6} = 5\dfrac{5}{6}$

31. $\left(\dfrac{5}{6}\right)^3 = \dfrac{5 \times 5 \times 5}{6 \times 6 \times 6} = \dfrac{125}{216}$

33. $\dfrac{12}{25} \times \dfrac{11}{18} = \dfrac{\overset{2}{\cancel{12}} \times 11}{25 \times \underset{3}{\cancel{18}}} = \dfrac{2 \times 11}{25 \times 3} = \dfrac{22}{75}$

35. $\dfrac{14}{15} \times \dfrac{10}{21} = \dfrac{\overset{2}{\cancel{14}} \times \overset{2}{\cancel{10}}}{\underset{3}{\cancel{15}} \times \underset{3}{\cancel{21}}} = \dfrac{2 \times 2}{3 \times 3} = \dfrac{4}{9}$

37. $\dfrac{18}{28} \times \dfrac{35}{22} = \dfrac{\overset{9}{\cancel{18}} \times \overset{5}{\cancel{35}}}{\underset{4}{\cancel{28}} \times \underset{11}{\cancel{22}}} = \dfrac{9 \times 5}{4 \times 11} = \dfrac{45}{44} = 1\dfrac{1}{44}$

39. $-\dfrac{3}{5} \times \left(-3\dfrac{1}{2}\right) = \dfrac{3}{5} \times \dfrac{7}{2} = \dfrac{21}{10} = 2\dfrac{1}{10}$

41. $-4\dfrac{3}{4} \bullet \left(2\dfrac{1}{5}\right) = -\dfrac{19}{4} \times \dfrac{11}{5} = -\dfrac{209}{20} = -10\dfrac{9}{20}$

43. $\dfrac{4}{9} \times 3\dfrac{3}{5} = \dfrac{4}{9} \times \dfrac{18}{5} = \dfrac{4 \times \overset{2}{\cancel{18}}}{\underset{1}{\cancel{9}} \times 5} = \dfrac{4 \times 2}{1 \times 5} = \dfrac{8}{5} = 1\dfrac{3}{5}$

45. $\dfrac{10}{27} \times 3\dfrac{3}{5} = \dfrac{10}{27} \times \dfrac{18}{5} = \dfrac{\overset{2}{\cancel{10}} \times \overset{2}{\cancel{18}}}{\underset{3}{\cancel{27}} \times \underset{1}{\cancel{5}}} = \dfrac{2 \times 2}{3 \times 1} = \dfrac{4}{3}$

$= 1\dfrac{1}{3}$

47. $2\dfrac{2}{5} \times 3\dfrac{3}{4} = \dfrac{12}{5} \times \dfrac{15}{4} = \dfrac{\overset{3}{\cancel{12}} \times \overset{3}{\cancel{15}}}{\underset{1}{\cancel{5}} \times \underset{1}{\cancel{4}}} = \dfrac{3 \times 3}{1 \times 1} = \dfrac{9}{1} = 9$

49. $4\frac{1}{5} \times \frac{10}{21} \times \frac{9}{20} = \frac{21}{5} \times \frac{10}{21} \times \frac{9}{20} = \frac{\overset{1}{\cancel{21}} \times \overset{1}{\cancel{10}} \times 9}{5 \times \cancel{21} \times \cancel{20}}$

$$= \frac{9}{10}$$

51. $3\frac{1}{3} \times \frac{4}{5} \times 1\frac{1}{8} = \frac{10}{3} \times \frac{4}{5} \times \frac{9}{8} = \frac{\overset{2}{\cancel{10}} \times \overset{1}{\cancel{4}} \times \overset{3}{\cancel{9}}}{\underset{1}{\cancel{3}} \times \underset{1}{\cancel{5}} \times \underset{2}{\cancel{8}}} = \frac{6}{2}$

$$= \frac{3}{1} = 3$$

53. $\frac{2}{3} \times \frac{3}{7} = \frac{2 \times \overset{1}{\cancel{3}}}{\underset{1}{\cancel{3}} \times 7} = \frac{2 \times 1}{1 \times 7} = \frac{2}{7}$

< Objective 3 >

55. $3\frac{1}{5} \to 3$

$4\frac{2}{3} \to 5$

Estimate of product: $3 \times 5 = 15$

57. $11\frac{3}{4} \to 12$

$5\frac{1}{4} \to 5$

Estimate of product: $12 \times 5 = 60$

59. $8\frac{2}{9} \to 8$

$7\frac{11}{12} \to 8$

Estimate of product: $8 \times 8 = 64$

61. $36 \text{ mi/hr} \times 4\text{hr} = \frac{36 \text{ mi}}{1 \text{ hr}} \times \frac{4 \cancel{\text{hr}}}{1} = 144 \text{ mi}$

63. $55 \text{ joules/s} \times 11 \text{ s} = \frac{55 \text{ joules}}{1 \cancel{\text{s}}} \times \frac{11 \cancel{\text{s}}}{1}$

$= 605 \text{ joules}$

65. $88 \text{ ft/s} \times 1 \text{ mi} / 5280 \text{ ft} \times 3,600 \text{ s/hr}$

$= \frac{88 \cancel{\text{ft}}}{1 \cancel{\text{s}}} \times \frac{1 \text{ mi}}{\underset{22}{\cancel{5280}} \cancel{\text{ft}}} \times \frac{\overset{15}{\cancel{3,600}} \cancel{\text{s}}}{1 \text{ hr}}$

$= \frac{\overset{4}{\cancel{88}} \times 1 \times 15}{1 \times \underset{1}{\cancel{22}} \times 1} \frac{\text{mi}}{\text{hr}} = \frac{60}{1} \frac{\text{mi}}{\text{hr}} = 60 \text{ mi/hr}$

< Objective 4 >

67. $\$11/\text{hr} \times 9 \text{ hr/day} \times 6 \text{ days}$

$= \frac{\$11}{1 \cancel{\text{hr}}} \times \frac{9 \cancel{\text{hr}}}{1 \cancel{\text{day}}} \times \frac{6 \cancel{\text{days}}}{1} = \$\frac{11 \times 9 \times 6}{1 \times 1 \times 1} = \$594$

69. $\frac{2}{3} \text{ cup/serving} \times 6 \text{ servings}$

$= \frac{2}{3} \frac{\text{cup}}{\cancel{\text{serving}}} \times \frac{6 \cancel{\text{servings}}}{1} = \frac{2 \times \overset{2}{\cancel{6}}}{\underset{1}{\cancel{3}} \times 1} \text{cup}$

$= \frac{4}{1} \text{cup} = 4 \text{ cup}$

71. $540 \text{ mi/hr} \times 4\frac{2}{3} \text{ hr} = \frac{540 \text{ mi}}{1 \cancel{\text{hr}}} \times \frac{14}{3} \cancel{\text{hr}}$

$= \frac{\overset{180}{\cancel{540}} \times 14}{1 \times \cancel{3}} \text{mi} = 2,520 \text{ mi}$

73. $2\frac{1}{4} \text{in.} \times 3\frac{7}{8} \text{in.} \times 4\frac{5}{6} \text{in.}$

$= \frac{\overset{3}{\cancel{9}}}{4} \text{in.} \times \frac{31}{8} \text{in.} \times \frac{29}{\underset{2}{\cancel{6}}} \text{in.} = \frac{2,697}{64} \text{in.}^3$

$= 42\frac{9}{64} \text{in.}^3$

75. $11 \times \frac{3}{4} = \frac{11}{1} \times \frac{3}{4} = \frac{33}{4} = 8\frac{1}{4} \text{ in.}$

77. $14\frac{3}{4} \times 7\frac{1}{2} = \frac{59}{4} \times \frac{15}{2} = \frac{885}{8} = 110\frac{5}{8}\cancel{c} \text{ or } \$1.11$

Exercises 3.6

< Objective 1 >

1. The reciprocal of $\dfrac{7}{8}$ is $\dfrac{8}{7}$.

3. The reciprocal of $\dfrac{5}{2}$ is $\dfrac{2}{5}$.

5. The reciprocal of $\dfrac{1}{2}$ is 2.

7. $2\dfrac{1}{3} = \dfrac{7}{3}$

The reciprocal of $2\dfrac{1}{3}$ is $\dfrac{3}{7}$.

9. $9\dfrac{3}{4} = \dfrac{39}{4}$

The reciprocal of $9\dfrac{3}{4}$ is $\dfrac{4}{39}$.

11. $6 = \dfrac{6}{1}$

The reciprocal of 6 is $\dfrac{1}{6}$.

13. The reciprocal of $-\dfrac{7}{16}$ is $-\dfrac{16}{7}$.

15. The reciprocal of $-\dfrac{1}{8}$ is -8.

17. $-3\dfrac{3}{8} = -\dfrac{27}{8}$

The reciprocal of $-3\dfrac{3}{8}$ is $-\dfrac{8}{27}$.

< Objective 2 >

19. $\dfrac{1}{5} \div \dfrac{3}{4} = \dfrac{1}{5} \times \dfrac{4}{3} = \dfrac{4}{15}$

21. $\dfrac{\frac{2}{5}}{\frac{3}{4}} = \dfrac{2}{5} \times \dfrac{4}{3} = \dfrac{8}{15}$

23. $\dfrac{8}{9} \div \dfrac{4}{3} = \dfrac{\overset{2}{\cancel{8}}}{\underset{3}{\cancel{9}}} \times \dfrac{\overset{1}{\cancel{3}}}{\underset{1}{\cancel{4}}} = \dfrac{2\times 1}{3\times 1} = \dfrac{2}{3}$

25. $\dfrac{7}{10} \div \dfrac{5}{9} = \dfrac{7}{10} \times \dfrac{9}{5} = \dfrac{63}{50} = 1\dfrac{13}{50}$

27. $\dfrac{8}{15} \div \dfrac{2}{5} = \dfrac{\overset{4}{\cancel{8}}}{\underset{3}{\cancel{15}}} \times \dfrac{\overset{1}{\cancel{5}}}{\underset{1}{\cancel{2}}} = \dfrac{4\times 1}{3\times 1} = \dfrac{4}{3} = 1\dfrac{1}{3}$

29. $\dfrac{\frac{5}{27}}{\frac{25}{36}} = \dfrac{\overset{1}{\cancel{5}}}{\underset{3}{\cancel{27}}} \times \dfrac{\overset{4}{\cancel{36}}}{\underset{5}{\cancel{25}}} = \dfrac{1\times 4}{3\times 5} = \dfrac{4}{15}$

31. $\dfrac{4}{5} \div 4 = \dfrac{4}{5} \div \dfrac{4}{1} = \dfrac{\overset{1}{\cancel{4}}}{5} \times \dfrac{1}{\underset{1}{\cancel{4}}} = \dfrac{1\times 1}{5\times 1} = \dfrac{1}{5}$

33. $12 \div \dfrac{2}{3} = \dfrac{12}{1} \div \dfrac{2}{3} = \dfrac{\overset{6}{\cancel{12}}}{1} \times \dfrac{3}{\underset{1}{\cancel{2}}} = \dfrac{18}{1} = 18$

35. $\dfrac{\frac{12}{17}}{\frac{6}{7}} = \dfrac{12}{17} \div \dfrac{6}{7} = \dfrac{\overset{2}{\cancel{12}}}{17} \times \dfrac{7}{\underset{1}{\cancel{6}}} = \dfrac{2\times 7}{17\times 1} = \dfrac{14}{17}$

37. $-\dfrac{3}{10} \div \dfrac{4}{5} = -\dfrac{3}{\underset{2}{\cancel{10}}} \times \dfrac{\overset{1}{\cancel{5}}}{4} = -\dfrac{3}{8}$

39. $-\dfrac{11}{16} \div \left(-\dfrac{5}{8}\right) = \dfrac{11}{\underset{2}{\cancel{16}}} \times \dfrac{\overset{1}{\cancel{8}}}{5} = \dfrac{11}{10} = 1\dfrac{1}{10}$

41. $\dfrac{-\dfrac{8}{15}}{6} = -\dfrac{8}{15} \div \dfrac{6}{1} = -\dfrac{\cancel{8}}{15} \times \dfrac{1}{\cancel{6}_3} = -\dfrac{4}{45}$

< Objective 3 >

43. $15 \div 3\dfrac{1}{3} = \dfrac{15}{1} \div \dfrac{10}{3} = \dfrac{\cancel{15}^3}{1} \times \dfrac{3}{\cancel{10}_2} = \dfrac{3 \times 3}{1 \times 2} = \dfrac{9}{2}$

$= 4\dfrac{1}{2}$

45. $1\dfrac{3}{5} \div \dfrac{4}{15} = \dfrac{8}{5} \div \dfrac{4}{15} = \dfrac{\cancel{8}^2}{\cancel{5}_1} \times \dfrac{\cancel{15}^3}{\cancel{4}_1} = \dfrac{6}{1} = 6$

47. $\dfrac{7}{12} \div 2\dfrac{1}{3} = \dfrac{7}{12} \div \dfrac{7}{3} = \dfrac{\cancel{7}^1}{\cancel{12}_4} \times \dfrac{\cancel{3}^1}{\cancel{7}_1} = \dfrac{1}{4}$

49. $5\dfrac{3}{5} \div \dfrac{7}{15} = \dfrac{28}{5} \div \dfrac{7}{15} = \dfrac{\cancel{28}^4}{\cancel{5}_1} \times \dfrac{\cancel{15}^3}{\cancel{7}_1} = \dfrac{12}{1} = 12$

51. $8 \div 5\dfrac{1}{4} = \dfrac{8}{1} \div \dfrac{21}{4} = \dfrac{8}{1} \times \dfrac{4}{21} = \dfrac{32}{21} = 1\dfrac{11}{21}$

53. $4\dfrac{3}{8} \div 6\dfrac{2}{3} = \dfrac{35}{8} \div \dfrac{20}{3} = \dfrac{\cancel{35}^7}{8} \times \dfrac{3}{\cancel{20}_4} = \dfrac{7}{8} \times \dfrac{3}{4}$

$= \dfrac{21}{32}$

55. $\left(\dfrac{2}{3}\right)^3 \div 4 = \dfrac{2}{3} \times \dfrac{2}{3} \times \dfrac{2}{3} \div 4 = \dfrac{8}{27} \div 4 = \dfrac{\cancel{8}^2}{27} \times \dfrac{1}{\cancel{4}_1}$

$= \dfrac{2}{27}$

57. $\dfrac{3}{5} \div \left(\dfrac{3}{10}\right)^2 = \dfrac{3}{5} \div \left(\dfrac{3}{10} \times \dfrac{3}{10}\right) = \dfrac{3}{5} \div \dfrac{9}{100}$

$= \dfrac{\cancel{3}^1}{\cancel{5}_1} \times \dfrac{\cancel{100}^{20}}{\cancel{9}_3} = \dfrac{20}{3} = 6\dfrac{2}{3}$

59. $-8\dfrac{1}{4} \div 6\dfrac{1}{2} = -\dfrac{33}{4} \div \dfrac{13}{2} = -\dfrac{33}{\cancel{4}_2} \times \dfrac{\cancel{2}^1}{13} = -\dfrac{33}{26}$

$= -1\dfrac{7}{26}$

61. $-12\dfrac{2}{5} \div \left(-2\dfrac{3}{4}\right) = -\dfrac{62}{5} \div \left(-\dfrac{11}{4}\right) = \dfrac{62}{5} \times \dfrac{4}{11}$

$= \dfrac{248}{55} = 4\dfrac{28}{55}$

63. $-15\dfrac{3}{8} \div 6 = -\dfrac{123}{8} \div \dfrac{6}{1} = -\dfrac{\cancel{123}^{41}}{8} \times \dfrac{1}{\cancel{6}_2} = -\dfrac{41}{16}$

$= -2\dfrac{9}{16}$

65. $900 \text{ mi} \div 15\dfrac{\text{mi}}{\text{gal}} = 900 \text{ mi} \div \dfrac{15 \text{ mi}}{1 \text{ gal}}$

$= 900 \cancel{\text{mi}} \times \dfrac{1 \text{ gal}}{15 \cancel{\text{mi}}} = \dfrac{\cancel{900}^{60} \text{ gal}}{\cancel{15}_1}$

$= 60 \text{ gal}$

67. $8,750 \text{ watts} \div 350\dfrac{\text{watts}}{\text{s}}$

$= 8,750 \text{ watts} \div \dfrac{350 \text{ watts}}{1 \text{ s}}$

$= \cancel{8,750}^{25} \cancel{\text{watts}} \times \dfrac{1 \text{ s}}{\cancel{350}_1 \cancel{\text{watts}}} = 25 \text{ s}$

< Objective 4 >

69. $5\dfrac{1}{4}\,\text{ft} \div 7 = \dfrac{21}{4}\,\text{ft} \div \dfrac{7}{1} = \dfrac{\overset{3}{\cancel{21}}}{4}\,\text{ft} \times \dfrac{1}{\underset{1}{\cancel{7}}} = \dfrac{3}{4}\,\text{ft}$

The length of each piece is $\dfrac{3}{4}$ ft.

71. $95\,\text{mi} \div 1\dfrac{1}{4}\,\text{hr} = \dfrac{95}{1}\,\text{mi} \div \dfrac{5}{4}\,\text{hr}$

$= \dfrac{\overset{19}{\cancel{95}}}{1}\,\text{mi} \times \dfrac{4}{\underset{1}{\cancel{5}}}\dfrac{1}{\text{hr}} = 76\,\text{mi/hr}$

Virginia's average speed was 76 mi/hr.

73. $3\dfrac{1}{4} \div \dfrac{1}{4} = \dfrac{13}{4} \div \dfrac{1}{4} = \dfrac{13}{\underset{1}{\cancel{4}}} \times \dfrac{\overset{1}{\cancel{4}}}{1} = 13$

There are 13 servings.

75. With a scientific calculator:

7 $\boxed{\textbf{a b / c}}$ 12 $\boxed{\times}$ 36 $\boxed{\textbf{a b / c}}$ 63 $\boxed{=}$

With a graphing calculator:

7 $\boxed{\div}$ 12 $\boxed{\times}$ 36 $\boxed{\div}$ 63 $\boxed{\textbf{MATH}}$

$\boxed{\textbf{1:▶ Frac}}$ $\boxed{\text{ENTER}}$

Display: $\dfrac{1}{3}$

77. With a scientific calculator:

12 $\boxed{\textbf{a b / c}}$ 45 $\boxed{\times}$ 27 $\boxed{\textbf{a b / c}}$ 72 $\boxed{=}$

With a graphing calculator:

12 $\boxed{\div}$ 45 $\boxed{\times}$ 27 $\boxed{\div}$ 72 $\boxed{\textbf{MATH}}$

$\boxed{\textbf{1:▶ Frac}}$ $\boxed{\text{ENTER}}$

Display: $\dfrac{1}{10}$

79. With a scientific calculator:

27 $\boxed{\textbf{a b / c}}$ 72 $\boxed{\times}$ 24 $\boxed{\textbf{a b / c}}$ 45 $\boxed{=}$

With a graphing calculator:

27 $\boxed{\div}$ 72 $\boxed{\times}$ 24 $\boxed{\div}$ 45 $\boxed{\textbf{MATH}}$

$\boxed{\textbf{1:▶ Frac}}$ $\boxed{\text{ENTER}}$

Display: $\dfrac{1}{5}$

81. With a scientific calculator:

$\boxed{-}$ 5 $\boxed{\textbf{a b / c}}$ 8 $\boxed{\times}$ 9 $\boxed{\textbf{a b / c}}$ 20 $\boxed{=}$

With a graphing calculator:

$\boxed{(}$ $\boxed{-}$ 5 $\boxed{\div}$ 8 $\boxed{)}$ $\boxed{\times}$ $\boxed{(}$ 9 $\boxed{\div}$ 20 $\boxed{)}$

$\boxed{\textbf{MATH}}$ $\boxed{\textbf{1:▶ Frac}}$ $\boxed{\text{ENTER}}$

Display: $-\dfrac{9}{32}$

83. With a scientific calculator:

3 $\boxed{\textbf{a b / c}}$ 64 $\boxed{+/-}$ $\boxed{\times}$ 16 $\boxed{\textbf{a b / c}}$ 27

$\boxed{+/-}$ $\boxed{=}$

With a graphing calculator:

$\boxed{(}$ $\boxed{(-)}$ 3 $\boxed{\div}$ 64 $\boxed{)}$ $\boxed{\times}$ $\boxed{(}$ $\boxed{(-)}$ 16 $\boxed{\div}$ 27 $\boxed{)}$

$\boxed{\textbf{MATH}}$ $\boxed{\textbf{1:▶ Frac}}$ $\boxed{\text{ENTER}}$

Display: $\dfrac{1}{36}$

85. With a scientific calculator:

1 $\boxed{\textbf{a b / c}}$ 5 $\boxed{\div}$ 2 $\boxed{\textbf{a b / c}}$ 15 $\boxed{=}$

With a graphing calculator:

1 $\boxed{\div}$ 5 $\boxed{\div}$ $\boxed{(}$ 2 $\boxed{\div}$ 15 $\boxed{)}$ $\boxed{\textbf{MATH}}$

$\boxed{\textbf{1:▶ Frac}}$ $\boxed{\text{ENTER}}$

Display: $\dfrac{3}{2}$ or $1\dfrac{1}{2}$

87. With a scientific calculator:

5 [a b / c] 7 [÷] 15 [a b / c] 28 [=]

With a graphing calculator:

5 [÷] 7 [÷] [(] 15 [÷] 28 [)] [**MATH**]

[**1:▶ Frac**] [ENTER]

Display: $\frac{4}{3}$ or $1\frac{1}{3}$

89. With a scientific calculator:

15 [a b / c] 18 [÷] 45 [a b / c] 27 [=]

With a graphing calculator:

15 [÷] 18 [÷] [(] 45 [÷] 27 [)] [**MATH**]

[**1:▶ Frac**] [ENTER]

Display: $\frac{1}{2}$

91. With a scientific calculator:

25 [a b / c] 45 [÷] 100 [a b / c] 135 [=]

With a graphing calculator:

25 [÷] 45 [÷] [(] 100 [÷] 135 [)] [**MATH**]

[**1:▶ Frac**] [ENTER]

Display: $\frac{3}{4}$

93. With a scientific calculator:

9 [a b / c] 3 [a b / c] 4 [×] 8 [a b / c] 2

[a b / c] 3 [=]

With a graphing calculator:

[(] 9 [+] 3 [÷] 4 [)] [×] [(] 8 [+] 2 [÷] 3 [)]

[**MATH**] [**1:▶ Frac**] [ENTER]

Display: $\frac{169}{2}$ or $84\frac{1}{2}$

95. With a scientific calculator:

6 [a b / c] 3 [a b / c] 4 [+/−] [×] 2

[a b / c] 2 [a b / c] 3 [+/−] [=]

With a graphing calculator:

[(] [(−)] 6 [+] 3 [÷] 4 [)] [×] [(] [(−)] 2 [+] 2

[÷] 3 [)] [**MATH**] [**1:▶ Frac**] [ENTER]

Display: 18

97. With a scientific calculator:

10 [÷] 6 [a b / c] 2 [a b / c] 3 [=]

With a graphing calculator:

10 [÷] [(] 6 [+] 2 [÷] 3 [)] [**MATH**]

[**1:▶ Frac**] [ENTER]

Display: $\frac{3}{2}$ or $1\frac{1}{2}$

99. With a scientific calculator:

2 [a b / c] 3 [÷] 4 [a b / c] 11 [a b / c] 16

[+/−] [=]

With a graphing calculator:

[(] 2 [÷] 3 [)] [÷] [(] [(−)] 4 [+] 11 [÷] 16 [)]

[**MATH**] [**1:▶ Frac**] [ENTER]

Display: $-\dfrac{32}{225}$

101. $8\dfrac{1}{4} \div 11 = \dfrac{33}{4} \times \dfrac{1}{11} = \dfrac{\overset{3}{\cancel{33}}}{\underset{4}{\cancel{44}}} = \dfrac{3}{4}$

$\dfrac{3}{4} \times 3 = \dfrac{9}{4} = 2\dfrac{1}{4}$ in.

103. $\dfrac{3}{4} \div \dfrac{3}{8} = \dfrac{\overset{1}{\cancel{3}}}{\underset{1}{\cancel{4}}} \times \dfrac{\overset{2}{\cancel{8}}}{\underset{1}{\cancel{3}}} = \dfrac{2}{1} = 2$ in.

105. $7\dfrac{1}{2} \div 1\dfrac{3}{4} = \dfrac{15}{2} \div \dfrac{7}{4} = \dfrac{15}{\cancel{2}} \times \dfrac{\overset{2}{\cancel{4}}}{7} = \dfrac{30}{7} = 4\dfrac{2}{7}$

Manuel will have 4 strips of cloth. He will

have $\dfrac{2}{7}$ of one $1\dfrac{3}{4}$ yd piece left.

$\dfrac{2}{7} \times 1\dfrac{3}{4}$ yd $= \dfrac{\overset{1}{\cancel{2}}}{\cancel{7}} \times \dfrac{\overset{1}{\cancel{7}}}{\cancel{4}}$ yd $= \dfrac{1}{2}$ yd

$\dfrac{1}{2}$ yd of cloth remains.

107. (a) $\dfrac{1}{3}$ c/3 oranges $\times$ 24 oranges

$= \dfrac{\dfrac{1}{3}\text{c}}{3 \text{ oranges}} \times \dfrac{24 \text{ oranges}}{1} = \dfrac{\dfrac{1}{3} \times 24}{3}\text{c}$

$= \dfrac{8}{3}\text{c} = 2\dfrac{2}{3}\text{cups}$

(b) Since one bag gives $2\dfrac{2}{3}$ cups, we would

divide 8 by $2\dfrac{2}{3}$ to find out how many bags

are needed.

$8 \div 2\dfrac{2}{3} = \dfrac{8}{1} \div \dfrac{8}{3} = \dfrac{\overset{1}{\cancel{8}}}{1} \times \dfrac{3}{\underset{1}{\cancel{8}}} = 3$

3 bags are needed.

109. Above and Beyond

111. Above and Beyond

113. Above and Beyond

Exercises 3.7

< Objectives 1 and 2 >

1. $5x = 20$ Check: $5(4) \overset{?}{=} 20$

$\dfrac{5x}{5} = \dfrac{20}{5}$ $20 = 20$

$x = 4$

The solution is 4.

3. $9x = 54$ Check: $9(6) \overset{?}{=} 54$

$\dfrac{9x}{9} = \dfrac{54}{9}$ $54 = 54$

$x = 6$

The solution is 6.

5. $63 = 9x$ Check: $63 \overset{?}{=} 9(7)$

$\dfrac{63}{9} = \dfrac{9x}{9}$ $63 = 63$

$7 = x$

The solution is 7.

7. $4x = -16$ Check: $4(-4) \overset{?}{=} -16$

$\dfrac{4x}{4} = \dfrac{-16}{4}$ $-16 = -16$

$x = -4$

The solution is −4.

9. $-9x = 72$ Check: $-9(-8) \overset{?}{=} 72$

$\dfrac{-9x}{-9} = \dfrac{72}{-9}$ $72 = 72$

$x = -8$

The solution is −8.

11. $6x = -54$ Check: $6(-9) \overset{?}{=} -54$

$\dfrac{6x}{6} = \dfrac{-54}{6}$ $-54 = -54$

$x = -9$

The solution is −9.

13. $-4x = -12$ Check: $-4(3) \overset{?}{=} -12$

$\dfrac{-4x}{-4} = \dfrac{-12}{-4}$ $-12 = -12$

$x = 3$

The solution is 3.

15. $-42 = 6x$ Check: $-42 \overset{?}{=} 6(-7)$

$\dfrac{-42}{6} = \dfrac{6x}{6}$ $-42 = -42$

$-7 = x$

The solution is -7.

17. $-6x = -54$ Check: $-6(9) \overset{?}{=} -54$

$\dfrac{-6x}{-6} = \dfrac{-54}{-6}$ $-54 = -54$

$x = 9$

The solution is 9.

19. $\dfrac{x}{2} = 4$ Check: $\dfrac{8}{2} \overset{?}{=} 4$

$(2)\dfrac{x}{2} = 4(2)$ $4 = 4$

$x = 8$

The solution is 8.

21. $\dfrac{x}{5} = 3$ Check: $\dfrac{15}{5} \overset{?}{=} 3$

$(5)\dfrac{x}{5} = 3(5)$ $3 = 3$

$x = 15$

The solution is 15.

23. $6 = \dfrac{x}{7}$ Check: $6 \overset{?}{=} \dfrac{42}{7}$

$(7)6 = \dfrac{x}{7}(7)$ $6 = 6$

$42 = x$

The solution is 42.

25. $\dfrac{x}{5} = -4$ Check: $-\dfrac{20}{5} \overset{?}{=} -4$

$(5)\dfrac{x}{5} = -4(5)$ $-4 = -4$

$x = -20$

The solution is -20.

27. $-\dfrac{x}{3} = 8$ Check: $-\dfrac{(-24)}{3} \overset{?}{=} 8$

$-3\left(-\dfrac{x}{3}\right) = 8(-3)$ $8 = 8$

$x = -24$

The solution is -24.

29. $\dfrac{2}{3}x = 6$ Check: $\dfrac{2}{3}(9) \overset{?}{=} 6$

$3\left(\dfrac{2}{3}x\right) = 6(3)$ $\dfrac{2}{\cancel{3}}\left(\overset{3}{\cancel{9}}\right) \overset{?}{=} 6$

$\dfrac{2x}{2} = \dfrac{18}{2}$ $6 = 6$

$x = 9$

The solution is 9.

31. $\dfrac{3}{4}x = -15$ Check: $\dfrac{3}{4}(-20) \overset{?}{=} -15$

$4\left(\dfrac{3}{4}x\right) = -15(4)$ $\dfrac{3}{\cancel{4}}\left(-\overset{5}{\cancel{20}}\right) \overset{?}{=} -15$

$\dfrac{3x}{3} = \dfrac{-60}{3}$ $-15 = -15$

$x = -20$

The solution is -20.

33.
$$-\frac{2}{5}x = 10$$
$$-5\left(-\frac{2}{5}x\right) = 10(-5)$$
$$\frac{2x}{2} = \frac{-50}{2}$$
$$x = -25$$

Check: $-\frac{2}{5}(-25)\overset{?}{=}10$
$$-\frac{2}{\overset{}{5}}\left(-\overset{5}{25}\right)\overset{?}{=}10$$
$$10 = 10$$

The solution is -25.

35. $5x + 4x = 36$ Check: $5(4)+4(4)\overset{?}{=}36$
$$9x = 36$$
$$\frac{9x}{9} = \frac{36}{9}$$ $20 + 16 \overset{?}{=} 36$
$$x = 4$$ $36 = 36$

The solution is 4.

37. $16x + (-9x) = -42$
$$7x = -42$$
$$\frac{7x}{7} = \frac{-42}{7}$$
$$x = -6$$

Check: $16(-6) + (-9(-6)) \overset{?}{=} -42$
$$-96 + 54 \overset{?}{=} -42$$
$$-42 = -42$$

The solution is -6.

39. $4x + (-2x) + 7x = 36$
$$9x = 36$$
$$\frac{9x}{9} = \frac{36}{9}$$
$$x = 4$$

Check: $4(4) + (-2(4)) + 7(4) \overset{?}{=} 36$
$$16 - 8 + 28 \overset{?}{=} 36$$
$$36 = 36$$

The solution is 4.

41. $8x = 5$ Check: $8\left(\frac{5}{8}\right)\overset{?}{=}5$
$$\frac{1}{8}(8x) = \frac{1}{8}(5)$$ $\overset{}{8}\left(\frac{5}{\overset{}{8}}\right)\overset{?}{=}5$
$$x = \frac{5}{8}$$ $5 = 5$

The solution is $\frac{5}{8}$.

43. $15x = -9$
$$\frac{1}{15}(15x) = \frac{1}{15}(-9)$$
$$x = -\frac{9}{15} = -\frac{3}{5}$$

Check: $15\left(-\frac{3}{5}\right)\overset{?}{=} -9$
$$\overset{3}{15}\left(-\frac{3}{\overset{}{5}}\right)\overset{?}{=} -9$$
$$-9 = -9$$

The solution is $-\frac{3}{5}$.

45. $\frac{4}{5}x = 10$ Check: $\frac{4}{5}\left(\frac{25}{2}\right)\overset{?}{=}10$
$$\frac{5}{4}\left(\frac{4}{5}x\right) = \frac{5}{4}(10)$$ $\frac{\overset{}{4}}{\overset{}{5}}\left(\frac{\overset{5}{25}}{\overset{}{2}}\right)\overset{?}{=}10$
$$x = \frac{5 \times 5}{2} = \frac{25}{2}$$ $10 = 10$

The solution is $\frac{25}{2}$.

47.
$$\frac{3x}{4} = -16$$
$$\frac{4}{3}\left(\frac{3}{4}x\right) = \frac{4}{3}(-16)$$
$$x = -\frac{64}{3}$$
Check: $\frac{3}{4}\left(-\frac{64}{3}\right) \overset{?}{=} -16$

$$\frac{\cancel{3}}{\cancel{4}}\left(-\frac{\overset{16}{\cancel{64}}}{\cancel{3}}\right) \overset{?}{=} -16$$
$$-16 = 16$$
The solution is $-\frac{64}{3}$.

49.
$$-x = 0 \qquad\qquad \text{Check: } -(0)\overset{?}{=}0$$
$$-(-x) = -(0) \qquad\qquad\qquad 0 = 0$$
$$x = 0$$
The solution is 0.

51.
$$-x = -4 \qquad\qquad \text{Check: } (4)\overset{?}{=}4$$
$$-(-x) = -(-4) \qquad\qquad\qquad 4 = 4$$
$$x = 4$$
The solution is 4.

53. Let x be the number.
$$6x = 72$$

55. Let x be the number.
$$\frac{x}{7} = 6$$

57. Let x be the number.
$$\frac{1}{3}x = 8$$

59. Let x be the number.
$$\frac{3}{4}x = 18$$

61. Let x be the number.
$$\frac{2x}{5} = 12$$

< Objective 3 >

63. Let x be the number of patients treated on Monday.
Number of patients treated on Tuesday $= 2x$
Total number of patients treated in the 2-day period $= x + 2x$.
$$48 = x + 2x$$
$$38 = 3x$$
$$x = \frac{48}{3} = 16$$
16 patients were treated on Monday.

65. Let x be the total theater audience.
$$\frac{3}{4}x = 87$$
$$\frac{4}{3}\left(\frac{3}{4}x\right) = \frac{4}{3}(87)$$
$$x = 116$$
Originally there were 116 patrons.

67.
$$200 = \frac{8}{5}x$$
$$\frac{5}{8} \times 200 = \frac{5}{8} \times \frac{8}{5}x$$
$$125 = x$$
The equivalent distance is 125 mi.

69. False

71. sometimes

73. $p = \dfrac{t}{d}$

75.
$$hp = \frac{P \cdot L \cdot A \cdot N}{33,000}$$
$$144 = \frac{P \times \dfrac{1}{3} \times 9 \times 8,000}{33,000}$$
$$4,752,000 = 24,000P$$
$$198 = P$$
The average pressure of the engine is 198 lb/in.^{2}

77. Above and Beyond

Summary Exercises

1. $4 \times 13 = 52$

$2 \times 26 = 52$

$1 \times 52 = 52$

1, 2, 4, 13, 26, and 52 are all the factors of 52.

3. Since each number has exactly two factors, 1 and itself, the numbers 2, 5, 7, 11, 17, 23, and 43 are prime numbers.

5. Since 2,350 has a last digit of 0, it is divisible by both 2 and 5. Since the sum of the digits, $2 + 3 + 5 + 0 = 10$, is not divisible by 3, the number 2,350 is not divisible by 3. Of the given numbers, 2,350 is divisible by 2 and 5, making 2 and 5 factors of 2,350.

7.

$$= \quad 8 \quad \times \quad 6$$
$$= \quad 4 \times 2 \times 2 \times 3$$
$$= 2 \times 2 \times 2 \times 2 \times 3$$
$$48 = 2 \times 2 \times 2 \times 2 \times 3$$

9.
$2\overline{)2,640}$
$2\overline{)1,320}$
$2\overline{)660}$
$2\overline{)330}$
$3\overline{)165}$
$5\overline{)55}$
11

$2640 = 2 \times 2 \times 2 \times 2 \times 3 \times 5 \times 11$

11. The factors of each of the two numbers are,

15: 1, 3, 5, 15

20: 1, 2, 4, 5, 10, 20

The GCF is 5.

13. The prime factorizations of the numbers are,

$24 = 2 \times 2 \times 2 \times 3$

$40 = 2 \times 2 \times 2 \times 5$

The GCF is $2 \times 2 \times 2 = 8$.

15. The prime factorizations of the numbers are,

$49 = 7 \times 7$

$84 = 2 \times 2 \times 3 \times 7$

$119 = 7 \times 17$

The GCF is 7.

17. $\dfrac{5}{9}$ $\leftarrow$ numerator $\leftarrow$ denominator

19. Three of the eight identical parts are shaded. The fraction $\dfrac{3}{8}$ names the shaded portion.

$\dfrac{3}{8}$ $\leftarrow$ numerator $\leftarrow$ denominator

21. Because the numerator is less than the denominator, $\dfrac{2}{3}$ and $\dfrac{7}{10}$ are proper fractions. Because the numerator is greater than or equal to the denominator, $\dfrac{5}{4}$, $\dfrac{45}{8}$, $\dfrac{7}{7}$, $\dfrac{9}{1}$, and $\dfrac{12}{5}$ are improper fractions. Because each is the sum of a whole number and a proper fraction, $2\dfrac{3}{7}$, $3\dfrac{4}{5}$, and $5\dfrac{2}{9}$ are mixed numbers.

23.
$6\overline{)41}$ with quotient 6
$\dfrac{36}{5}$

$\dfrac{41}{6} = 6\dfrac{5}{6}$ $\leftarrow$ Remainder $\leftarrow$ Original denominator, Quotient

25.
$3\overline{)23}$ with quotient 7
$\dfrac{21}{2}$

$\dfrac{23}{3} = 7\dfrac{2}{3}$ $\leftarrow$ Remainder $\leftarrow$ Original denominator, Quotient

27. $-\dfrac{14}{3}=-\left(\dfrac{14}{3}\right)$

$3\overline{)14}$ $\dfrac{14}{3}=4\dfrac{2}{3}$ $\begin{array}{l}\leftarrow \text{Remainder}\\ \\ \leftarrow \text{Original denominator}\end{array}$

$\quad\dfrac{4}{3\overline{)14}}$

$\quad\underline{12}$

$\quad\;\;2$ $\qquad\qquad$ Quotient

$-\dfrac{14}{3}=-\left(\dfrac{14}{3}\right)=-4\dfrac{2}{3}$

29. $7\dfrac{5}{8}=\dfrac{(8\times7)+5}{8}=\dfrac{61}{8}$

31. $5\dfrac{2}{7}=\dfrac{(7\times5)+2}{7}=\dfrac{37}{7}$

33. $-12\dfrac{3}{8}=-\left(12\dfrac{3}{8}\right)=-\left[\dfrac{(12\times8)+3}{8}\right]=-\dfrac{99}{8}$

35. Cross products $5\times12=60$ and $8\times7=56$ are not equal. The fractions are not equivalent.

37. $\dfrac{24}{36}=\dfrac{\cancel{2}\times\cancel{2}\times2\times\cancel{3}}{\cancel{2}\times\cancel{2}\times\cancel{3}\times3}=\dfrac{2}{3}$

39. $\dfrac{140}{180}=\dfrac{\cancel{2}\times\cancel{2}\times\cancel{5}\times7}{\cancel{2}\times\cancel{2}\times3\times3\times\cancel{5}}=\dfrac{7}{9}$

41. $-\dfrac{30}{105}=-\dfrac{2\times\cancel{3}\times\cancel{5}}{\cancel{3}\times\cancel{5}\times7}=-\dfrac{2}{7}$

43. $\dfrac{15}{25}=\dfrac{3\times\cancel{5}}{5\times\cancel{5}}=\dfrac{3}{5}$

Therefore, the statement is true.

45. $\dfrac{7}{15}\times\dfrac{5}{21}=\dfrac{\cancel{7}\times\cancel{5}}{\cancel{15}\times\cancel{21}}=\dfrac{1\times1}{3\times3}=\dfrac{1}{9}$

47. $4\times\dfrac{3}{8}=\dfrac{4}{1}\times\dfrac{3}{8}=\dfrac{\cancel{4}\times3}{1\times\cancel{8}}=\dfrac{3}{2}=1\dfrac{1}{2}$

49. $5\dfrac{1}{3}\times1\dfrac{4}{5}=\dfrac{16}{3}\times\dfrac{9}{5}=\dfrac{16\times\cancel{9}}{\cancel{3}\times5}=\dfrac{48}{5}=9\dfrac{3}{5}$

51. $3\dfrac{1}{5}\times\dfrac{7}{8}\times2\dfrac{6}{7}=\dfrac{16}{5}\times\dfrac{7}{8}\times\dfrac{20}{7}=\dfrac{\cancel{16}\times\cancel{7}\times\cancel{20}}{\cancel{5}\times\cancel{8}\times\cancel{7}}$

$=\dfrac{8}{1}=8$

53. $-\dfrac{5}{9}\times\dfrac{6}{7}=-\dfrac{5\times\cancel{6}}{\cancel{9}\times7}=-\dfrac{10}{21}$

55. $\left(-\dfrac{8}{15}\right)\bullet\left(-\dfrac{3}{4}\right)=\dfrac{\cancel{8}\times\cancel{3}}{\cancel{15}\times\cancel{4}}=\dfrac{2}{5}$

57. $8\dfrac{1}{3}\times\left(-3\dfrac{3}{4}\right)=-\dfrac{25}{3}\times\dfrac{15}{4}=-\dfrac{25\times\cancel{15}}{\cancel{3}\times4}$

$=-\dfrac{125}{4}=-31\dfrac{1}{4}$

59. $2\dfrac{3}{4}\text{ in.}\times\dfrac{80\text{ mi}}{1\text{ in.}}=\dfrac{11}{4}\cancel{\text{ in.}}\times\dfrac{80\text{ mi}}{1\cancel{\text{ in.}}}$

$=\dfrac{11\times\cancel{80}}{\cancel{4}\times1}\text{ mi}=220\text{ mi}$

The actual distance between the cities is 220 mi.

61. $\left(6\dfrac{2}{3}\,\text{yd} \times 4\dfrac{1}{2}\,\text{yd}\right) \times \dfrac{\$18}{1\,\text{yd}^2}$

$= \left(\dfrac{\overset{10}{\cancel{20}}}{\cancel{3}_{1}}\,\text{yd} \times \dfrac{\overset{3}{\cancel{9}}}{\cancel{2}_{1}}\,\text{yd}\right) \times \dfrac{\$18}{1\,\text{yd}^2} = 30\,\cancel{\text{yd}^2} \times \dfrac{\$18}{1\,\cancel{\text{yd}^2}}$

$= \$540$

The cost to carpet the room is \$540.

63. $3\dfrac{2}{5}\,\text{in.} \times \dfrac{120\,\text{mi}}{1\,\text{in.}} = \dfrac{17}{5}\,\cancel{\text{in.}} \times \dfrac{120\,\text{mi}}{1\,\cancel{\text{in.}}}$

$= \dfrac{17 \times \overset{24}{\cancel{120}}}{\cancel{5}_{1} \times 1}\,\text{mi} = 408\,\text{mi}$

The actual distance is 408 mi.

65. $\dfrac{3}{4} \times \dfrac{5}{6} = \dfrac{\overset{1}{\cancel{3}} \times 5}{4 \times \cancel{6}_{2}} = \dfrac{5}{8}$

$\dfrac{5}{8}$ of the students surveyed work more than 20 hr per week.

67. $\dfrac{5}{12} \div \dfrac{5}{8} = \dfrac{5}{12} \times \dfrac{8}{5} = \dfrac{\overset{1}{\cancel{5}} \times \overset{2}{\cancel{8}}}{\underset{3}{\cancel{12}} \times \underset{1}{\cancel{5}}} = \dfrac{2}{3}$

69. $\dfrac{\frac{9}{20}}{2\frac{2}{5}} = \dfrac{\frac{9}{20}}{\frac{12}{5}} = \dfrac{9}{20} \times \dfrac{5}{12} = \dfrac{\overset{3}{\cancel{9}} \times \overset{1}{\cancel{5}}}{\underset{4}{\cancel{20}} \times \underset{4}{\cancel{12}}} = \dfrac{3}{16}$

71. $3\dfrac{3}{7} \div 8 = \dfrac{24}{7} \div \dfrac{8}{1} = \dfrac{\overset{3}{\cancel{24}}}{7} \times \dfrac{1}{\cancel{8}_{1}} = \dfrac{3}{7}$

73. $\dfrac{9}{22} \div \left(-\dfrac{6}{55}\right) = -\dfrac{\overset{3}{\cancel{9}}}{\underset{2}{\cancel{22}}} \times \dfrac{\overset{5}{\cancel{55}}}{\underset{2}{\cancel{6}}} = -\dfrac{15}{4} = -3\dfrac{3}{4}$

75. $\dfrac{-\frac{5}{6}}{-\frac{3}{10}} = \dfrac{\overset{5}{\cancel{5}}}{\cancel{6}} \times \dfrac{\overset{5}{\cancel{10}}}{3} = \dfrac{25}{9} = 2\dfrac{7}{9}$

77. $-8 \div \left(-\dfrac{4}{5}\right) = \overset{2}{\cancel{8}} \times \dfrac{5}{\cancel{4}} = 10$

79. $3\dfrac{3}{4}\,\text{ft} \div 5 = \dfrac{15}{4}\,\text{ft} \div \dfrac{5}{1} = \dfrac{\overset{3}{\cancel{15}}}{4}\,\text{ft} \times \dfrac{1}{\cancel{5}_{1}} = \dfrac{3}{4}\,\text{ft}$

81. $126\,\text{mi} \div 2\dfrac{1}{4}\,\text{hr} = 126\,\text{mi} \div \dfrac{9}{4}\,\text{hr}$

$= \overset{14}{\cancel{126}}\,\text{mi} \times \dfrac{4}{\cancel{9}_{1}}\dfrac{1}{\text{hr}} = 56\,\text{mi/hr}$

83. $5x = 35$ Check: $5(7)\overset{?}{=}35$

$\dfrac{5x}{5} = \dfrac{35}{5}$ $35 = 35$

$x = 7$

The solution is 7.

85. $-6x = 24$ Check: $-6(-4)\overset{?}{=}24$

$\dfrac{-6x}{-6} = \dfrac{24}{-6}$ $24 = 24$

$x = -4$

The solution is −4.

87. $\dfrac{x}{4} = 8$ Check: $\dfrac{32}{4}\overset{?}{=}8$

$4\left(\dfrac{x}{4}\right) = 8(4)$ $8 = 8$

$x = 32$

The solution is 32.

89.
$$\frac{2}{3}x = 18 \qquad \text{Check: } \frac{2}{3}(27) \overset{?}{=} 18$$
$$18 = 18$$
$$3\left(\frac{2}{3}x\right) = 18(3)$$
$$2x = 54$$
$$\frac{2x}{2} = \frac{54}{2}$$
$$x = 27$$
The solution is 27.

91. Let x be the regular price of Holiday pack.
$$\frac{3}{4}x = 57$$
$$\frac{4}{3} \times \frac{3}{4}x = \frac{4}{\cancel{3}} \times \cancel{57}^{19}$$
$$x = 76$$
The regular price of the Holiday pack was $76.

Chapter Test 3

1. Since each has exactly two factors, 1 and itself, the numbers 5, 13, 17, and 31 are prime numbers.
1, 3, and 9 are factors of 9.
1, 2, 11, and 22 are factors of 22.
1, 3, 9, and 27 are factors of 27.
1, 3, 5, 9, 15, and 45 are factors of 45.
Since each has more than two factors, the numbers 9, 22, 27, and 45 are composite numbers.

3. Since 54,204 has a last digit of 4, the number is divisible by 2. Since the sum of the digits, $5 + 4 + 2 + 0 + 4 = 15$, is divisible by 3, the number 54,204 is divisible by 3. Since 54,204 does not have a last digit of 0 or 5, the number is not divisible by 5. Of the given numbers, 54,204 is divisible by 2 and 3, making 2 and 3 factors of 54,204

5. The prime factorizations of the numbers are,
$16 = 2 \times 2 \times 2 \times 2$
$24 = 2 \times 2 \times 2 \times 3$
$72 = 2 \times 2 \times 2 \times 3 \times 3$
The GCF is $2 \times 2 \times 2 = 8$.

7. Five of the eight identical parts are shaded. The fraction $\frac{5}{8}$ names the shaded part.
$$\frac{5}{8} \begin{array}{l} \leftarrow \text{numerator} \\ \leftarrow \text{denominator} \end{array}$$

9. The fully shaded circles represent the whole number 4. For the remaining circle, one of the four identical parts is shaded. The fraction $\frac{1}{4}$ names this shaded part. So the mixed number $4\frac{1}{4}$ names the shaded portion of the diagram.

11. $\dfrac{36}{84} = \dfrac{\overset{1}{\cancel{2}} \times \overset{1}{\cancel{2}} \times \overset{1}{\cancel{3}} \times 3}{\underset{1}{\cancel{2}} \times \underset{1}{\cancel{2}} \times \underset{1}{\cancel{3}} \times 7} = \dfrac{3}{7}$

13. Cross products $2 \times 28 = 56$ and $7 \times 8 = 56$ are equal. The fractions are equivalent.

15. Cross products $3 \times 15 = 45$ and $20 \times 2 = 40$ are not equal. The fractions are not equivalent.

17. $5\dfrac{2}{7} = \dfrac{(7 \times 5) + 2}{7} = \dfrac{37}{7}$

19. $8\dfrac{2}{9} = \dfrac{(9 \times 8) + 2}{9} = \dfrac{74}{9}$

21.
$$\begin{array}{r} 4 \\ 4\overline{)17} \\ \underline{16} \\ 1 \end{array} \qquad \frac{17}{4} = 4\frac{1}{4} \begin{array}{l} \leftarrow \text{Remainder} \\ \downarrow \leftarrow \text{Original denominator} \\ \text{Quotient} \end{array}$$

23.

$$8\overline{)74}$$

$\quad\quad\quad\quad$ Remainder

$$\frac{72}{2}$$

$$\frac{74}{8} = 9\frac{2}{8} = 9\frac{1}{4}$$

Quotient$\quad$Original denominator

25. $-\dfrac{82}{64} = -\left(\dfrac{82}{64}\right)$

$$64\overline{)82}$$

$$\frac{82}{64} = 1\frac{18}{64} = 1\frac{9}{32}$$

$$\frac{64}{18}$$

$$-\frac{82}{64} = -\left(\frac{82}{64}\right) = -1\frac{9}{32}$$

27. $\dfrac{2}{3} \times \dfrac{5}{7} = \dfrac{2\times 5}{3\times 7} = \dfrac{10}{21}$

29. $\dfrac{7}{12} \div \dfrac{14}{15} = \dfrac{7}{12} \times \dfrac{15}{14} = \dfrac{\overset{1}{\cancel{7}} \times \overset{5}{\cancel{15}}}{\underset{4}{\cancel{12}} \times \underset{2}{\cancel{14}}} = \dfrac{5}{8}$

31. $\dfrac{16}{35} \times \dfrac{14}{24} = \dfrac{\overset{2}{\cancel{16}} \times \overset{2}{\cancel{14}}}{\underset{5}{\cancel{35}} \times \underset{3}{\cancel{24}}} = \dfrac{2\times 2}{5\times 3} = \dfrac{4}{15}$

33. $\dfrac{9}{10} \times \dfrac{5}{8} = \dfrac{9\times \overset{1}{\cancel{5}}}{\underset{2}{\cancel{10}} \times 8} = \dfrac{9\times 1}{2\times 8} = \dfrac{9}{16}$

35. $3\dfrac{5}{6} \times 2\dfrac{2}{5} = \dfrac{23}{6} \times \dfrac{12}{5} = \dfrac{23\times \overset{2}{\cancel{12}}}{\underset{1}{\cancel{6}} \times 5} = \dfrac{23\times 2}{1\times 5} = \dfrac{46}{5}$

$$= 9\frac{1}{5}$$

37. $-14\left(-\dfrac{5}{6}\right) = \overset{7}{\cancel{14}} \times \dfrac{5}{\underset{3}{\cancel{6}}} = \dfrac{35}{3} = 11\dfrac{2}{3}$

39. $\dfrac{-\dfrac{7}{12}}{-\dfrac{1}{6}} = \dfrac{7}{\underset{2}{\cancel{12}}} \times \dfrac{\overset{1}{\cancel{6}}}{1} = \dfrac{7}{2} = 3\dfrac{1}{2}$

41. $48\cancel{c}/lb \times 2\dfrac{3}{4}\,lb = \dfrac{48}{1}\dfrac{\cancel{c}}{\cancel{lb}} \times \dfrac{11}{4}\,\cancel{lb} = \dfrac{\overset{12}{\cancel{48}}\times 11}{1\times \underset{1}{\cancel{4}}}$

$$= 132c \text{ or } \$1.32$$

The cost of $2\dfrac{3}{4}$ lb of apples is \$1.32

43. $31\dfrac{1}{3} \div \dfrac{2}{3} = \dfrac{94}{3} \div \dfrac{2}{3} = \dfrac{\overset{47}{\cancel{94}}}{\underset{1}{\cancel{3}}} \times \dfrac{\overset{1}{\cancel{3}}}{\underset{1}{\cancel{2}}} = 47$

47 homes can be built.

45. $2\dfrac{3}{8}\,in. \times \dfrac{80\ mi}{1\ in.} = \dfrac{19}{8}\,\cancel{in.} \times \dfrac{80\ mi}{1\ \cancel{in.}}$

$$= \dfrac{19 \times \overset{10}{\cancel{80}}}{\underset{1}{\cancel{8}} \times 1}\,mi = 190\ mi$$

The actual distance between the towns is 190 mi.

Cumulative Review: Chapters 1–3

1. 3,738,500
0 ones, 0 tens, 5 hundreds, 8 thousands,
$\quad$ 3 ten thousands, 7 hundred thousands,
$\quad$ 3 millions
The place value of 7 is hundred thousands.

3. Two million, four hundred thirty thousand as a numeral is 2,430,000.

5. The sum of 9 and 0 is 9 by the additive identity property.

7.
$$
\begin{array}{r}
{}^{2\,1}\\
593\\
275\\
+\ 98\\
\hline
966
\end{array}
$$

9. The hundreds digit is 8. The digit to the right, 7, is 5 or more. So round up. 5,873 is rounded up to 5,900.

11. Estimate:
$$
\begin{array}{r}
{}^{2}\\
900\\
3,300\\
800\\
2,100\\
+\ 600\\
\hline
7,700
\end{array}
$$

13. Since 80 lies to the left of 90 on the number line, $80 < 90$.

15.
$$
\begin{array}{r}
{}^{4}\,2\slashed{8},{}^{1}000\\
-\ 7,535\\
\end{array}
$$

$$
\begin{array}{r}
{}^{9}\\
{}^{4}\slashed{}\,{}^{\slashed{10}}\\
2\slashed{8},\slashed{0}\,{}^{1}00\\
-\ 7,5\,35\\
\end{array}
$$

$$
\begin{array}{r}
{}^{9\ 9}\\
{}^{4\ \slashed{10}\ \slashed{10}}\\
2\slashed{8},\slashed{0}\slashed{0}\,{}^{1}0\\
-\ 7,5\,3\,5\\
\hline
4\,6\,5
\end{array}
$$

$$
\begin{array}{r}
{}^{9\ 9}\\
{}^{1\ 14\ \slashed{10}\ \slashed{10}}\\
\slashed{2}\,\slashed{8},\slashed{0}\slashed{0}\,{}^{1}0\\
-\ 7,5\,3\,5\\
\hline
1\,7,4\,6\,5
\end{array}
$$

17. First, find the total amount invested in the car using addition. Write $18,975 + 439 + 615 = 20,029$. Alan spent \$20,029 to purchase the car. To find the balance, subtract the down payment made. Write $20,029 - 2450 = 17,579$. A balance of \$17,579 remained on the car.

19. The order of the factors, 3 and 4, was changed. The product remains the same by the commutative property of multiplication.

21.
$$
\begin{array}{r}
{}^{2\ 5}\\
{}^{1\ 2}\\
538\\
\times\ 703\\
\hline
1614\\
376600\\
\hline
378,214
\end{array}
$$

23. First, find the area of the classroom.
$8\text{ yd} \times 9\text{ yd} = 72\text{ yd}^2$
Then multiply by the cost per square yard.
$72 \times \$14 = \$1,008$
The carpeting will cost \$1,008.

25.
$$
\begin{array}{r}
103\\
458\overline{)47,350}\\
458\\
\hline
155\\
0\\
\hline
1550\\
1374\\
\hline
176
\end{array}
$$
We have $47,350 \div 458 = 103\text{ r}176$.

27. $(3+5) \times 7 = 8 \times 7 = 56$

29. $2 + 8 \times 3 \div 4 = 2 + 24 \div 4 = 2 + 6 = 8$

31. $32 + (-45) = -13$

33. $40 - 67 = 40 + (-67) = -27$

35. $8(-14) = -112$

37. $2x - 3y + z = 2(-4) - 3(2) + 5 = -8 - 6 + 5$
$= -9$

39.
$$
\begin{aligned}
x + 6 &= 18\\
x + 6 + (-6) &= 18 + (-6)\\
x &= 12
\end{aligned}
$$

41. Since each has exactly two factors, 1 and itself, the numbers 5, 13, 17, and 31 are prime numbers.

1, 3, and 9 are factors of 9.

1, 2, 11, and 22 are factors of 22.

1, 3, 9, and 27 are factors of 27.

1, 3, 5, 9, 15, and 45 are factors of 45.

Since each has more than two factors, the numbers 9, 22, 27, and 45 are composite numbers.

43. $2\overline{)264} \searrow 2\overline{)132} \searrow 2\overline{)66} \searrow 3\overline{)33}$ with quotients $132, 66, 33, 11$

$$264 = 2 \times 2 \times 2 \times 3 \times 11$$

45. The factors of each of the two numbers are,

16: 1, 2, 4, 8, 16

40: 1, 2, 4, 5, 8, 10, 20, 40

72: 1, 2, 3, 4, 6, 8, 9, 12, 18, 24, 36, 72

The GCF is 8.

47. $5\overline{)14}$ quotient 2, $\frac{14}{5} = 2\frac{4}{5}$ ← Remainder, ↓5 ← Original denominator, Quotient

$\frac{10}{4}$

49. $4\frac{1}{3} = \frac{(3 \times 4) + 1}{3} = \frac{13}{3}$

51. $-5\frac{3}{20} = -\left(5\frac{3}{20}\right) = -\left[\frac{(20 \times 5) + 3}{20}\right] = -\frac{103}{20}$

53. Cross products $7 \times 24 = 168$ and $21 \times 8 = 168$ are equal. The fractions are equivalent.

55. $\frac{28}{42} = \frac{\cancel{2} \times 2 \times \cancel{7}}{\cancel{2} \times 3 \times \cancel{7}} = \frac{2}{3}$

57. $-\frac{12}{20} = -\frac{\cancel{2} \times \cancel{2} \times 3}{\cancel{2} \times \cancel{2} \times 5} = -\frac{3}{5}$

59. $\frac{5}{9} \times \frac{8}{15} = \frac{\cancel{5} \times 8}{9 \times \cancel{15}} = \frac{1 \times 8}{9 \times 3} = \frac{8}{27}$

61. $1\frac{1}{8} \cdot 4\frac{4}{5} = \frac{9}{8} \times \frac{24}{5} = \frac{9 \times \cancel{24}}{\cancel{8} \times 5} = \frac{9 \times 3}{1 \times 5} = \frac{27}{5}$

$$= 5\frac{2}{5}$$

63. $\frac{2}{3} \times 1\frac{4}{5} \times \frac{5}{8} = \frac{2}{3} \times \frac{9}{5} \times \frac{5}{8} = \frac{\cancel{2} \times \cancel{9} \times \cancel{5}}{\cancel{3} \times \cancel{5} \times \cancel{8}} = \frac{3}{4}$

65. $\frac{5}{8} \div \frac{15}{32} = \frac{5}{8} \times \frac{32}{15} = \frac{\cancel{5} \times \cancel{32}}{\cancel{8} \times \cancel{15}} = \frac{4}{3} = 1\frac{1}{3}$

67. $4\frac{1}{6} \div 5 = \frac{25}{6} \div \frac{5}{1} = \frac{\cancel{25}}{6} \times \frac{1}{\cancel{5}} = \frac{5}{6}$

69. $\left(6\frac{2}{3}\text{yd} \times 4\frac{1}{2}\text{yd}\right) \times \frac{\$18}{1\text{ yd}^2}$

$$= \left(\frac{\cancel{20}}{\cancel{3}}\text{yd} \times \frac{\cancel{9}}{\cancel{2}}\text{yd}\right) \times \frac{\$18}{1\text{ yd}^2} = 30\cancel{\text{ yd}^2} \times \frac{\$18}{1\cancel{\text{ yd}^2}}$$

$$= \$540$$

The cost to carpet the room is $540

71. $4x = -24$ Check: $4(-6) \overset{?}{=} -24$

$\frac{4x}{4} = \frac{-24}{4}$ $-24 = -24$

$x = -6$

The solution is −6.

Chapter 4
Adding and Subtracting Fractions

Prerequisite Check

1. $2\overline{)24}^{\,12} \searrow 2\overline{)12}^{\,6} \searrow 2\overline{)6}^{\,3}$

   $24 = 2 \times 2 \times 2 \times 3$

3. $2\overline{)90}^{\,45} \searrow 3\overline{)45}^{\,15} \searrow 3\overline{)15}^{\,5}$

   $24 = 2 \times 3 \times 3 \times 5$

5. $-6\dfrac{2}{5} = -\left[\dfrac{(5 \times 6) + 2}{5}\right] = -\dfrac{32}{5}$

7. $4\overline{)29}^{\,7}$ $\dfrac{29}{4} = 7\dfrac{1}{4}$ $\begin{array}{l}\leftarrow \text{Remainder}\\ \leftarrow \text{Original denominator}\end{array}$

   $\dfrac{28}{1}$ Quotient

9. $-\dfrac{8}{8} = -1$

11. To simplify $\dfrac{8}{12}$, factor.

    $\dfrac{8}{12} = \dfrac{\cancel{2}^{\,1} \times \cancel{2}^{\,1} \times 2}{\cancel{2}_{\,1} \times \cancel{2}_{\,1} \times 3} = \dfrac{2}{3}$

13. To simplify $\dfrac{21}{35}$, factor.

    $\dfrac{21}{35} = \dfrac{\cancel{7}^{\,1} \times 3}{\cancel{7}_{\,1} \times 5} = \dfrac{3}{5}$

Exercises 4.1

< Objective 1 >

1. $\dfrac{3}{5} + \dfrac{1}{5} = \dfrac{4}{5}$

3. $\dfrac{4}{11} + \dfrac{6}{11} = \dfrac{10}{11}$

5. $\dfrac{2}{10} + \dfrac{3}{10} = \dfrac{5}{10} = \dfrac{1}{2}$

7. $\dfrac{3}{7} + \dfrac{4}{7} = \dfrac{7}{7} = 1$

9. $\dfrac{29}{30} + \dfrac{11}{30} = \dfrac{40}{30} = \dfrac{4}{3} = 1\dfrac{1}{3}$

11. $\dfrac{13}{48} + \dfrac{23}{48} = \dfrac{36}{48} = \dfrac{3}{4}$

13. $\dfrac{3}{7} + \dfrac{6}{7} = \dfrac{9}{7} = 1\dfrac{2}{7}$

15. $\dfrac{7}{10} + \dfrac{9}{10} = \dfrac{16}{10} = \dfrac{8}{5} = 1\dfrac{3}{5}$

17. $\dfrac{11}{12} + \dfrac{10}{12} = \dfrac{21}{12} = \dfrac{7}{4} = 1\dfrac{3}{4}$

< Objective 2 >

19. $\dfrac{1}{8} + \dfrac{1}{8} + \dfrac{3}{8} = \dfrac{5}{8}$

75

21. $\dfrac{1}{9} + \dfrac{4}{9} + \dfrac{5}{9} = \dfrac{10}{9} = 1\dfrac{1}{9}$

< Objective 3 >

23. $\dfrac{3}{5} - \dfrac{1}{5} = \dfrac{3-1}{5} = \dfrac{2}{5}$

25. $\dfrac{7}{9} - \dfrac{4}{9} = \dfrac{7-4}{9} = \dfrac{3}{9} = \dfrac{1}{3}$

27. $\dfrac{13}{20} - \dfrac{3}{20} = \dfrac{13-3}{20} = \dfrac{10}{20} = \dfrac{1}{2}$

29. $\dfrac{19}{24} - \dfrac{5}{24} = \dfrac{19-5}{24} = \dfrac{14}{24} = \dfrac{7}{12}$

31. $\dfrac{7}{12} - \dfrac{11}{12} = \dfrac{7-11}{12} = -\dfrac{4}{12} = -\dfrac{1}{3}$

33. $\dfrac{3}{9} - \dfrac{8}{9} = \dfrac{3-8}{9} = -\dfrac{5}{9}$

< Objective 4 >

35. $-\dfrac{12}{25} + \dfrac{18}{25} = \dfrac{-12+18}{25} = \dfrac{6}{25}$

37. $-\dfrac{4}{5} + \dfrac{3}{5} = \dfrac{-4+3}{5} = -\dfrac{1}{5}$

39. $\dfrac{23}{32} + \left(-\dfrac{5}{32}\right) = \dfrac{23}{32} - \dfrac{5}{32} = \dfrac{23-5}{32} = \dfrac{\overset{9}{\cancel{18}}}{\underset{16}{\cancel{32}}} = \dfrac{9}{16}$

41. $\dfrac{17}{64} + \left(-\dfrac{35}{64}\right) = \dfrac{17}{64} - \dfrac{35}{64} = \dfrac{17-35}{64} = -\dfrac{\overset{9}{\cancel{18}}}{\underset{32}{\cancel{64}}}$

$\qquad = -\dfrac{9}{32}$

43. $-\dfrac{40}{43} + \left(-\dfrac{8}{43}\right) = -\dfrac{40}{43} - \dfrac{8}{43} = \dfrac{-40-8}{43}$

$\qquad = -\dfrac{48}{43} \text{ or } -1\dfrac{5}{43}$

45. $\dfrac{3}{4} - \left(-\dfrac{3}{4}\right) = \dfrac{3}{4} + \dfrac{3}{4} = \dfrac{3+3}{4} = \dfrac{\overset{3}{\cancel{6}}}{\underset{2}{\cancel{4}}} = \dfrac{3}{2} \text{ or } 1\dfrac{1}{2}$

47. $-\dfrac{3}{16} - \dfrac{9}{16} = \dfrac{-3-9}{16} = -\dfrac{\overset{3}{\cancel{12}}}{\underset{4}{\cancel{16}}} = -\dfrac{3}{4}$

49. $-\dfrac{7}{10} - \left(-\dfrac{7}{10}\right) = -\dfrac{7}{10} + \dfrac{7}{10} = 0$

51. $-\dfrac{3}{4} - \left(-\dfrac{1}{4}\right) = -\dfrac{3}{4} + \dfrac{1}{4} = \dfrac{-3+1}{4} = -\dfrac{2}{4} = -\dfrac{1}{2}$

53. $-\dfrac{1}{9} - \left(-\dfrac{5}{9}\right) = -\dfrac{1}{9} + \dfrac{5}{9} = \dfrac{-1+5}{9} = \dfrac{4}{9}$

55. $\dfrac{7}{12} - \dfrac{4}{12} + \dfrac{3}{12} = \dfrac{7-4+3}{12} = \dfrac{6}{12} = \dfrac{1}{2}$

57. $\dfrac{6}{13} - \dfrac{3}{13} + \dfrac{11}{13} = \dfrac{6-3+11}{13} = \dfrac{14}{13} = 1\dfrac{1}{13}$

59. $\dfrac{18}{23} - \dfrac{13}{23} - \dfrac{3}{23} = \dfrac{18-13-3}{23} = \dfrac{2}{23}$

61. 1 hour $= \dfrac{1}{24}$ of a day.

$\dfrac{7}{24} + \dfrac{5}{24} + \dfrac{6}{24} = \dfrac{18}{24} = \dfrac{3}{4}$ of a day.

63. $\dfrac{7}{10} + \dfrac{2}{10} + \dfrac{7}{10} + \dfrac{2}{10} = \dfrac{18}{10} = \dfrac{9}{5} = 1\dfrac{4}{5}$

The perimeter is $\dfrac{9}{5}$ in. or $1\dfrac{4}{5}$ in.

65. $2 - \dfrac{4}{9} - \dfrac{7}{9} = \dfrac{18}{9} - \dfrac{4}{9} - \dfrac{7}{9} = \dfrac{18-4-7}{9} = \dfrac{7}{9}$

He should spend $\dfrac{7}{9}$ hr on Sunday.

67. $\dfrac{3}{4} + \dfrac{5}{4} + \dfrac{7}{4} = \dfrac{3+5+7}{4} = \dfrac{15}{4} = 3\dfrac{3}{4}$

The perimeter is $3\dfrac{3}{4}$ in.

69. $\dfrac{7}{8}+\dfrac{9}{8}+\dfrac{9}{8}=\dfrac{7+9+9}{8}=\dfrac{25}{8}=3\dfrac{1}{8}$

The perimeter is $3\dfrac{1}{8}$ in.

71. $\dfrac{7}{8}+\dfrac{11}{8}+\dfrac{15}{8}+\dfrac{5}{8}+\dfrac{7}{8}+\dfrac{15}{8}$

$=\dfrac{7+11+15+5+7+15}{8}=\dfrac{60}{8}=\dfrac{15}{2}=7\dfrac{1}{2}$

The perimeter is $7\dfrac{1}{2}$ in.

73. $\dfrac{1}{8}+\dfrac{1}{8}+\dfrac{3}{8}=\dfrac{1+1+3}{8}=\dfrac{5}{8}$

Carla will receive $\dfrac{5}{8}$ g of medication in one day.

75. $\dfrac{5}{8}-\dfrac{1}{8}=\dfrac{4}{8}=\dfrac{1}{2}$

$\dfrac{1}{2}$ ton of steel has yet to arrive.

Exercises 4.2

< Objective 1 >

1. $2=2$
$\underline{3=\quad 3}$
$\quad 2\times3$ Bring down the factors.
$2\times3=6$
So 6 is the LCM of 2 and 3.

3. $4=2\times2$
$\underline{6=2\quad\times3}$
$\quad 2\times2\times3$ Bring down the factors.
$2\times2\times3=12$
So 12 is the LCM of 4 and 6.

5. $10=2\quad\times5$
$\underline{20=2\times2\times5}$
$\quad 2\times2\times5$ Bring down the factors.
$2\times2\times5=20$
So 20 is the LCM of 10 and 20.

7. $9=\quad\quad 3\times3$
$\underline{12=2\times2\times3}$
$\quad 2\times2\times3\times3$ Bring down the factors.
$2\times2\times3\times3=36$
So 36 is the LCM of 9 and 12.

9. $12=2\times2\quad\times3$
$\underline{16=2\times2\times2\times2}$
$\quad 2\times2\times2\times2\times3$ Bring down the factors.
$2\times2\times2\times2\times3=48$
So 48 is the LCM of 12 and 16.

11. $12=2\times2\times3$
$\underline{15=\quad\quad 3\times5}$
$\quad 2\times2\times3\times5$ Bring down the factors.
$2\times2\times3\times5=60$
So 60 is the LCM of 12 and 15.

13. $18=2\quad\times3\times3$
$\underline{36=2\times2\times3\times3}$
$\quad 2\times2\times3\times3$ Bring down the factors.
$2\times2\times3\times3=36$
So 36 is the LCM of 18 and 36.

15. $25=\quad\quad 5\times5$
$\underline{40=2\times2\times2\times5}$
$\quad 2\times2\times2\times5\times5$ Bring down the factors.
$2\times2\times2\times5\times5=200$
So 200 is the LCM of 25 and 40.

< Objective 2 >

17. $3=\quad 3$
$5=\quad\quad 5$
$\underline{6=2\times3}$
$\quad 2\times3\times5$ Bring down the factors.
$2\times3\times5=30$
So 30 is the LCM of 3, 5, and 6.

19. $18 = 2 \quad \times 3 \times 3$

$21 = \quad\quad 3 \quad \times 7$

$28 = \underline{2 \times 2 \quad\quad \times 7}$

$\quad\quad 2 \times 2 \times 3 \times 3 \times 7$ Bring down the factors.

$2 \times 2 \times 3 \times 3 \times 7 = 252$

So 252 is the LCM of 18, 21, and 28.

21. $20 = 2 \times 2 \quad\quad \times 5$

$30 = 2 \quad \times 3 \quad \times 5$

$45 = \underline{\quad\quad 3 \times 3 \times 5}$

$\quad\quad 2 \times 2 \times 3 \times 3 \times 5$ Bring down the factors.

$2 \times 2 \times 3 \times 3 \times 5 = 180$

So 180 is the LCM of 20, 30, and 45.

23. To find the LCM of 8 and 3, we factor each number.

$8 = 2 \times 2 \times 2$

$3 = \underline{\quad\quad\quad 3}$

$\quad\quad 2 \times 2 \times 2 \times 3$ Bring down the factors.

$2 \times 2 \times 2 \times 3 = 24$

So 24 is the LCM of 8 and 3.

25. To find the LCM of 2 and 8, we factor each number.

$2 = 2$

$8 = \underline{2 \times 2 \times 2}$

$\quad\quad 2 \times 2 \times 2$ Bring down the factors.

$2 \times 2 \times 2 = 8$

So 8 is the LCM of 2 and 8.

27. To find the LCM of 4 and 5, we factor each number.

$4 = 2 \times 2$

$5 = \underline{\quad\quad 5}$

$\quad\quad 2 \times 2 \times 5$ Bring down the factors.

$2 \times 2 \times 5 = 20$

So 20 is the LCM of 4 and 5.

29. To find the LCM of 6 and 9, we factor each number.

$6 = 2 \times 3$

$9 = \underline{\quad 3 \times 3}$

$\quad\quad 2 \times 3 \times 3$ Bring down the factors.

$2 \times 3 \times 3 = 18$

So 18 is the LCM of 6 and 9.

31. Find a form of one to multiply by.

$$\frac{4}{5} = \frac{}{25}$$

$$\frac{4}{5}\left(\frac{5}{5}\right) = \frac{\boxed{20}}{25}$$

33. Find a form of one to multiply by.

$$\frac{5}{6} = \frac{25}{}$$

$$\frac{5}{6}\left(\frac{5}{5}\right) = \frac{25}{\boxed{30}}$$

35. Find a form of one to multiply by.

$$\frac{11}{37} = \frac{}{111}$$

$$\frac{11}{37}\left(\frac{3}{3}\right) = \frac{\boxed{33}}{111}$$

37. Find a form of one to multiply by.

$$\frac{1}{2} = \frac{}{16}$$

$$\frac{1}{2}\left(\frac{8}{8}\right) = \frac{\boxed{8}}{16}$$

39. Find a form of one to multiply by.

$$\frac{5}{12} = \frac{25}{}$$

$$\frac{5}{12}\left(\frac{5}{5}\right) = \frac{25}{\boxed{60}}$$

41. Find a form of one to multiply by.

$$\frac{17}{10} = \frac{}{30}$$

$$\frac{17}{10}\left(\frac{3}{3}\right) = \frac{\boxed{51}}{30}$$

< Objective 3 >

43. Because 170 is a common multiple of 17 and 10, let's use 170 as our common denominator. Because $\dfrac{12}{17} = \dfrac{120}{170}$ and $\dfrac{9}{10} = \dfrac{153}{170}$, we see that $\dfrac{12}{17}$ is smaller than $\dfrac{9}{10}$. The order, from smaller to larger, is $\dfrac{12}{17}$, $\dfrac{9}{10}$.

45. Because 40 is a common multiple of 8 and 5, let's use 40 as our common denominator. Because $\dfrac{5}{8} = \dfrac{25}{40}$ and $\dfrac{3}{5} = \dfrac{24}{40}$, we see that $\dfrac{3}{5}$ is smaller than $\dfrac{5}{8}$. The order, from smaller to larger, is $\dfrac{3}{5}$, $\dfrac{5}{8}$.

47. Because 24 is a common multiple of 8, 3, and 4, let's use 24 as our common denominator. Because $\dfrac{3}{8} = \dfrac{9}{24}$, $\dfrac{1}{3} = \dfrac{8}{24}$, and $\dfrac{1}{4} = \dfrac{6}{24}$, we see that $\dfrac{1}{4}$ is smaller than $\dfrac{1}{3}$ and $\dfrac{1}{3}$ is smaller than $\dfrac{3}{8}$. The order, from smallest to largest, is $\dfrac{1}{4}$, $\dfrac{1}{3}$, $\dfrac{3}{8}$.

49. Because 60 is a common multiple of 12, 5, and 6, let's use 60 as our common denominator. Because $\dfrac{11}{12} = \dfrac{55}{60}$, $\dfrac{4}{5} = \dfrac{48}{60}$, and $\dfrac{5}{6} = \dfrac{50}{60}$, we see that $\dfrac{4}{5}$ is smaller than $\dfrac{5}{6}$ and $\dfrac{5}{6}$ is smaller than $\dfrac{11}{12}$. The order, from smallest to largest, is $\dfrac{4}{5}$, $\dfrac{5}{6}$, $\dfrac{11}{12}$.

51. Because $\dfrac{5}{6}$ $\left(\text{or } \dfrac{25}{30}\right)$ is larger than $\dfrac{2}{5}$ $\left(\text{or } \dfrac{12}{30}\right)$, we write $\dfrac{5}{6} > \dfrac{2}{5}$.

53. Because $\dfrac{4}{9}$ $\left(\text{or } \dfrac{28}{63}\right)$ is larger than $\dfrac{3}{7}$ $\left(\text{or } \dfrac{27}{63}\right)$, we write $\dfrac{4}{9} > \dfrac{3}{7}$.

55. Because $\dfrac{7}{20}$ $\left(\text{or } \dfrac{35}{100}\right)$ is smaller than $\dfrac{9}{25}$ $\left(\text{or } \dfrac{36}{100}\right)$, we write $\dfrac{7}{20} < \dfrac{9}{25}$.

57. Because $\dfrac{5}{16}$ $\left(\text{or } \dfrac{25}{80}\right)$ is smaller than $\dfrac{7}{20}$ $\left(\text{or } \dfrac{28}{80}\right)$, we write $\dfrac{5}{16} < \dfrac{7}{20}$.

59. $\dfrac{3}{8} = \dfrac{12}{32}$, $\dfrac{5}{16} = \dfrac{10}{32}$, $\dfrac{11}{32}$

So the drill bit marked $\dfrac{3}{8}$ is the largest.

61. $\dfrac{5}{8}$, $\dfrac{3}{4} = \dfrac{6}{8}$, $\dfrac{1}{2} = \dfrac{4}{8}$, $\dfrac{3}{8}$

So the plywood of thickness $\dfrac{3}{4}$ in. is the thickest.

63. Elian added 3 to both the numerator and denominator.
$$\dfrac{1}{4} \neq \dfrac{1+3}{4+3} = \dfrac{4}{7}$$
A correct answer would be to multiply both the numerator and the denominator by 3 (or any other nonzero factor).
$$\dfrac{1}{4} = \dfrac{1 \times 3}{4 \times 3} = \dfrac{3}{12}$$

65. **(b)** (1,760 is a multiple of 440 but 771 is not.)

67. Above and Beyond

69. Across

2. $11 = 11$

    $13 = \underline{\quad 13}$

    $\quad 11 \times 13$ Bring down the factors.

    $11 \times 13 = 143$
    So 143 is the LCM of 11 and 13.

4. $120 = 2 \times 2 \times 2 \times 3 \times 5$

    $300 = 2 \times 2 \times 3 \times 5 \times 5$

    $2 \times 2 \times 3 \times 5 = 60$
    So 60 is the GCF of 120 and 300.

7. $13 = 13$

    $52 = 2 \times 2 \times 13$
    So 13 is the GCF of 120 and 300.

8. $360 = 2 \times 2 \times 2 \times 3 \times 3 \times 5$

    $540 = 2 \times 2 \times 3 \times 3 \times 3 \times 5$

    $2 \times 2 \times 3 \times 3 \times 5 = 180$
    Of 180 is the GCF of 360 and 540.

Down

1. $18 = 2 \times 2 \times 2$

    $14 = 2 \qquad\quad \times 7$

    $21 = \underline{\qquad\qquad 3 \times 7}$

    $\quad 2 \times 2 \times 2 \times 3 \times 7$ Bring down the factors.

    $2 \times 5 \times 13 = 130$

    $2 \times 2 \times 2 \times 3 \times 7 = 168$
    So 168 is the LCM of 8, 14, and 21.

3. $16 = 2 \times 2 \times 2 \times 2$

    $12 = \underline{2 \times 2 \qquad\quad \times 3}$

    $\quad 2 \times 2 \times 2 \times 2 \times 3$ Bring down the factors.

    $2 \times 2 \times 2 \times 2 \times 3 = 48$
    So 48 is the LCM of 16 and 12.

5. $2 = 2$

    $5 = \qquad 5$

    $13 = \underline{\qquad\quad 13}$

    $\quad 2 \times 5 \times 13$ Bring down the factors.

    $2 \times 5 \times 13 = 130$
    So 130 is the LCM of 2, 5, and 13.

6. $54 = 2 \times 3 \times 3 \times 3$

    $90 = 2 \times 3 \times 3 \times 5$

    $2 \times 3 \times 3 = 18$
    So 18 is the GCF of 54 and 90.

Exercises 4.3

1. $3 = \qquad 3$

    $4 = \underline{2 \times 2 \qquad}$

    $\quad 2 \times 2 \times 3$ Bring down the factors.

    $2 \times 2 \times 3 = 12$
    So 12 is the LCD for the fractions with denominators of 3 and 4.

3. $4 = 2 \times 2$

    $8 = \underline{2 \times 2 \times 2}$

    $\quad 2 \times 2 \times 2$ Bring down the factors.

    $2 \times 2 \times 2 = 8$
    So 8 is the LCD for the fractions with denominators of 4 and 8.

5. $9 = 3 \times 3$

    $27 = \underline{3 \times 3 \times 3}$

    $\quad 3 \times 3 \times 3$ Bring down the factors.

    $3 \times 3 \times 3 = 27$
    So 27 is the LCD for the fractions with denominators of 9 and 27.

7. $8 = 2 \times 2 \times 2$

    $12 = \underline{2 \times 2 \quad\; \times 3}$

    $\quad 2 \times 2 \times 2 \times 3$ Bring down the factors.

    $2 \times 2 \times 2 \times 3 = 24$
    So 24 is the LCD for the fractions with denominators of 8 and 12.

9. $14 = 2 \quad \times 7$

$21 = \underline{\quad 3 \times 7}$

$\quad 2 \times 3 \times 7 \qquad$ Bring down the factors.

$2 \times 3 \times 7 = 42$

So 42 is the LCD for the fractions with denominators of 14 and 21.

11. $48 = 2 \times 2 \times 2 \times 2 \times 3$

$80 = \underline{2 \times 2 \times 2 \times 2 \quad \times 5}$

$\quad 2 \times 2 \times 2 \times 2 \times 3 \times 5 \qquad$ Bring down the factors.

$2 \times 2 \times 2 \times 2 \times 3 \times 5 = 240$

So 240 is the LCD for the fractions with denominators of 48 and 80.

13. $2 = 2$

$3 = \underline{\quad 3}$

$\quad 2 \times 3 \qquad$ Bring down the factors.

$2 \times 3 = 6$

So 6 is the LCD for the fraction with denominators 2 and 3.

15. $3 = \qquad 3$

$8 = \underline{2 \times 2 \times 2 \quad}$

$\quad 2 \times 2 \times 2 \times 3 \qquad$ Bring down the factors.

$2 \times 2 \times 2 \times 3 = 24$

So 24 is the LCD for the fraction with denominators 3 and 8.

17. $8 = 2 \times 2 \times 2$

$12 = \underline{2 \times 2 \quad \times 3}$

$\quad 2 \times 2 \times 2 \times 3 \qquad$ Bring down the factors.

$2 \times 2 \times 2 \times 3 = 24$

So 24 is the LCD for the fraction with denominators 8 and 12.

19. $12 = 2 \times 2 \qquad \times 3$

$32 = \underline{2 \times 2 \times 2 \times 2 \times 2 \quad}$

$\quad 2 \times 2 \times 2 \times 2 \times 2 \times 3 \qquad$ Bring down the factors.

$2 \times 2 \times 2 \times 2 \times 2 \times 3 = 96$

So 96 is the LCD for the fraction with denominators 12 and 32.

21. $3 = 3$

$9 = \underline{3 \times 3}$

$\quad 3 \times 3 \qquad$ Bring down the factors.

$3 \times 3 = 9$

So 9 is the LCD for the fraction with denominators 3 and 9.

23. $9 = 3 \times 3$

$12 = \underline{\quad 3 \times 2 \times 2}$

$\quad 3 \times 3 \times 2 \times 2 \qquad$ Bring down the factors.

$3 \times 3 \times 2 \times 2 = 36$

So 36 is the LCD for the fraction with denominators 9 and 12.

25. $3 = 3$

$4 = \quad 2 \times 2$

$5 = \underline{\qquad \quad 5}$

$\quad 3 \times 2 \times 2 \times 5 \qquad$ Bring down the factors.

$3 \times 2 \times 2 \times 5 = 60$

So 60 is the LCD for the fraction with denominators 3, 4, and 5.

27. $8 = 2 \times 2 \times 2$

$10 = 2 \qquad \times 5$

$15 = \underline{\qquad \quad 5 \times 3}$

$\quad 2 \times 2 \times 2 \times 5 \times 3 \qquad$ Bring down the factors.

$2 \times 2 \times 2 \times 5 \times 3 = 120$

So 120 is the LCD for the fraction with denominators 8, 10, and 15.

29. $5 = 5$

$10 = \quad 2 \times 5$

$25 = \underline{5 \quad \times 5}$

$\quad 5 \times 2 \times 5 \qquad$ Bring down the factors.

$5 \times 2 \times 5 = 50$

So 50 is the LCD for the fraction with denominators 5, 10, and 25.

< Objectives 1 and 2 >

31. Step 1: The LCD is 12.

Step 2: $\dfrac{2}{3} = \dfrac{8}{12}$

$\dfrac{1}{4} = \dfrac{3}{12}$

Step 3: $\dfrac{2}{3} + \dfrac{1}{4} = \dfrac{8}{12} + \dfrac{3}{12} = \dfrac{11}{12}$

33. Step 1: The LCD is 10.

Step 2: $\dfrac{1}{5} = \dfrac{2}{10}$

$\dfrac{3}{10} = \dfrac{3}{10}$

Step 3: $\dfrac{1}{5} + \dfrac{3}{10} = \dfrac{2}{10} + \dfrac{3}{10} = \dfrac{5}{10} = \dfrac{1}{2}$

35. Step 1: The LCD is 8.

Step 2: $\dfrac{3}{4} = \dfrac{6}{8}$

$\dfrac{1}{8} = \dfrac{1}{8}$

Step 3: $\dfrac{3}{4} + \dfrac{1}{8} = \dfrac{6}{8} + \dfrac{1}{8} = \dfrac{7}{8}$

37. Step 1: The LCD is 35.

Step 2: $\dfrac{1}{7} = \dfrac{5}{35}$

$\dfrac{3}{5} = \dfrac{21}{35}$

Step 3: $\dfrac{1}{7} + \dfrac{3}{5} = \dfrac{5}{35} + \dfrac{21}{35} = \dfrac{26}{35}$

39. Step 1: The LCD is 14.

Step 2: $\dfrac{3}{7} = \dfrac{6}{14}$

$\dfrac{3}{14} = \dfrac{3}{14}$

Step 3: $\dfrac{3}{7} + \dfrac{3}{14} = \dfrac{6}{14} + \dfrac{3}{14} = \dfrac{9}{14}$

41. Step 1: The LCD is 105.

Step 2: $\dfrac{7}{15} = \dfrac{49}{105}$

$\dfrac{2}{35} = \dfrac{6}{105}$

Step 3: $\dfrac{7}{15} + \dfrac{2}{35} = \dfrac{49}{105} + \dfrac{6}{105} = \dfrac{55}{105} = \dfrac{11}{21}$

43. Step 1: The LCD is 24.

Step 2: $\dfrac{5}{8} = \dfrac{15}{24}$

$\dfrac{1}{12} = \dfrac{2}{24}$

Step 3: $\dfrac{5}{8} + \dfrac{1}{12} = \dfrac{15}{24} + \dfrac{2}{24} = \dfrac{17}{24}$

45. Step 1: The LCD is 30.

Step 2: $\dfrac{1}{5} = \dfrac{6}{30}$

$\dfrac{7}{10} = \dfrac{21}{30}$

$\dfrac{4}{15} = \dfrac{8}{30}$

Step 3: $\dfrac{1}{5} + \dfrac{7}{10} + \dfrac{4}{15} = \dfrac{6}{30} + \dfrac{21}{30} + \dfrac{8}{30} = \dfrac{35}{30}$

$= 1\dfrac{5}{30} = 1\dfrac{1}{6}$

47. Step 1: The LCD is 72.

Step 2: $\dfrac{1}{9} = \dfrac{8}{72}$

$\dfrac{7}{12} = \dfrac{42}{72}$

$\dfrac{5}{8} = \dfrac{45}{72}$

Step 3: $\dfrac{1}{9} + \dfrac{7}{12} + \dfrac{5}{8} = \dfrac{8}{72} + \dfrac{42}{72} + \dfrac{45}{72} = \dfrac{95}{72}$

$= 1\dfrac{23}{72}$

49. Step 1: The LCD is 50.

Step 2: $\dfrac{12}{25} = \dfrac{24}{50}$

$\dfrac{3}{10} = \dfrac{15}{50}$

Step 3: $-\dfrac{12}{25} + \dfrac{3}{10} = -\dfrac{24}{50} + \dfrac{15}{50} = \dfrac{-24 + 15}{50}$

$= -\dfrac{9}{50}$

51. Step 1: The LCD is 40.

Step 2: $\dfrac{4}{5} = \dfrac{32}{40}$

$\dfrac{3}{8} = \dfrac{15}{40}$

Step 3: $\dfrac{4}{5} + \left(-\dfrac{3}{8}\right) = \dfrac{32}{40} - \dfrac{15}{40} = \dfrac{32 - 15}{40}$

$= \dfrac{17}{40}$

53. Step 1: The LCD is 32.

Step 2: $\dfrac{3}{8} = \dfrac{12}{32}$

$\dfrac{5}{32} = \dfrac{5}{32}$

Step 3: $-\dfrac{3}{8} + \left(-\dfrac{5}{32}\right) = -\dfrac{12}{32} - \dfrac{5}{32} = \dfrac{-12 - 5}{32}$

$= -\dfrac{17}{32}$

< Objective 3 >

55. Step 1: The LCD is 15.

Step 2: $\dfrac{4}{5} = \dfrac{12}{15}$

$\dfrac{1}{3} = \dfrac{5}{15}$

Step 3: $\dfrac{4}{5} - \dfrac{1}{3} = \dfrac{12}{15} - \dfrac{5}{15} = \dfrac{7}{15}$

57. Step 1: The LCD is 15.

Step 2: $\dfrac{11}{15} = \dfrac{11}{15}$

$\dfrac{3}{5} = \dfrac{9}{15}$

Step 3: $\dfrac{11}{15} - \dfrac{3}{5} = \dfrac{11}{15} - \dfrac{9}{15} = \dfrac{2}{15}$

59. Step 1: The LCD is 8.

Step 2: $\dfrac{3}{8} = \dfrac{3}{8}$

$\dfrac{1}{4} = \dfrac{2}{8}$

Step 3: $\dfrac{3}{8} - \dfrac{1}{4} = \dfrac{3}{8} - \dfrac{2}{8} = \dfrac{1}{8}$

61. Step 1: The LCD is 24.

Step 2: $\dfrac{5}{12} = \dfrac{10}{24}$

$\dfrac{3}{8} = \dfrac{9}{24}$

Step 3: $\dfrac{5}{12} - \dfrac{3}{8} = \dfrac{10}{24} - \dfrac{9}{24} = \dfrac{1}{24}$

63. Step 1: The LCD is 60.

Step 2: $\dfrac{8}{15} = \dfrac{32}{60}$

$\dfrac{3}{4} = \dfrac{45}{60}$

Step 3: $\dfrac{8}{15} - \dfrac{3}{4} = \dfrac{32}{60} - \dfrac{45}{60} = -\dfrac{13}{60}$

65. Step 1: The LCD is 10.

Step 2: $\dfrac{1}{2} = \dfrac{5}{10}$

$\dfrac{7}{10} = \dfrac{7}{10}$

Step 3: $\dfrac{1}{2} - \dfrac{7}{10} = \dfrac{5}{10} - \dfrac{7}{10} = -\dfrac{2}{10} = -\dfrac{1}{5}$

67. Step 1: The LCD is 42.

Step 2: $\dfrac{5}{14} = \dfrac{15}{42}$

$\dfrac{10}{21} = \dfrac{20}{42}$

Step 3: $-\dfrac{5}{14} - \dfrac{10}{21} = -\dfrac{15}{42} - \dfrac{20}{42} = -\dfrac{35}{42} = -\dfrac{5}{6}$

69. Step 1: The LCD is 45.

Step 2: $\dfrac{4}{9} = \dfrac{20}{45}$

$\dfrac{2}{15} = \dfrac{4}{45}$

Step 3: $\dfrac{4}{9} - \left(-\dfrac{2}{15}\right) = \dfrac{4}{9} + \dfrac{2}{15} = \dfrac{20}{45} + \dfrac{4}{45}$

$= -\dfrac{24}{45} = -\dfrac{8}{15}$

71. Step 1: The LCD is 20.

Step 2: $\dfrac{3}{4} = \dfrac{15}{20}$

$\dfrac{1}{5} = \dfrac{4}{20}$

Step 3: $-\dfrac{3}{4} - \left(-\dfrac{1}{5}\right) = -\dfrac{15}{20} + \dfrac{4}{20} = \dfrac{-15+4}{20}$

$= -\dfrac{11}{20}$

73. Step 1: The LCD is 144.

Step 2: $\dfrac{5}{16} = \dfrac{45}{144}$

$\dfrac{11}{18} = \dfrac{88}{144}$

Step 3: $-\dfrac{5}{16} - \left(-\dfrac{11}{18}\right) = -\dfrac{45}{144} + \dfrac{88}{144}$

$= \dfrac{-45+88}{144} = \dfrac{43}{144}$

75. Step 1: The LCD is 120.

Step 2: $\dfrac{33}{40} = \dfrac{99}{120}$

$\dfrac{7}{24} = \dfrac{35}{120}$

$\dfrac{11}{30} = \dfrac{44}{120}$

Step 3: $\dfrac{33}{40} - \dfrac{7}{24} + \dfrac{11}{30} = \dfrac{99}{120} - \dfrac{35}{120} + \dfrac{44}{120}$

$= \dfrac{108}{120} = \dfrac{9}{10}$

77. Step 1: The LCD is 16.

Step 2: $\dfrac{15}{16} = \dfrac{15}{16}$

$\dfrac{5}{8} = \dfrac{10}{16}$

$\dfrac{1}{4} = \dfrac{4}{16}$

Step 3: $\dfrac{15}{16} + \dfrac{5}{8} - \dfrac{1}{4} = \dfrac{15}{16} + \dfrac{10}{16} - \dfrac{4}{16} = \dfrac{21}{16}$

$= 1\dfrac{5}{16}$

79. $\dfrac{1}{2} + \dfrac{3}{8} = \dfrac{4}{8} + \dfrac{3}{8} = \dfrac{7}{8}$

Paul bought $\dfrac{7}{8}$ lb of nuts.

81. $\dfrac{2}{5} + \dfrac{1}{6} = \dfrac{12}{30} + \dfrac{5}{30} = \dfrac{17}{30}$

$\dfrac{17}{30}$ of Amy's income is used for housing and food.

$\dfrac{30}{30} - \dfrac{17}{30} = \dfrac{13}{30}$

$\dfrac{13}{30}$ of her income remains.

83. $\dfrac{3}{4} + \dfrac{1}{2} + \dfrac{2}{3} = \dfrac{9}{12} + \dfrac{6}{12} + \dfrac{8}{12} = \dfrac{23}{12} = 1\dfrac{11}{12}$

Jose walked $1\dfrac{11}{12}$ mi.

85. $\dfrac{1}{4} + \dfrac{3}{16} + \dfrac{1}{16} + \dfrac{1}{8} = \dfrac{4}{16} + \dfrac{3}{16} + \dfrac{1}{16} + \dfrac{2}{16}$

$= \dfrac{10}{16} = \dfrac{5}{8}$

$\dfrac{5}{8}$ of your salary should be used for housing, food, clothing, and transportation.

87. $\dfrac{3}{4} - \dfrac{7}{16} = \dfrac{12}{16} - \dfrac{7}{16} = \dfrac{5}{16}$

The missing dimension is $\dfrac{5}{16}$ in.

89. always

91. sometimes

93. With a scientific calculator:

1 | **a b / c** | 10 | + | 7 | **a b / c** | 12 | = |

Display: $\dfrac{41}{60}$

With a graphing calculator:

1 | ÷ | 10 | + | 7 | ÷ | 12 | ▶ **Frac** | ENTER |

Display: $\dfrac{41}{60}$

95. With a scientific calculator:

8 | **a b / c** | 9 | + | 6 | **a b / c** | 7 | = |

Display: $1\dfrac{47}{63}$

With a graphing calculator:

8 | ÷ | 9 | + | 6 | ÷ | 7 | ▶ **Frac** | ENTER |

Display: $\dfrac{110}{63}$

97. With a scientific calculator:

11 | **a b / c** | 18 | + | 5 | **a b / c** | 12 | = |

Display: $1\dfrac{1}{36}$

With a graphing calculator:

11 | ÷ | 18 | + | 5 | ÷ | 12 | ▶ **Frac** | ENTER |

Display: $\dfrac{37}{36}$

99. With a scientific calculator:

2 | **a b / c** | 5 | − | 4 | **a b / c** | 9 | = |

Display: $-\dfrac{2}{45}$

With a graphing calculator:

2 | ÷ | 5 | − | 4 | ÷ | 9 | ▶ **Frac** | ENTER |

Display: $-\dfrac{2}{45}$

101. With a scientific calculator:

3 | **a b / c** | 14 | +/− | + | 9 | **a b / c** | 20 | = |

Display: $\dfrac{33}{140}$

With a graphing calculator:

(−) | 3 | ÷ | 14 | + | 9 | ÷ | 20 | ▶ **Frac** | ENTER |

Display: $\dfrac{33}{140}$

103. $\dfrac{5}{16} - \left(-\dfrac{5}{42}\right) = \dfrac{5}{16} + \dfrac{5}{42}$

With a scientific calculator:

5 | **a b / c** | 16 | + | 5 | **a b / c** | 42 | = |

Display: $\dfrac{145}{336}$

With a graphing calculator:

5 | ÷ | 16 | + | 5 | ÷ | 42 | ▶ **Frac** | ENTER |

Display: $\dfrac{145}{336}$

105. $\dfrac{5}{6} - \dfrac{1}{4} = \dfrac{10}{12} - \dfrac{3}{12} = \dfrac{7}{12}$

The tumor weighed $\dfrac{7}{12}$ lb at the end of the week.

107. $\dfrac{3}{16} + \dfrac{3}{16} + \dfrac{1}{4} + \dfrac{1}{8} = \dfrac{3}{16} + \dfrac{3}{16} + \dfrac{4}{16} + \dfrac{2}{16} = \dfrac{12}{16}$

$= \dfrac{3}{4}$

The total thickness of the part is $\dfrac{3}{4}$ in.

109. $\dfrac{1}{R_{eq}} = \dfrac{1}{10} + \dfrac{1}{20} + \dfrac{1}{40} = \dfrac{4}{40} + \dfrac{2}{40} + \dfrac{1}{40} = \dfrac{7}{40}$

$R_{eq} = \dfrac{1}{\left(\dfrac{7}{40}\right)} = 1 \div \dfrac{7}{40} = 1 \times \dfrac{40}{7} = \dfrac{40}{7} = 5\dfrac{5}{7}$

111. Let us consider one hook to be located at a distance of x feet from one edge.

$1 \text{ in.} = \dfrac{1}{12} \text{ ft}$

$\dfrac{3}{2} \text{ in} = \dfrac{3}{2} \times \dfrac{1}{12} \text{ ft} = \dfrac{1}{8} \text{ ft}$

$2x + \dfrac{1}{8} = 4\dfrac{1}{4}$

$2x = 4\dfrac{1}{4} - \dfrac{1}{8} = \dfrac{17}{4} - \dfrac{1}{8}$

$x = \dfrac{33}{8} \times \dfrac{1}{2} = \dfrac{33}{16} = 2\dfrac{1}{16} \text{ ft}$

Each hook is located at a $2\dfrac{1}{16}$ ft from the edge of the door.

Exercises 4.4

< Objective 1 >

1. $4\dfrac{2}{9} + 5\dfrac{5}{9} = (4+5) + \left(\dfrac{2}{9} + \dfrac{5}{9}\right) = 9 + \dfrac{7}{9} = 9\dfrac{7}{9}$

3. $3\dfrac{3}{8} + 7\dfrac{3}{8} = (3+7) + \dfrac{3}{8} + \dfrac{3}{8} = 10 + \dfrac{6}{8} = 10\dfrac{3}{4}$

5. $8\dfrac{1}{6} + 8\dfrac{5}{6} = (8+8) + \dfrac{1}{6} + \dfrac{5}{6} = 16 + \dfrac{6}{6} = 16+1$

$= 17$

7. $9\dfrac{5}{8} + 12\dfrac{7}{8} = (9+12) + \dfrac{5}{8} + \dfrac{7}{8} = 21\dfrac{12}{8}$

$= 21 + 1\dfrac{4}{8} = 22\dfrac{1}{2}$

9. $3\dfrac{1}{3} + 6\dfrac{3}{5} = (3+6) + \dfrac{5}{15} + \dfrac{9}{15} = 9 + \dfrac{14}{15} = 9\dfrac{14}{15}$

11. $6\dfrac{1}{2} + 7\dfrac{1}{8} = (6+7) + \dfrac{4}{8} + \dfrac{1}{8} = 13 + \dfrac{5}{8} = 13\dfrac{5}{8}$

13. $11\dfrac{3}{10} + 4\dfrac{5}{6} = (11+4) + \dfrac{18}{60} + \dfrac{50}{60} = 15 + \dfrac{68}{60}$

$= 15 + 1\dfrac{8}{60} = 16\dfrac{2}{15}$

15. $2\dfrac{7}{12} + 6\dfrac{7}{9} = (2+6) + \dfrac{21}{36} + \dfrac{28}{36} = 8 + \dfrac{49}{36}$

$= 8 + 1\dfrac{13}{36} = 9\dfrac{13}{36}$

17. $9\dfrac{1}{2} + \dfrac{3}{4} = 9 + \dfrac{2}{4} + \dfrac{3}{4} = 9 + \dfrac{5}{4} = 9 + 1\dfrac{1}{4} = 10\dfrac{1}{4}$

19. $2\dfrac{1}{4} + 3\dfrac{5}{8} + 1\dfrac{1}{6} = (2+3+1) + \dfrac{6}{24} + \dfrac{15}{24} + \dfrac{4}{24}$

$= 6 + \dfrac{25}{24} = 6 + 1\dfrac{1}{24} = 7\dfrac{1}{24}$

21. $3\dfrac{3}{5} + 4\dfrac{1}{4} + 5\dfrac{3}{10} = (3+4+5) + \dfrac{12}{20} + \dfrac{5}{20} + \dfrac{6}{20}$

$= 12 + \dfrac{23}{20} = 12 + 1\dfrac{3}{20}$

$= 13\dfrac{3}{20}$

< Objective 2 >

23. $11\dfrac{7}{8} - 4\dfrac{3}{8} = (11-4) + \left(\dfrac{7}{8} - \dfrac{3}{8}\right) = 7 + \dfrac{4}{8} = 7\dfrac{1}{2}$

25. $6\dfrac{1}{4} - 1\dfrac{3}{4} = (5-1) + \left(\dfrac{5}{4} - \dfrac{3}{4}\right) = 4 + \dfrac{2}{4} = 4\dfrac{1}{2}$

27. $3\dfrac{2}{3} - 2\dfrac{1}{4} = (3-2) + \left(\dfrac{8}{12} - \dfrac{3}{12}\right) = 1 + \dfrac{5}{12}$

$= 1\dfrac{5}{12}$

29. $7\dfrac{5}{12} - 3\dfrac{11}{18} = (6-3) + \left(\dfrac{51}{36} - \dfrac{22}{36}\right) = 3 + \dfrac{29}{36}$

$= 3\dfrac{29}{36}$

31. $4\dfrac{1}{4} - 3\dfrac{2}{3} = (3-3) + \left(\dfrac{15}{12} - \dfrac{8}{12}\right) = 0 + \dfrac{7}{12} = \dfrac{7}{12}$

33. $1\dfrac{5}{12} - \dfrac{11}{18} = \dfrac{17}{12} - \dfrac{11}{18} = \dfrac{51}{36} - \dfrac{22}{36} = \dfrac{51-22}{36}$

$= \dfrac{29}{36}$

35. $5 - 2\dfrac{1}{4} = \dfrac{20}{4} - \dfrac{9}{4} = \dfrac{11}{4} = 2\dfrac{3}{4}$

37. $17 - 8\dfrac{3}{4} = \dfrac{68}{4} - \dfrac{35}{4} = \dfrac{33}{4} = 8\dfrac{1}{4}$

39. $3\dfrac{3}{4} + 5\dfrac{1}{2} - 2\dfrac{3}{8} = \dfrac{15}{4} + \dfrac{11}{2} - \dfrac{19}{8}$

$= \dfrac{30}{8} + \dfrac{44}{8} - \dfrac{19}{8} = \dfrac{55}{8} = 6\dfrac{7}{8}$

41. $2\dfrac{3}{8} + 2\dfrac{1}{4} - 1\dfrac{5}{6} = \dfrac{19}{8} + \dfrac{9}{4} - \dfrac{11}{6} = \dfrac{57}{24} + \dfrac{54}{24} - \dfrac{44}{24}$

$= \dfrac{67}{24} = 2\dfrac{19}{24}$

43. $4\dfrac{1}{8} + \dfrac{3}{7} - 2\dfrac{23}{28} = 4\underbrace{\dfrac{7}{56} + \dfrac{24}{56}}_{\text{add}} - 2\dfrac{46}{56}$

$= 4\dfrac{31}{56} - 2\dfrac{46}{56} = 3\dfrac{87}{56} - 2\dfrac{46}{56}$

$= 1\dfrac{41}{56}$

45. $6\dfrac{1}{11} + \dfrac{2}{3} - 2\dfrac{1}{6} = 6\underbrace{\dfrac{6}{66} + \dfrac{44}{66}}_{\text{add}} - 2\dfrac{11}{66}$

$= 6\dfrac{50}{66} - 2\dfrac{11}{66} = 4\dfrac{39}{66} = 4\dfrac{13}{22}$

47. $6\dfrac{1}{11} - \dfrac{2}{3} + 2\dfrac{1}{6} = 6\underbrace{\dfrac{6}{66} - \dfrac{44}{66}}_{\text{borrow}} + 2\dfrac{11}{66}$

$= 5\underbrace{\dfrac{72}{66} - \dfrac{44}{66}}_{\text{subtract}} + 2\dfrac{11}{66}$

$= 5\dfrac{28}{66} + 2\dfrac{11}{66} = 7\dfrac{39}{66} = 7\dfrac{13}{22}$

49. $\dfrac{9}{4} + \dfrac{3}{2} = \dfrac{9}{4} + \dfrac{6}{4} = \dfrac{15}{4} \text{ or } 3\dfrac{3}{4}$

51. $2\dfrac{1}{2} + \dfrac{11}{6} = 2\underbrace{\dfrac{3}{6}}_{\text{borrow}} + \dfrac{11}{6} = 2\underbrace{\dfrac{3}{6} + \dfrac{11}{6}}_{\text{add}} = \dfrac{13}{3} \text{ or } 4\dfrac{1}{3}$

53. $\dfrac{9}{4} - \dfrac{3}{2} = \dfrac{9}{4} - \dfrac{6}{4} = \dfrac{3}{4}$

55. $5\dfrac{2}{3} - \dfrac{15}{4} = 5\underbrace{\dfrac{8}{12}}_{\text{borrow}} - \dfrac{45}{12} = 5\underbrace{\dfrac{8}{12} - \dfrac{45}{12}}_{\text{subtract}}$

$= \dfrac{23}{12} \text{ or } 1\dfrac{11}{12}$

57. $1\dfrac{3}{4} - 4\dfrac{1}{4} = \dfrac{7}{4} - \dfrac{17}{4} = -\dfrac{10}{4} = -\dfrac{5}{2} = -2\dfrac{1}{2}$

59. $3\dfrac{1}{2} - 9\dfrac{3}{4} = \dfrac{7}{2} - \dfrac{39}{4} = \dfrac{14}{4} - \dfrac{39}{4} = -\dfrac{25}{4} = -6\dfrac{1}{4}$

< Objective 3 >

61. $-2\dfrac{5}{8}+6\dfrac{3}{10}=-\dfrac{21}{8}+\dfrac{63}{10}=-\dfrac{105}{40}+\dfrac{252}{40}$

$$=\dfrac{147}{40}=3\dfrac{27}{40}$$

63. $3\dfrac{2}{5}+\left(-11\dfrac{5}{6}\right)=\dfrac{17}{5}-\dfrac{71}{6}=\dfrac{102}{30}-\dfrac{355}{40}$

$$=-\dfrac{253}{30}=-8\dfrac{13}{30}$$

65. $-4\dfrac{1}{4}+\left(-3\dfrac{3}{5}\right)=-\dfrac{17}{4}-\dfrac{18}{5}=-\dfrac{85}{20}-\dfrac{72}{20}$

$$=-\dfrac{157}{20}=-7\dfrac{17}{20}$$

67. $-2\dfrac{11}{16}-4\dfrac{3}{5}=-\dfrac{43}{16}-\dfrac{23}{5}=-\dfrac{215}{80}-\dfrac{368}{80}$

$$=-\dfrac{583}{80}=-7\dfrac{23}{80}$$

69. $4\dfrac{2}{9}-\left(-3\dfrac{1}{2}\right)=\dfrac{38}{9}+\dfrac{7}{2}=\dfrac{76}{18}+\dfrac{63}{18}=\dfrac{139}{18}$

$$=7\dfrac{13}{18}$$

< Objective 4 >

71. $\dfrac{3}{4}+1\dfrac{1}{4}+\dfrac{5}{8}+\dfrac{1}{8}=\dfrac{3}{4}+\dfrac{5}{4}+\dfrac{5}{8}+\dfrac{1}{8}$

$$=\dfrac{6}{8}+\dfrac{10}{8}+\dfrac{5}{8}+\dfrac{1}{8}=\dfrac{22}{8}=2\dfrac{6}{8}$$

$$=2\dfrac{3}{4}$$

Senta should buy $2\dfrac{3}{4}$ yd of fabric.

73. $34\dfrac{3}{8}-28\dfrac{3}{4}=\dfrac{275}{8}-\dfrac{115}{4}=\dfrac{275}{8}-\dfrac{230}{8}=\dfrac{45}{8}$

$$=5\dfrac{5}{8}$$

The stock dropped $5\dfrac{5}{8}$ points.

75. $30\dfrac{1}{4}-16\dfrac{7}{8}=\dfrac{121}{4}-\dfrac{135}{8}=\dfrac{242}{8}-\dfrac{135}{8}=\dfrac{107}{8}$

$$=13\dfrac{3}{8}$$

$13\dfrac{3}{8}$ yd of the paper remains.

77. $4\dfrac{1}{4}-3\dfrac{1}{2}=\dfrac{17}{4}-\dfrac{7}{2}=\dfrac{17}{4}-\dfrac{14}{4}=\dfrac{3}{4}$

The bolt extends $\dfrac{3}{4}$ in. beyond the board.

79. $5\dfrac{1}{4}-\dfrac{5}{8}-\dfrac{5}{8}=\dfrac{21}{4}-\dfrac{5}{8}-\dfrac{5}{8}=\dfrac{42}{8}-\dfrac{5}{8}-\dfrac{5}{8}=\dfrac{32}{8}$

$$=4$$

The missing dimension is 4 in.

81. False

83. False

85. With a scientific calculator:

4 $\boxed{\textbf{a b / c}}$ 7 $\boxed{\textbf{a b / c}}$ 9 $\boxed{-}$ 2 $\boxed{\textbf{a b / c}}$ 11
$\boxed{\textbf{a b / c}}$ 18 $\boxed{=}$

Display: $2\dfrac{1}{6}$

With a graphing calculator:

4 $\boxed{+}$ 7 $\boxed{\div}$ 9 $\boxed{-}$ $\boxed{(}$ 2 $\boxed{+}$ 11 $\boxed{\div}$ 18 $\boxed{)}$
$\boxed{\blacktriangleright\textbf{Frac}}$ $\boxed{\textbf{ENTER}}$

Display: $\dfrac{13}{6}$

87. With a scientific calculator:

5 [ab/c] 11 [ab/c] 16 [−] 2 [ab/c] 5 [ab/c] 12 [=]

Display: $3\dfrac{13}{48}$

With a graphing calculator:

5 [+] 11 [÷] 16 [−] [(] 2 [+] 5 [÷] 12 [)]

[▶Frac] [ENTER]

Display: $\dfrac{157}{48}$

89. With a scientific calculator:

6 [ab/c] 2 [ab/c] 3 [−] 1 [ab/c] 5 [ab/c] 6 [=]

Display: $4\dfrac{5}{6}$

With a graphing calculator:

6 [+] 2 [÷] 3 [−] [(] 1 [+] 5 [÷] 6 [)]

[▶Frac] [ENTER]

Display: $\dfrac{29}{6}$

91. With a scientific calculator:

10 [ab/c] 2 [ab/c] 3 [+] 4 [ab/c] 1 [ab/c] 5 [+] 7 [ab/c] 2 [ab/c] 15 [=]

Display: 22

With a graphing calculator:

10 [+] 2 [÷] 3 [+] 4 [+] 1 [÷] 5 [+] 7 [+] 2 [÷] 15 [▶Frac] [ENTER]

Display: 22

93. With a scientific calculator:

11 [ab/c] 3 [ab/c] 8 [−] 20 [ab/c] 4 [ab/c] 5 [=]

Display: $-\dfrac{377}{40} = -9\dfrac{17}{40}$

With a graphing calculator:

11 [+] 3 [÷] 8 [−] [(] 20 [+] 4 [÷] 5 [)]

[▶Frac] [ENTER]

Display: $-\dfrac{377}{40} = -9\dfrac{17}{40}$

95. $10\dfrac{3}{4} + \left(-4\dfrac{7}{8}\right) = 10\dfrac{3}{4} - 4\dfrac{7}{8}$

With a scientific calculator:

10 [ab/c] 3 [ab/c] 4 [−] 4 [ab/c] 7 [ab/c] 8 [=]

Display: $\dfrac{47}{8} = 5\dfrac{7}{8}$

With a graphing calculator:

10 [+] 3 [÷] 4 [−] [(] 4 [+] 7 [÷] 8 [)]

[▶Frac] [ENTER]

Display: $\dfrac{47}{8} = 5\dfrac{7}{8}$

97. $-3\dfrac{2}{7} + \left(-15\dfrac{1}{2}\right) = -3\dfrac{2}{7} - 15\dfrac{1}{2}$

With a scientific calculator:

3 [ab/c] 2 [ab/c] 7 [+/−] [−] 15 [ab/c] 1 [ab/c] 2 [=]

Display: $-\dfrac{263}{14} = -18\dfrac{11}{14}$

With a graphing calculator:

[(−)] 8 [+] 2 [÷] 7 [−] [(] 15 [+] 1 [÷] 2 [)]

[▶Frac] [ENTER]

Display: $-\dfrac{263}{14} = -18\dfrac{11}{14}$

99. $2\dfrac{7}{12}-\left(-3\dfrac{2}{3}\right)=2\dfrac{7}{12}+3\dfrac{2}{3}$

With a scientific calculator:

2 | **a b / c** | 7 | **a b / c** | 12 | **+** | 3 | **a b / c** | 2

| **a b / c** | 3 | **=**

Display: $\dfrac{25}{4}=6\dfrac{1}{4}$

With a graphing calculator:

2 | **+** | 7 | **÷** | 12 | **+** | **(** | 3 | **+** | 2 | **÷** | 3 | **)**

| ▶ **Frac** | **ENTER**

Display: $\dfrac{25}{4}=6\dfrac{1}{4}$

101. $-9\dfrac{1}{2}-\left(-15\dfrac{1}{3}\right)=-9\dfrac{1}{2}+15\dfrac{1}{3}$

With a scientific calculator:

9 | **a b / c** | 1 | **a b / c** | 2 | **+/−** | **+** | 15 | **a b / c** | 1

| **a b / c** | 3 | **=**

Display: $\dfrac{35}{6}=5\dfrac{5}{6}$

With a graphing calculator:

(−) | 9 | **+** | 1 | **÷** | 2 | **+** | **(** | 15 | **+** | 1 | **÷** | 3 | **)**

| ▶ **Frac** | **ENTER**

Display: $\dfrac{35}{6}=5\dfrac{5}{6}$

103. $\dfrac{1}{4}+\dfrac{3}{8}+11\dfrac{3}{4}+\dfrac{13}{16}=\dfrac{4}{16}+\dfrac{6}{16}+11\dfrac{12}{16}+\dfrac{13}{16}$

$$=11\dfrac{35}{16}=13\dfrac{3}{16}\ \text{in.}$$

The total thickness of the floor is $13\dfrac{3}{16}$ in.

105. $100-60\dfrac{1}{4}=99\dfrac{4}{4}-60\dfrac{1}{4}=39\dfrac{3}{4}$

Joy needs $39\dfrac{3}{4}$ ft more of cable.

Exercises 4.5

< Objective 1 >

1. Use Order of Operations

$\dfrac{1}{3}-\underbrace{\left(\dfrac{1}{2}-\dfrac{1}{4}\right)}_{\text{parentheses}}=\dfrac{1}{3}-\left(\dfrac{2}{4}-\dfrac{1}{4}\right)=\dfrac{1}{3}-\dfrac{1}{4}$

$$=\dfrac{4}{12}-\dfrac{3}{12}=\dfrac{1}{12}$$

3. Use Order of Operations

$\dfrac{3}{4}-\underbrace{\left(\dfrac{1}{2}\right)\left(\dfrac{1}{3}\right)}_{\text{multiply}}=\dfrac{3}{4}-\dfrac{1}{6}=\dfrac{9}{12}-\dfrac{2}{12}=\dfrac{7}{12}$

5. Use Order of Operations

$\left(\dfrac{1}{2}\right)^2-\underbrace{\left(\dfrac{1}{4}-\dfrac{1}{5}\right)}_{\text{parentheses}}=\left(\dfrac{1}{2}\right)^2-\underbrace{\left(\dfrac{5}{20}-\dfrac{4}{20}\right)}_{\text{parentheses}}$

$=\underbrace{\left(\dfrac{1}{2}\right)^2}_{\text{exponent}}-\left(\dfrac{1}{20}\right)=\dfrac{1}{4}-\dfrac{1}{20}$

$=\dfrac{5}{20}-\dfrac{1}{20}=\dfrac{4}{20}=\dfrac{1}{5}$

7. Use Order of Operations

$\dfrac{1}{2}+\underbrace{\dfrac{2}{3}\cdot\dfrac{9}{16}}_{\text{multiply}}=\dfrac{1}{2}+\dfrac{\overset{1}{\cancel{2}}}{\underset{1}{\cancel{3}}}\cdot\dfrac{\overset{3}{\cancel{9}}}{\underset{8}{\cancel{16}}}=\dfrac{1}{2}+\dfrac{3}{8}=\dfrac{4}{8}+\dfrac{3}{8}$

$$=\dfrac{7}{8}$$

9. Use Order of Operations

$\left(\dfrac{1}{3}\right)\underbrace{\left(\dfrac{1}{2}+\dfrac{1}{4}\right)}_{\text{parentheses}}=\left(\dfrac{1}{3}\right)\underbrace{\left(\dfrac{2}{4}+\dfrac{1}{4}\right)}_{\text{parentheses}}=\left(\dfrac{1}{3}\right)\left(\dfrac{3}{4}\right)$

$=\left(\dfrac{1}{\underset{1}{\cancel{3}}}\right)\left(\dfrac{\overset{1}{\cancel{3}}}{4}\right)=\dfrac{1}{4}$

11. Use Order of Operations

$$5\left(\underbrace{\frac{2}{3}+\frac{5}{6}}_{\text{parentheses}}\right)=5\left(\underbrace{\frac{4}{6}+\frac{5}{6}}_{\text{parentheses}}\right)=5\left(\frac{9}{6}\right)=5\left(\frac{\cancel{9}^{3}}{\cancel{6}_{2}}\right)$$

$$=\frac{15}{2}\ \text{or}\ 7\frac{1}{2}$$

13. Use Order of Operations

$$\left(4\frac{1}{2}\right)\left(\underbrace{\frac{7}{8}+2\frac{1}{4}}_{\text{parentheses}}\right)=\left(\frac{9}{2}\right)\left(\underbrace{\frac{7}{8}+\frac{9}{4}}_{\text{parentheses}}\right)$$

$$=\left(\frac{9}{2}\right)\left(\frac{7}{8}+\frac{18}{8}\right)$$

$$=\left(\frac{9}{2}\right)\left(\frac{7+18}{8}\right)$$

$$=\left(\frac{9}{2}\right)\left(\frac{25}{8}\right)=\frac{225}{16}\ \text{or}\ 14\frac{1}{16}$$

15. Use Order of Operations

$$\left(\underbrace{4-1\frac{1}{2}}_{\text{parentheses}}\right)\left(\underbrace{2\frac{1}{4}-1\frac{2}{3}}_{\text{parentheses}}\right)=\left(\underbrace{\frac{4}{1}-\frac{3}{2}}_{\text{parentheses}}\right)\left(\underbrace{\frac{9}{4}-\frac{5}{3}}_{\text{parentheses}}\right)$$

$$=\left(\frac{8}{2}-\frac{3}{2}\right)\left(\frac{27}{12}-\frac{20}{12}\right)$$

$$=\left(\frac{8-3}{2}\right)\left(\frac{27-20}{12}\right)$$

$$=\left(\frac{5}{2}\right)\left(\frac{7}{12}\right)$$

$$=\frac{35}{24}\ \text{or}\ 1\frac{11}{24}$$

17. Use Order of Operations

$$\left(\frac{1}{2}\right)^{3}+\left(\frac{1}{3}\right)\bullet\left(\underbrace{\frac{1}{2}+\frac{1}{4}}_{\text{parentheses}}\right)$$

$$=\left(\frac{1}{2}\right)^{3}+\left(\frac{1}{3}\right)\bullet\left(\underbrace{\frac{2}{4}+\frac{1}{4}}_{\text{parentheses}}\right)=\left(\underbrace{\frac{1}{2}}_{\text{exponent}}\right)^{3}+\left(\frac{1}{3}\right)\bullet\left(\frac{3}{4}\right)$$

$$=\frac{1}{8}+\underbrace{\left(\frac{1}{\cancel{3}}\right)\bullet\left(\frac{\cancel{3}^{1}}{4}\right)}_{\text{multiply}}=\frac{1}{8}+\frac{1}{4}=\frac{1}{8}+\frac{2}{8}=\frac{3}{8}$$

19. Use Order of Operations

$$\left(\frac{3}{4}\right)^{2}+\left(\frac{1}{2}\right)^{3}\bullet\left(\underbrace{\frac{1}{2}+\frac{3}{4}}_{\text{parentheses}}\right)$$

$$=\left(\frac{3}{4}\right)^{2}+\left(\frac{1}{2}\right)^{3}\bullet\left(\underbrace{\frac{2}{4}+\frac{3}{4}}_{\text{parentheses}}\right)=\left(\underbrace{\frac{3}{4}}_{\text{exponent}}\right)^{2}+\left(\underbrace{\frac{1}{2}}_{\text{exponent}}\right)^{3}\bullet\frac{5}{4}$$

$$=\frac{9}{16}+\frac{1}{8}\bullet\frac{5}{4}=\frac{9}{16}+\frac{5}{32}=\frac{18}{32}+\frac{5}{32}=\frac{23}{32}$$

21. Use Order of Operations

$$\left(\frac{2}{3}\right)^{2}\bullet\left(\underbrace{\frac{2}{15}+\frac{1}{2}}_{\text{parentheses}}\right)\div\frac{3}{5}$$

$$=\left(\frac{2}{3}\right)^{2}\bullet\left(\underbrace{\frac{4}{30}+\frac{15}{30}}_{\text{parentheses}}\right)\div\frac{3}{5}=\left(\underbrace{\frac{2}{3}}_{\text{exponent}}\right)^{2}\bullet\frac{19}{30}\div\frac{3}{5}$$

$$=\frac{4}{9}\bullet\frac{19}{30}\div\frac{3}{5}=\frac{4}{9}\bullet\frac{19}{30}\bullet\frac{5}{3}=\frac{380}{810}=\frac{38}{81}$$

23. Use Order of Operations

$$2\frac{1}{5}-\frac{1}{2}\bullet\frac{1}{4}=\frac{11}{5}-\underbrace{\frac{1}{2}\bullet\frac{1}{4}}_{\text{multiply}}=\frac{11}{5}-\frac{1}{8}=\frac{88}{40}-\frac{5}{40}$$

$$=\frac{83}{40}=2\frac{3}{40}$$

25. Use Order of Operations

$$11\frac{1}{10}-\frac{1}{2}\bullet\left(6\frac{2}{3}\right)=11\frac{1}{10}-\frac{1}{2}\bullet\frac{20}{3}$$

$$=11\frac{1}{10}-\frac{1}{\cancel{2}_{1}}\bullet\frac{\cancel{20}^{10}}{3}$$

$$=\frac{111}{10}-\frac{10}{3}=\frac{333}{30}-\frac{100}{30}$$

$$=\frac{233}{30}=7\frac{23}{30}$$

27. $\dfrac{8+12}{4}=\dfrac{\cancel{20}^{5}}{\cancel{4}_{1}}=5$

29. $\dfrac{3(8-3)}{7}=\dfrac{3\times5}{7}=\dfrac{15}{7}\ \text{or}\ 2\frac{1}{7}$

31. $6\left(\dfrac{4+9}{2(8+5)}\right) - \left(\dfrac{3}{4} + \dfrac{3}{4}\right)^2$

$= 6\left(\dfrac{\cancel{13}^{\,1}}{2 \times \cancel{13}_{\,1}}\right) - \left(\dfrac{\cancel{6}^{\,3}}{\cancel{4}_{\,2}}\right)^2 = 6 \times \dfrac{1}{2} - \left(\dfrac{3}{2}\right)^2$

$= \dfrac{6}{2} - \dfrac{9}{4} = \dfrac{12}{4} - \dfrac{9}{4} = \dfrac{3}{4}$

33. $\dfrac{4}{5} + \left(\dfrac{2+3}{1+2}\right)^2 - \left(4\dfrac{1}{2}\right) \div \left(\dfrac{3(5-2)}{2^2}\right)$

$= \dfrac{4}{5} + \left(\dfrac{5}{3}\right)^2 - \left(\dfrac{9}{2}\right) \div \left(\dfrac{3 \times 3}{4}\right)$

$= \dfrac{4}{5} + \dfrac{25}{9} - \dfrac{9}{2} \times \dfrac{4}{9} = \dfrac{4}{5} + \dfrac{25}{9} - \dfrac{36}{18}$

$= \dfrac{72 + 250 - 180}{90} = \dfrac{\overset{71}{\cancel{142}}}{\underset{45}{\cancel{90}}} = \dfrac{71}{45} \text{ or } 1\dfrac{26}{45}$

< Objective 2 >

35. $\dfrac{\frac{2}{3}}{\frac{6}{8}} = \dfrac{2}{3} \div \dfrac{6}{8} = \dfrac{\cancel{2}}{3} \cdot \dfrac{8}{\cancel{6}_{3}} = \dfrac{8}{9}$

37. $\dfrac{\frac{1}{2}}{\frac{1}{4}} = \dfrac{1}{2} \div \dfrac{1}{4} = \dfrac{1}{\cancel{2}_{1}} \cdot \dfrac{\cancel{4}^{2}}{1} = 2$

39. $\dfrac{\frac{4}{5}}{\frac{9}{10}} = \dfrac{4}{5} \div \dfrac{9}{10} = \dfrac{4}{\cancel{5}} \cdot \dfrac{\cancel{10}^{2}}{9} = \dfrac{8}{9}$

41. $\dfrac{\frac{6}{19}}{\frac{5}{38}} = \dfrac{6}{19} \div \dfrac{5}{38} = \dfrac{6}{\cancel{19}} \cdot \dfrac{\cancel{38}^{2}}{5} = \dfrac{12}{5}$

43. $\dfrac{6\frac{1}{2}}{\frac{2}{3}} = \dfrac{\frac{13}{2}}{\frac{2}{3}} = \dfrac{13}{2} \div \dfrac{2}{3} = \dfrac{13}{2} \cdot \dfrac{3}{2} = \dfrac{39}{4}$

45. $\dfrac{4\frac{1}{2}}{5\frac{1}{4}} = \dfrac{\frac{9}{2}}{\frac{21}{4}} = \dfrac{9}{2} \div \dfrac{21}{4} = \dfrac{\cancel{9}^{3}}{\cancel{2}_{1}} \cdot \dfrac{\cancel{4}^{2}}{\cancel{21}_{7}} = \dfrac{6}{7}$

47. $\dfrac{2 - \frac{1}{2}}{2 + \frac{1}{4}}$

The LCD of the fractions is 4.

$\dfrac{2 - \frac{1}{2}}{2 + \frac{1}{4}} = \dfrac{4 \cdot \left(2 - \frac{1}{2}\right)}{4 \cdot \left(2 + \frac{1}{4}\right)} = \dfrac{4 \cdot 2 - 4 \cdot \frac{1}{2}}{4 \cdot 2 + 4 \cdot \frac{1}{4}} = \dfrac{8 - 2}{8 + 1}$

$= \dfrac{6}{9} = \dfrac{2}{3}$

49. $\dfrac{3 + \frac{1}{8}}{4 - \frac{1}{4}}$

The LCD of the fractions is 8.

$\dfrac{3 + \frac{1}{8}}{4 - \frac{1}{4}} = \dfrac{8 \cdot \left(3 + \frac{1}{8}\right)}{8 \cdot \left(4 - \frac{1}{4}\right)} = \dfrac{8 \cdot 3 + 8 \cdot \frac{1}{8}}{8 \cdot 4 - 8 \cdot \frac{1}{4}} = \dfrac{24 + 1}{32 - 2}$

$= \dfrac{25}{30} = \dfrac{5}{6}$

51. $\dfrac{2 + \frac{1}{6}}{5 - \frac{1}{3}}$

The LCD of the fractions is 6.

$\dfrac{2 + \frac{1}{6}}{5 - \frac{1}{3}} = \dfrac{6 \cdot \left(2 + \frac{1}{6}\right)}{6 \cdot \left(5 - \frac{1}{3}\right)} = \dfrac{6 \cdot 2 + 6 \cdot \frac{1}{6}}{6 \cdot 5 - 6 \cdot \frac{1}{3}} = \dfrac{12 + 1}{30 - 2}$

$= \dfrac{13}{28}$

53. $\dfrac{8 + \frac{1}{2}}{4 - \frac{1}{4}}$

The LCD of the fractions is 4.

$\dfrac{8 + \frac{1}{2}}{4 - \frac{1}{4}} = \dfrac{4 \cdot \left(8 + \frac{1}{2}\right)}{4 \cdot \left(4 - \frac{1}{4}\right)} = \dfrac{4 \cdot 8 + 4 \cdot \frac{1}{2}}{4 \cdot 4 - 4 \cdot \frac{1}{4}} = \dfrac{32 + 2}{16 - 1}$

$= \dfrac{34}{15}$

55. $\dfrac{4-\dfrac{1}{3}}{3+\dfrac{1}{2}}$

The LCD of the fractions is 6.

$$\dfrac{4-\dfrac{1}{3}}{3+\dfrac{1}{2}}=\dfrac{6\cdot\left(4-\dfrac{1}{3}\right)}{6\cdot\left(3+\dfrac{1}{2}\right)}=\dfrac{6\cdot4-6\cdot\dfrac{1}{3}}{6\cdot3+6\cdot\dfrac{1}{2}}=\dfrac{24-2}{18+3}$$

$$=\dfrac{22}{21}$$

57. $\dfrac{5+\dfrac{3}{4}}{7-\dfrac{1}{6}}$

The LCD of the fractions is 12.

$$\dfrac{5+\dfrac{3}{4}}{7-\dfrac{1}{6}}=\dfrac{12\cdot\left(5+\dfrac{3}{4}\right)}{12\cdot\left(7-\dfrac{1}{6}\right)}=\dfrac{12\cdot5+12\cdot\dfrac{3}{4}}{12\cdot7-12\cdot\dfrac{1}{6}}$$

$$=\dfrac{60+9}{84-2}=\dfrac{69}{82}$$

< Objective 3 >

59. Use Order of Operations

$$8-\left(1\dfrac{1}{2}+\dfrac{3}{4}+3\dfrac{1}{3}\right)=8-\left(1\dfrac{6}{12}+\dfrac{9}{12}+3\dfrac{4}{12}\right)$$

$$=8-\left(4\dfrac{19}{12}\right)=8-5\dfrac{7}{12}$$

$$=7\dfrac{12}{12}-5\dfrac{7}{12}=2\dfrac{5}{12}$$

They can take on $2\dfrac{5}{12}$ mi more work in July.

61. Use Order of Operations

$$20-\left(5\dfrac{1}{2}+4\dfrac{1}{4}+4\dfrac{3}{4}+2\dfrac{1}{8}\right)$$

$$=20-\left(5\dfrac{4}{8}+4\dfrac{2}{8}+4\dfrac{6}{8}+2\dfrac{1}{8}\right)=20-\left(15\dfrac{13}{8}\right)$$

$$=20-\left(16\dfrac{5}{8}\right)=19\dfrac{8}{8}-16\dfrac{5}{8}=3\dfrac{3}{8}$$

She must run $3\dfrac{3}{8}$ mi on Saturday to reach her goal.

63. Use Order of Operations

$$1-\left(\dfrac{1}{2}+\dfrac{1}{8}\right)=1-\left(\dfrac{4}{8}+\dfrac{1}{8}\right)=1-\dfrac{5}{8}=\dfrac{8}{8}-\dfrac{5}{8}=\dfrac{3}{8}$$

$\dfrac{3}{8}$ of the landfill is used for other materials.

65. $\dfrac{2}{3}+4\dfrac{3}{10}\times8\dfrac{5}{6}$

The calculator will display $\dfrac{773}{20}=38\dfrac{13}{20}$.

67. $4-\dfrac{2\dfrac{4}{5}-3\times\dfrac{3}{4}}{2\times6}$

The calculator will display $\dfrac{949}{240}=3\dfrac{229}{240}$.

69. Multiply

$$\left(2\dfrac{1}{2}\right)\left(15\dfrac{1}{3}\right)=\left(\dfrac{5}{2}\right)\left(\dfrac{46}{3}\right)=\left(\dfrac{5}{\underset{1}{\cancel{2}}}\right)\left(\dfrac{\overset{23}{\cancel{46}}}{3}\right)$$

$$=\left(\dfrac{115}{3}\right)=38\dfrac{1}{3}$$

$38\dfrac{1}{3}$ units should be prescribed.

71. Divide

$$3\dfrac{3}{8}\div\dfrac{3}{4}=\dfrac{27}{8}\div\dfrac{3}{4}=\dfrac{27}{8}\cdot\dfrac{4}{3}=\dfrac{\overset{9}{\cancel{27}}}{\underset{2}{\cancel{8}}}\cdot\dfrac{\overset{1}{\cancel{4}}}{\underset{1}{\cancel{3}}}=\dfrac{9}{2}=4\dfrac{1}{2}$$

It takes $4\dfrac{1}{2}$ min to make the cut.

73. Amara: $8\cdot18\dfrac{1}{4}=8\cdot\dfrac{73}{4}=\overset{2}{\cancel{8}}\cdot\dfrac{73}{\underset{1}{\cancel{4}}}=146$

Burt: $4\cdot34\dfrac{2}{3}=4\cdot\dfrac{104}{3}=\dfrac{416}{3}=138\dfrac{2}{3}$

Both: $146+138\dfrac{2}{3}=284\dfrac{2}{3}$

284 completed boards can be populated in 8 hours.

75. Since both fraction bars are the same length, the complex fraction could be interpreted as:

$$\frac{2}{3} \div 5 = \frac{2}{3} \div \frac{5}{1} = \frac{2}{3} \cdot \frac{1}{5} = \frac{2}{15}$$

or

$$2 \div \frac{3}{5} = \frac{2}{1} \div \frac{3}{5} = \frac{2}{1} \cdot \frac{5}{3} = \frac{10}{3}$$

Exercises 4.6

< Objectives 1 and 2 >

1.
$$2x + 1 = 9 \qquad \text{Check: } 2(4) + 1 \overset{?}{=} 9$$
$$2x + 1 + (-1) = 9 + (-1) \qquad\qquad 9 = 9$$
$$2x = 8$$
$$x = 4$$
The solution is 4.

3.
$$3x + (-2) = 7$$
$$3x + (-2) + 2 = 7 + 2$$
$$3x = 9$$
$$\frac{3x}{3} = \frac{9}{3}$$
$$x = 3$$
Check: $3(3) + (-2) \overset{?}{=} 7$
$$7 = 7$$
The solution is 3.

5.
$$4x + 7 = 35$$
$$4x + 7 + (-7) = 35 + (-7)$$
$$4x = 28$$
$$\frac{4x}{4} = \frac{28}{4}$$
$$x = 7$$
Check: $4(7) + 7 \overset{?}{=} 35$
$$35 = 35$$
The solution is 7.

7.
$$2x + 9 = 5$$
$$2x + 9 + (-9) = 5 + (-9)$$
$$2x = -4$$
$$\frac{2x}{2} = \frac{-4}{2}$$
$$x = -2$$
Check: $2(-2) + 9 \overset{?}{=} 5$
$$5 = 5$$
The solution is –2.

9.
$$4 + (-7x) = 18$$
$$4 + (-4) + (-7x) = 18 + (-4)$$
$$-7x = 14$$
$$\frac{-7x}{-7} = \frac{14}{-7}$$
$$x = -2$$
Check: $4 + (-7(-2)) \overset{?}{=} 18$
$$18 = 18$$
The solution is –2.

11.
$$3 + (-4x) = -9$$
$$3 + (-3) + (-4x) = -9 + (-3)$$
$$\frac{-4x}{-4} = \frac{-12}{-4}$$
$$x = 3$$
Check: $3 + (-4(3)) \overset{?}{=} -9$
$$-9 = -9$$
The solution is 3.

13.
$$\frac{x}{2} + 1 = 5 \qquad \text{Check: } \frac{8}{2} + 1 \overset{?}{=} 5$$
$$\frac{x}{2} + 1 + (-1) = 5 + (-1) \qquad\qquad 5 = 5$$
$$\frac{x}{2} = 4$$
$$x = 8$$
The solution is 8.

15. $\dfrac{x}{4}+(-5)=3$ Check: $\dfrac{32}{4}+(-5)\overset{?}{=}3$

$\dfrac{x}{4}+(-5)+5=3+5$ $\qquad 3=3$

$\dfrac{x}{4}=8$

$x=32$

The solution is 32.

17. $\dfrac{2}{3}x+5=17$

$\dfrac{2}{3}x+5+(-5)=17+(-5)$

$\dfrac{2}{3}x=12$

$x=18$

Check: $\dfrac{2}{3}(18)+5\overset{?}{=}17$

$12+5\overset{?}{=}17$

$17=17$

The solution is 18.

19. $\dfrac{4}{5}x+(-3)=13$

$\dfrac{4}{5}x+(-3)+3=13+3$

$\dfrac{4}{5}x=16$

$x=20$

Check: $\dfrac{4}{5}(20)+(-3)\overset{?}{=}13$

$16-3\overset{?}{=}13$

$13=13$

The solution is 20.

21. $5x=2x+9$

$5x+(-2x)=2x+(-2x)+9$

$3x=9$

$\dfrac{1}{3}(3x)=\dfrac{1}{3}(9)$

$x=3$

Check: $5(3)\overset{?}{=}2(3)+9$

$15=15$

The solution is 3.

23. $3x=10+(-2x)$

$3x+2x=10+(-2x)+2x$

$5x=10$

$\dfrac{1}{5}(5x)=\dfrac{1}{5}(10)$

$x=2$

Check: $3(2)\overset{?}{=}10+(-2(2))$

$6=6$

The solution is 2.

25. $9x+2=3x+38$

$9x+(-3x)+2=3x+(-3x)+38$

$6x+2=38$

$6x=36$

$\dfrac{1}{6}(6x)=\dfrac{1}{6}(36)$

$x=6$

Check: $9(6)+2\overset{?}{=}3(6)+38$

$54+2\overset{?}{=}18+38$

$56=56$

The solution is 6.

27.
$$4x + (-8) = x + (-14)$$
$$4x + (-x) + (-8) = x + (-x) + (-14)$$
$$3x + (-8) + 8 = -14 + 8$$
$$3x = -6$$
$$\frac{1}{3}(3x) = \frac{1}{3}(-6)$$
$$x = -2$$
Check: $4(-2) + (-8) \overset{?}{=} -2 + (-14)$
$$-16 = -16$$
The solution is –2.

29.
$$5x + 7 = 2x + (-3)$$
$$5x + (-2x) + 7 = 2x + (-2x) + (-3)$$
$$3x + 7 + (-7) = -3 + (-7)$$
$$3x = -10$$
$$\frac{1}{3}(3x) = \frac{1}{3}(-10)$$
$$x = -\frac{10}{3}$$
Check: $5\left(-\frac{10}{3}\right) + 7 \overset{?}{=} 2\left(-\frac{10}{3}\right) + (-3)$
$$3\left(-\frac{10}{3}\right) + 7 \overset{?}{=} (-3)$$
$$-10 + 7 \overset{?}{=} (-3)$$
$$-3 = -3$$
The solution is $-\frac{10}{3}$.

31.
$$7x + (-3) = 9x + 5$$
$$7x + (-7x) + (-3) = 9x + (-7x) + 5$$
$$-3 = 2x + 5$$
$$-8 = 2x$$
$$\frac{1}{4}(-8) = \frac{1}{4}(2x)$$
$$-4 = x$$
Check: $7(-4) + (-3) \overset{?}{=} 9(-4) + 5$
$$-31 = -31$$
The solution is –4.

33.
$$5x + 4 = 7x + (-8)$$
$$5x + (-5x) + 4 = 7x + (-5x) + (-8)$$
$$4 = 2x + (-8)$$
$$12 = 2x$$
$$\frac{1}{2}(12) = \frac{1}{2}(2x)$$
$$6 = x$$
Check: $5(6) + 4 \overset{?}{=} 7(6) + (-8)$
$$30 + 4 = 42 - 8$$
$$34 = 34$$
The solution is 6.

35. $2x + (-3) + 5x = 7 + 4x + 2$
$$7x + (-3) = 9 + 4x$$
$$7x + (-3) - 4x = 9 + 4x - 4x$$
$$3x = 12$$
$$\frac{1}{3}(3x) = \frac{1}{3}(12)$$
$$x = 4$$
Check: $2(4) + (-3) + 5(4) \overset{?}{=} 7 + 4(4) + 2$
$$8 - 3 + 20 \overset{?}{=} 7 + 16 + 2$$
$$25 = 25$$
The solution is 4.

37.
$$6x + 7 + (-4x) = 8 + 7x + (-26)$$
$$2x + 7 = 7x + (-18)$$
$$7 = 5x + (-18)$$
$$25 = 5x$$
$$\frac{1}{5}(25) = \frac{1}{5}(5x)$$
$$5 = x$$

Check:
$$6(5) + 7 + (-4(5)) \overset{?}{=} 8 + 7(5) + (-26)$$
$$30 + 7 + (-20) \overset{?}{=} 8 + 35 + (-26)$$
$$17 = 17$$
The solution is 5.

39.
$$9x + (-2) + 7x + 13 = 10x + (-13)$$
$$16x + 11 = 10x + (-13)$$
$$6x + 11 = -13$$
$$6x = -24$$
$$\frac{1}{6}(6x) = \frac{1}{6}(-24)$$
$$x = -4$$

Check:
$$9(-4) + (-2) + 7(-4) + 13 \overset{?}{=} 10(-4) + (-13)$$
$$-36 + (-2) + (-28) + 13 \overset{?}{=} -40 + (-13)$$
$$-53 = -53$$
The solution is –4.

41.
$$8x + (-7) + 5x - 10 = 10x + (-12)$$
$$13x + (-17) = 10x + (-12)$$
$$3x + (-17) = -12$$
$$3x = 5$$
$$x = \frac{5}{3}$$

Check:
$$8\left(\frac{5}{3}\right) + (-7) + 5\left(\frac{5}{3}\right) - 10 \overset{?}{=} 10\left(\frac{5}{3}\right) + (-12)$$
$$13\left(\frac{5}{3}\right) + (-17) \overset{?}{=} 10\left(\frac{5}{3}\right) + (-12)$$
$$65 - 51 \overset{?}{=} 50 - 36$$
$$14 = 14$$

The solution is $\frac{5}{3}$.

43.
$$7\left[2x + (-1)\right] + (-5x) = x + 25$$
$$14x + (-7) + (-5x) = x + 25$$
$$9x - 7 = x + 25$$
$$8x = 32$$
$$\frac{1}{8}(8x) = \frac{1}{8}(32)$$
$$x = 4$$

Check:
$$7(2(4) + (-1)) + (-5(4)) \overset{?}{=} 4 + 25$$
$$49 + (-20) \overset{?}{=} 29$$
$$29 = 29$$
The solution is 4.

45. $3x + 2(4x + (-3)) = 6x + (-9)$

$$3x + 8x + (-6) = 6x + (-9)$$

$$3x + 8x + (-6) - 6x = 6x + (-9) - 6x$$

$$5x + (-6) = -9$$

$$5x = -3$$

$$x = -\frac{3}{5}$$

Check:

$$3\left(-\frac{3}{5}\right) + 2\left(4\left(-\frac{3}{5}\right) + (-3)\right) \overset{?}{=} 6\left(-\frac{3}{5}\right) + (-9)$$

$$-\frac{9}{5} + 2\left(-\frac{27}{5}\right) \overset{?}{=} -\frac{18}{5} + (-9)$$

$$-\frac{63}{5} = -\frac{63}{5}$$

The solution is $-\dfrac{3}{5}$.

47. $\dfrac{8}{3}x + (-3) = \dfrac{2}{3}x + 15$

$$\frac{8}{3}x + \left(-\frac{2}{3}x\right) + (-3) = \frac{2}{3}x + \left(-\frac{2}{3}x\right) + 15$$

$$\frac{6}{3}x + (-3) = 15$$

$$2x = 18$$

$$\frac{1}{2}(2x) = \frac{1}{2}(18)$$

$$x = 9$$

Check: $\dfrac{8}{3}(9) + (-3) \overset{?}{=} \dfrac{2}{3}(9) + 15$

$$24 + (-3) \overset{?}{=} 6 + 15$$

$$21 = 21$$

The solution is 9.

49. $\dfrac{2}{5}x + (-5) = \dfrac{12}{5}x + 8$

$$\frac{2}{5}x + \left(-\frac{2}{5}x\right) + (-5) = \frac{12}{5}x + \left(-\frac{2}{5}x\right) + 8$$

$$-5 - 8 = \frac{10}{5}x$$

$$-13 = 2x$$

$$-\frac{13}{2} = x$$

Check: $\dfrac{2}{5}\left(-\dfrac{13}{2}\right) + (-5) \overset{?}{=} \dfrac{12}{5}\left(-\dfrac{13}{2}\right) + 8$

$$-\frac{13}{5} + (-5) \overset{?}{=} -\frac{78}{5} + 8$$

$$-\frac{38}{5} = -\frac{38}{5}$$

The solution is $-\dfrac{13}{2}$.

51.
$$\frac{3}{4}x + 4\frac{2}{3} = 9$$

$$\frac{3}{4}x + 4\frac{2}{3} + \left(-4\frac{2}{3}\right) = 9 + \left(-4\frac{2}{3}\right)$$

$$\frac{3}{4}x = 9 - \frac{14}{3} = \frac{27-14}{3} = \frac{13}{3}$$

$$\frac{4}{3} \cdot \frac{3}{4}x = \frac{4}{3} \cdot \frac{13}{3}$$

$$x = \frac{52}{9} = 5\frac{7}{9}$$

Check:

$$\frac{3}{4}x + 4\frac{2}{3} \overset{?}{=} 9$$

$$\frac{\cancel{3}}{\cancel{4}}\left(\frac{\overset{13}{\cancel{52}}}{\underset{3}{\cancel{9}}}\right) + \frac{14}{3} \overset{?}{=} 9$$

$$\frac{13}{3} + \frac{14}{3} \overset{?}{=} 9$$

$$\frac{27}{3} \overset{?}{=} 9$$

$$9 = 9$$

The solution is $\frac{52}{9}$ or $5\frac{7}{9}$.

< Objective 3 >

53. Let x represent the number of "no" votes on the election measure and $x + 55$ represent the number of "yes" votes. Since there were a total of 735 votes cast in all, our equation will be:

$$x + x + 55 = 735$$

$$2x + 55 = 735$$

$$2x + 55 + (-55) = 735 + (-55)$$

$$2x = 680$$

$$\frac{2x}{2} = \frac{680}{2}$$

$$x = 340$$

There were 340 "no" votes. Therefore, there were $x + 55 = 340 + 55 = 395$ "yes" votes.

55. Let x represent Rob's salary and $x + 120$ represent Francine's salary. Since they earn a total of $2,680 per month, our equation will be:

$$x + x + 120 = 2,680$$

$$2x + 120 = 2,680$$

$$2x + 120 + (-120) = 2,680 + (-120)$$

$$2x = 2,560$$

$$\frac{2x}{2} = \frac{2,560}{2}$$

$$x = 1,280$$

Thus, Rob earns $1,280 per month and so Francine earns

$x + \$120 = \$1,280 + \$120 = \$1,400$ per month.

57. Let Frank earn $\$x$ per month.

Jody's salary is $\$(x + 280)$.

Sum of their monthly salaries

$$= (x) + (x + 280)$$

$$\$5,520 = (x) + (x + 280)$$

$$\$5,520 = 2x + 280$$

$$x = \frac{(5,520 - 280)}{2} = \frac{5,240}{2} = 2,620$$

Frank's monthly salary is $2,620 and Jody's monthly salary is $2,900.

59. Let x be the age of Yan Ling's sister. Then, $2x - 1$ is Yan Ling's age.

$$x + (2x - 1) = 14$$

$$3x - 1 = 14$$

$$3x = 15$$

$$x = 5$$

Thus, $2x - 1 = 9$

Yan Ling is 9 years old.

61. Le x be Toyna's weekly pay before deductions.

Deduction $= \dfrac{7}{25}x$

Her take home pay $= x - \dfrac{7}{25}x = \dfrac{18}{25}x$

Now,

$\dfrac{18}{25}x = 1,080$

$x = 1,500$

Her weekly pay before deductions is $1,500.

63. $95 = \dfrac{9}{5}C + 32$

$63 = \dfrac{9}{5}C$

$35 = C$

The corresponding temperature is 35°C.

65. Let x be the number.

$2x + 16 = 24$

$2x = 8$

$x = 4$

The number is 4.

67. Let x be the number.

$5x - 12 = 78$

$5x = 90$

$x = 18$

The number is 18.

69. sometimes

71. $Y = 16r - 15$

$159 = 16r - 15$

$174 = 16r$

$10\dfrac{7}{8} = r$

Therefore, $10\dfrac{7}{8}$ in. of rainfall is needed.

73. $D = \dfrac{t+16}{4}$

$7 = \dfrac{t+16}{4}$

$4 \times 7 = t + 16$

$28 = t + 16$

$12 = t$

The child is 12 years old.

75. Let x be the size of the folder before compressing.

$x - \dfrac{9}{25}x = 11\dfrac{1}{5}$

$\dfrac{25-9}{25}x = \dfrac{56}{5}$

$\dfrac{16}{25}x = \dfrac{56}{5}$

$x = \dfrac{\overset{7}{\cancel{56}}}{\cancel{5}} \times \dfrac{\overset{5}{\cancel{25}}}{\underset{2}{\cancel{16}}}$

$x = \dfrac{35}{2} = 17\dfrac{1}{2}$

The size of the folder before compressing was $17\dfrac{1}{2}$ MB.

77. Above and Beyond

79. Above and Beyond

81. Above and Beyond

83. A solution to an equation is a value which, when substituted into the original equation for the variable, gives a true statement.

85. (a) $2x + 3 = 0$

$$2x = -3$$

$$x = -\frac{3}{2}$$

(b) $4x + 7 = 0$

$$4x = -7$$

$$x = -\frac{7}{4}$$

(c) $6x - 1 = 0$

$$6x = 1$$

$$x = \frac{1}{6}$$

(d) $5x - 2 = 0$

$$5x = 2$$

$$x = \frac{2}{5}$$

(e) $-3x + 8 = 0$

$$-3x = -8$$

$$x = \frac{8}{3}$$

(f) $-5x - 9 = 0$

$$-5x = 9$$

$$x = -\frac{9}{5}$$

(g) $ax + b = 0$

$$ax = -b$$

$$x = -\frac{b}{a}$$

Summary Exercises

1. $\dfrac{8}{15} + \dfrac{2}{15} = \dfrac{10}{15} = \dfrac{2}{3}$

3. $\dfrac{8}{13} + \dfrac{7}{13} = \dfrac{15}{13} = 1\dfrac{2}{13}$

5. $\dfrac{19}{24} + \dfrac{13}{24} = \dfrac{32}{24} = \dfrac{4}{3} = 1\dfrac{1}{3}$

7. $\dfrac{2}{9} + \dfrac{5}{9} + \dfrac{4}{9} = \dfrac{11}{9} = 1\dfrac{2}{9}$

9. $\dfrac{4}{9} - \dfrac{5}{9} = -\dfrac{1}{9}$

11. $-\dfrac{5}{6} + \dfrac{1}{6} = -\dfrac{4}{6} = -\dfrac{2}{3}$

13. $-\dfrac{6}{7} + \left(-\dfrac{5}{7}\right) = -\dfrac{6}{7} - \dfrac{5}{7} = -\dfrac{11}{7} = -1\dfrac{4}{7}$

15. $-\dfrac{3}{4} - \dfrac{1}{4} = -\dfrac{4}{4} = -1$

17. $4 = 2 \times 2$

$\underline{12 = 2 \times 2 \times 3}$

$\qquad 2 \times 2 \times 3$ Bring down the factors.

$2 \times 2 \times 3 = 12$
So 12 is the LCM of 4 and 12.

19. $18 = 2 \qquad \times 3 \times 3$

$\underline{24 = 2 \times 2 \times 2 \times 3 \qquad}$

$\qquad 2 \times 2 \times 2 \times 3 \times 3$ Bring down the factors.

$2 \times 2 \times 2 \times 3 \times 3 = 72$
So 72 is the LCM of 18 and 24.

21. $15 = \qquad 3 \times 5$

$\underline{20 = 2 \times 2 \qquad \times 5}$

$\qquad 2 \times 2 \times 3 \times 5$ Bring down the factors.

$2 \times 2 \times 3 \times 5 = 60$
So 60 is the LCM of 15 and 20.

23. $9 = \qquad 3 \times 3$

$12 = 2 \times 2 \qquad \times 3$

$\underline{24 = 2 \times 2 \times 2 \times 3 \qquad}$

$\qquad 2 \times 2 \times 2 \times 3 \times 3$ Bring down the factors.

$2 \times 2 \times 2 \times 3 \times 3 = 72$
So 72 is the LCM of 9, 12, and 24.

25. Because 48 is a common multiple of 8 and 12, let's use 48 as our common denominator. Because $\dfrac{5}{8} = \dfrac{30}{48}$ and $\dfrac{7}{12} = \dfrac{28}{48}$, we see that $\dfrac{7}{12}$ is smaller than $\dfrac{5}{8}$. The order, from smaller to larger, is $\dfrac{7}{12}, \dfrac{5}{8}$.

27. Because 12 is a common multiple of 6, 4 and 3, let's use 3 as our common denominator. Because $-\dfrac{3}{4} = -\dfrac{9}{12}$, $-\dfrac{5}{6} = -\dfrac{10}{12}$, and $-\dfrac{1}{3} = -\dfrac{4}{12}$, we see that $-\dfrac{10}{12}$ is smaller than $-\dfrac{9}{12}$ and $-\dfrac{9}{12}$ is smaller than $-\dfrac{4}{12}$. The order, from smallest to largest, is $-\dfrac{5}{6}, -\dfrac{3}{4}, -\dfrac{1}{3}$.

29. Because $\dfrac{3}{7}$ is the same as $\dfrac{9}{21}$, we write $\dfrac{3}{7} = \dfrac{9}{21}$.

31. Because $-\dfrac{7}{10}$ $\left(\text{or } -\dfrac{21}{30}\right)$ is smaller than $-\dfrac{2}{3}\left(\text{or } -\dfrac{20}{30}\right)$, we write $-\dfrac{7}{10} < -\dfrac{2}{3}$.

33. 15 is the LCD of 3 and 5. Then,
$$\dfrac{2}{3} = \left(\dfrac{2 \times 5}{3 \times 5}\right) = \dfrac{10}{15}$$
$$\dfrac{4}{5} = \left(\dfrac{4 \times 3}{5 \times 3}\right) = \dfrac{12}{15}$$

35. $12 = 2 \times 2 \times 3$
$\underline{18 = 2 \times \quad 3 \times 3}$
$\quad\quad 2 \times 2 \times 3 \times 3$ Bring down the factors.
$2 \times 2 \times 3 \times 3 = 36$
So 36 is the LCD for the fractions with denominators of 12 and 18.

37. $25 = \quad\quad\quad 5 \times 5$
$\underline{40 = 2 \times 2 \times 2 \times 5 \quad\quad}$
$\quad\quad 2 \times 2 \times 2 \times 5 \times 5$ Bring down the factors.
$2 \times 2 \times 2 \times 5 \times 5 = 200$
So 200 is the LCD for the fractions with denominators of 25 and 40.

39. $3 = \quad\quad 3$
$4 = 2 \times 2$
$\underline{11 = \quad\quad\quad\quad 11}$
$\quad\quad 2 \times 2 \times 3 \times 11$ Bring down the factors.
$2 \times 2 \times 3 \times 11 = 132$
So 132 is the LCD for the fractions with denominators of 3, 4, and 11.

41. $3 = \quad\quad\quad 3$
$6 = 2 \quad\quad \times 3$
$\underline{8 = 2 \times 2 \times 2 \quad\quad}$
$\quad\quad 2 \times 2 \times 2 \times 3$ Bring down the factors.
$2 \times 2 \times 2 \times 3 = 24$
So 24 is the LCD for the fractions with denominators of 3, 6, and 8.

43. Step 1: The LCD is 24.

Step 2: $\dfrac{3}{8} = \dfrac{9}{24}$

$\dfrac{5}{12} = \dfrac{10}{24}$

Step 3: $\dfrac{3}{8} + \dfrac{5}{12} = \dfrac{9}{24} + \dfrac{10}{24} = \dfrac{19}{24}$

45. Step 1: The LCD is 60.

Step 2: $\dfrac{2}{15} = \dfrac{8}{60}$

$\dfrac{9}{20} = \dfrac{27}{60}$

Step 3: $\dfrac{2}{15} + \dfrac{9}{20} = \dfrac{8}{60} + \dfrac{27}{60} = \dfrac{35}{60} = \dfrac{7}{12}$

47. Step 1: The LCD is 90.

Step 2: $\dfrac{7}{15} = \dfrac{42}{90}$

$\dfrac{13}{18} = \dfrac{65}{90}$

Step 3: $\dfrac{7}{15} + \dfrac{13}{18} = \dfrac{42}{90} + \dfrac{65}{90} = \dfrac{107}{90} = 1\dfrac{17}{90}$

49. Step 1: The LCD is 8.

Step 2: $\dfrac{1}{2} = \dfrac{4}{8}$

$\dfrac{1}{4} = \dfrac{2}{8}$

$\dfrac{1}{8} = \dfrac{1}{8}$

Step 3: $\dfrac{1}{2} + \dfrac{1}{4} + \dfrac{1}{8} = \dfrac{4}{8} + \dfrac{2}{8} + \dfrac{1}{8} = \dfrac{7}{8}$

51. Step 1: The LCD is 72.

Step 2: $\dfrac{3}{8} = \dfrac{27}{72}$

$\dfrac{5}{12} = \dfrac{30}{72}$

$\dfrac{7}{18} = \dfrac{28}{72}$

Step 3: $\dfrac{3}{8} + \dfrac{5}{12} + \dfrac{7}{18} = \dfrac{27}{72} + \dfrac{30}{72} + \dfrac{28}{72} = \dfrac{85}{72}$

$= 1\dfrac{13}{72}$

53. Step 1: The LCD is 48.

Step 2: $\dfrac{5}{12} = \dfrac{20}{48}$

$\dfrac{5}{16} = \dfrac{15}{48}$

Step 3: $-\dfrac{5}{12} + \dfrac{6}{16} = -\dfrac{20}{48} + \dfrac{15}{48} = -\dfrac{5}{48}$

55. Step 1: The LCD is 45.

Step 2: $\dfrac{7}{9} = \dfrac{35}{45}$

$\dfrac{2}{5} = \dfrac{18}{45}$

Step 3: $\dfrac{7}{9} + \left(-\dfrac{2}{5}\right) = \dfrac{35}{45} - \dfrac{18}{45} = \dfrac{17}{45}$

57. Step 1: The LCD is 24.

Step 2: $\dfrac{7}{8} = \dfrac{21}{24}$

$\dfrac{2}{3} = \dfrac{16}{24}$

Step 3: $\dfrac{7}{8} - \dfrac{2}{3} = \dfrac{21}{24} - \dfrac{16}{24} = \dfrac{5}{24}$

59. Step 1: The LCD is 18.

Step 2: $\dfrac{11}{18} = \dfrac{11}{18}$

$\dfrac{2}{9} = \dfrac{4}{18}$

Step 3: $\dfrac{11}{18} - \dfrac{2}{9} = \dfrac{11}{18} - \dfrac{4}{18} = \dfrac{7}{18}$

61. Step 1: The LCD is 24.

Step 2: $\dfrac{5}{8} = \dfrac{15}{24}$

$\dfrac{1}{6} = \dfrac{4}{24}$

Step 3: $\dfrac{5}{8} - \dfrac{1}{6} = \dfrac{15}{24} - \dfrac{4}{24} = \dfrac{11}{24}$

63. Step 1: The LCD is 42.

Step 2: $\dfrac{8}{21} = \dfrac{16}{42}$

$\dfrac{1}{14} = \dfrac{3}{42}$

Step 3: $\dfrac{8}{21} - \dfrac{1}{14} = \dfrac{16}{42} - \dfrac{3}{42} = \dfrac{13}{42}$

65. Step 1: The LCD is 12.

Step 2: $\dfrac{11}{12}=\dfrac{11}{12}$

$\dfrac{1}{4}=\dfrac{3}{12}$

$\dfrac{1}{3}=\dfrac{4}{12}$

Step 3: $\dfrac{11}{12}-\dfrac{1}{4}-\dfrac{1}{3}=\dfrac{11}{12}-\dfrac{3}{12}-\dfrac{4}{12}=\dfrac{4}{12}=\dfrac{1}{3}$

67. Step 1: The LCD is 20.

Step 2: $\dfrac{1}{4}=\dfrac{5}{20}$

$\dfrac{7}{10}=\dfrac{14}{20}$

Step 3: $\dfrac{1}{4}-\dfrac{7}{10}=\dfrac{5}{20}-\dfrac{14}{20}=-\dfrac{9}{20}$

69. Step 1: The LCD is 144.

Step 2: $\dfrac{7}{16}=\dfrac{63}{144}$

$\dfrac{4}{9}=\dfrac{64}{144}$

Step 3: $-\dfrac{7}{16}-\dfrac{4}{8}=-\dfrac{63}{144}-\dfrac{64}{144}=-\dfrac{127}{144}$

71. Step 1: The LCD is 160.

Step 2: $\dfrac{5}{32}=\dfrac{25}{160}$

$\dfrac{17}{20}=\dfrac{136}{160}$

Step 3: $-\dfrac{5}{32}-\left(-\dfrac{17}{20}\right)=-\dfrac{25}{160}+\dfrac{136}{160}=\dfrac{111}{160}$

73. $6\dfrac{5}{7}+3\dfrac{4}{7}=(6+3)+\dfrac{5}{7}+\dfrac{4}{7}=9+\dfrac{9}{7}=9+1\dfrac{2}{7}$

$=10\dfrac{2}{7}$

75. $5\dfrac{7}{10}+3\dfrac{11}{12}=(5+3)+\dfrac{42}{60}+\dfrac{55}{60}=8+\dfrac{97}{60}$

$=8+1\dfrac{37}{60}=9\dfrac{37}{60}$

77. $7\dfrac{4}{9}-3\dfrac{7}{9}=(6-3)+\left(\dfrac{13}{9}-\dfrac{7}{9}\right)=3+\dfrac{6}{9}=3\dfrac{2}{3}$

79. $4\dfrac{3}{10}-2\dfrac{7}{12}=\dfrac{43}{10}-\dfrac{31}{12}=\dfrac{258}{60}-\dfrac{155}{60}$

$=\dfrac{258-155}{60}=\dfrac{103}{60}=1\dfrac{43}{60}$

81. $11\dfrac{3}{5}-2\dfrac{4}{5}=(10-2)+\left(\dfrac{8}{5}-\dfrac{4}{5}\right)=8+\dfrac{4}{5}=8\dfrac{4}{5}$

83. $8-4\dfrac{3}{4}=\dfrac{8}{1}-\dfrac{19}{4}=\dfrac{32}{4}-\dfrac{19}{4}=\dfrac{32-19}{4}=\dfrac{13}{4}$

$=3\dfrac{1}{4}$

85. $2\dfrac{1}{2}+3\dfrac{5}{6}+3\dfrac{3}{8}=(2+3+3)+\dfrac{12}{24}+\dfrac{20}{24}+\dfrac{9}{24}$

$=8+\dfrac{41}{24}=8+1\dfrac{17}{24}=9\dfrac{17}{24}$

87. $-4\dfrac{1}{2}+6\dfrac{3}{4}=-\dfrac{9}{2}+\dfrac{27}{4}=-\dfrac{18}{4}+\dfrac{27}{4}=-\dfrac{9}{4}$

$=-2\dfrac{1}{4}$

89. $5\dfrac{1}{3}+\left(-8\dfrac{9}{10}\right)=\dfrac{16}{3}-\dfrac{89}{10}=\dfrac{160}{30}-\dfrac{267}{30}$

$=-\dfrac{107}{30}=-3\dfrac{17}{30}$

91. $2\dfrac{1}{4}-7\dfrac{1}{3}=\dfrac{9}{4}-\dfrac{22}{3}=\dfrac{27}{12}-\dfrac{88}{12}=-\dfrac{61}{12}=-5\dfrac{1}{12}$

93. $-6\dfrac{1}{2}-9\dfrac{1}{3}=-\dfrac{13}{2}-\dfrac{28}{3}=-\dfrac{39}{6}-\dfrac{56}{6}=-\dfrac{95}{6}$

$=-15\dfrac{5}{6}$

95. $-17\dfrac{31}{100}-\left(-12\dfrac{5}{16}\right)=-\dfrac{1,731}{100}+\dfrac{197}{16}$

$$=-\dfrac{6,924}{400}+\dfrac{4,925}{400}$$

$$=-\dfrac{1999}{400}=-4\dfrac{399}{400}$$

97. The amount of milk left over is the difference between the amount of milk you have and the amount of milk that the recipe calls for.

$$\dfrac{3}{4}-\dfrac{1}{3}=\dfrac{9}{12}-\dfrac{4}{12}=\dfrac{5}{12}$$

You will have $\dfrac{5}{12}$ cup of milk left over.

99. The perimeter of the triangle is the sum of the three sides.

$$5\dfrac{3}{8}+6\dfrac{7}{16}+7\dfrac{3}{4}=\dfrac{43}{8}+\dfrac{103}{16}+\dfrac{31}{4}$$

$$=\dfrac{86}{16}+\dfrac{103}{16}+\dfrac{124}{16}=\dfrac{313}{16}$$

$$=19\dfrac{9}{16}$$

The perimeter of the triangle is $19\dfrac{9}{16}$ in.

101. First, convert 8 ft to inches:

$$8\text{ ft}\times\dfrac{12\text{ in.}}{1\text{ ft}}=96\text{ in.}$$

$$96-42\dfrac{5}{16}-\dfrac{1}{8}=96-\dfrac{677}{16}-\dfrac{1}{8}$$

$$=\dfrac{1536}{16}-\dfrac{677}{16}-\dfrac{2}{16}=\dfrac{857}{16}$$

$$=53\dfrac{9}{16}$$

The length of the board remaining is $53\dfrac{9}{16}$ in.

103. To find the amount of paint he used, we compute the sum of the amount of the paint he used in the three rooms.

$$1\dfrac{3}{4}+1\dfrac{1}{3}+\dfrac{1}{2}=(1+1)+\dfrac{9}{12}+\dfrac{4}{12}+\dfrac{6}{12}=2+\dfrac{19}{12}$$

$$=2+1\dfrac{7}{12}=3\dfrac{7}{12}$$

He used $3\dfrac{7}{12}$ gal of paint.

105. Use Order of Operations:

$$\dfrac{3}{4}+\underbrace{\left(\dfrac{1}{2}\right)\left(\dfrac{1}{3}\right)}_{\text{multiply first}}=\dfrac{3}{4}+\dfrac{1}{6}$$

The LCD is 12.

So, $\dfrac{3}{4}=\dfrac{9}{12}$ and $\dfrac{1}{6}=\dfrac{2}{12}$.

$$\dfrac{3}{4}+\dfrac{1}{6}=\dfrac{9}{12}+\dfrac{2}{12}=\dfrac{11}{12}$$

107. Use Order of Operations:

$$\left(4\dfrac{3}{8}\right)\left(1\dfrac{1}{2}\right)\div\underbrace{\left(\dfrac{3}{4}+\dfrac{2}{4}\right)}_{\text{parentheses}}$$

$$\underbrace{\left(4\dfrac{3}{8}\right)\left(1\dfrac{1}{2}\right)}_{\text{multiply}}\div\left(\dfrac{5}{4}\right)$$

(make mixed numbers improper fractions)

$$\underbrace{\left(\dfrac{35}{8}\right)\left(\dfrac{3}{2}\right)}_{\text{multiply}}\div\left(\dfrac{5}{4}\right)$$

$$\dfrac{105}{16}\div\dfrac{5}{4}$$

(change to multiplication and take the reciprocal)

$$\dfrac{105}{16}\cdot\dfrac{4}{5}=\dfrac{\overset{21}{\cancel{105}}}{\underset{4}{\cancel{16}}}\cdot\dfrac{\overset{1}{\cancel{4}}}{\underset{1}{\cancel{5}}}=\dfrac{21}{4}=5\dfrac{1}{4}$$

109. $\underbrace{\left(\dfrac{2}{3}\right)^{3}}_{\text{exponent}}-\underbrace{\dfrac{1}{2}\cdot\dfrac{1}{9}}_{\text{multiply}}=\dfrac{8}{27}-\dfrac{1}{2}\cdot\dfrac{1}{9}=\dfrac{8}{27}-\dfrac{1}{18}$

$$=\dfrac{16}{54}-\dfrac{3}{54}=\dfrac{13}{54}$$

111. $\dfrac{\frac{2}{3}}{\frac{5}{6}} = \dfrac{2}{\cancel{3}} \times \dfrac{\cancel{6}^{\,2}}{5} = \dfrac{4}{5}$

113.
$$5x + (-3) = 12$$
$$5x + (-3) + 3 = 12 + 3$$
$$5x = 15$$
$$\frac{5x}{5} = \frac{15}{5}$$
$$x = 3$$

Check: $5(3) + (-3) \overset{?}{=} 12$

$\qquad 15 + (-3) \overset{?}{=} 12$

$\qquad\qquad 12 = 12$

The solution is 3.

115.
$$7x + 8 = 3x$$
$$7x + (-3x) + 8 = 3x + (-3x)$$
$$4x + 8 + (-8) = 0 + (-8)$$
$$\frac{4x}{4} = \frac{-8}{4}$$
$$x = -2$$

Check: $7(-2) + 8 \overset{?}{=} 3(-2)$

$\qquad -14 + 8 \overset{?}{=} -6$

$\qquad\qquad -6 = -6$

The solution is –2.

117.
$$3x + (-7) = x$$
$$3x + (-x) + (-7) = x + (-x)$$
$$2x + (-7) + 7 = 0 + 7$$
$$2x = 7$$
$$\frac{2x}{2} = \frac{7}{2}$$
$$x = \frac{7}{2}$$

Check: $3\left(\dfrac{7}{2}\right) + (-7) \overset{?}{=} \dfrac{7}{2}$

$\qquad \dfrac{21}{2} + (-7) \overset{?}{=} \left(\dfrac{7}{2}\right)$

$\qquad\qquad \dfrac{7}{2} = \dfrac{7}{2}$

The solution is $\dfrac{7}{2}$.

119. $\quad \dfrac{x}{3} + (-5) = 1 \qquad$ Check: $\dfrac{18}{3} + (-5) \overset{?}{=} 1$

$\qquad \dfrac{x}{3} + (-5) + 5 = 1 + 5 \qquad\qquad 6 + (-5) \overset{?}{=} 1$

$\qquad\qquad \dfrac{x}{3} = 6 \qquad\qquad\qquad\qquad 1 = 1$

$\qquad\qquad x = 18$

The solution is 18.

121.
$$6x + (-5) = 3x + 13$$
$$6x + (-3x) + (-5) = 3x + (-3x) + 13$$
$$3x + (-5) = 13$$
$$x = 6$$

Check: $6(6) + (-5) \overset{?}{=} 3(6) + 13$

$\qquad 36 + (-5) \overset{?}{=} 18 + 13$

$\qquad\qquad 31 = 31$

The solution is 6.

123.

$$7x + 4 = 2x + 6$$

$$7x + (-2x) + 4 = 2x + (-2x) + 6$$

$$5x + 4 = 6$$

$$5x + 4 + (-4) = 6 + (-4)$$

$$5x = 2$$

$$\frac{5x}{5} = \frac{2}{5}$$

$$x = \frac{2}{5}$$

Check: $7\left(\dfrac{2}{5}\right) + 4 \overset{?}{=} 2\left(\dfrac{2}{5}\right) + 6$

$$\frac{14}{5} + 4 \overset{?}{=} \frac{4}{5} + 6$$

$$\frac{34}{5} = \frac{34}{5}$$

The solution is $\dfrac{2}{5}$.

125.

$$2x + 7 = 4x + (-5)$$

$$2x + (-2x) + 7 = 4x + (-2x) + (-5)$$

$$7 = 2x + (-5)$$

$$7 + 5 = 2x + (-5) + 5$$

$$12 = 2x$$

$$\frac{12}{2} = \frac{2x}{2}$$

$$6 = x$$

Check: $2(6) + 7 \overset{?}{=} 4(6) + (-5)$

$$12 + 7 \overset{?}{=} 24 + -5$$

$$19 = 19$$

The solution is 6.

127.

$$\frac{10}{3}x + (-5) = \frac{4}{3}x + 7$$

$$\frac{10}{3}x + \left(-\frac{4}{3}x\right) + (-5) = \frac{4}{3}x + \left(-\frac{4}{3}x\right) + 7$$

$$2x + (-5) = 7$$

$$2x + (-5) + 5 = 7 + 5$$

$$2x = 12$$

$$\frac{2x}{2} = \frac{12}{2}$$

$$x = 6$$

Check: $\dfrac{10}{3}(6) + (-5) \overset{?}{=} \dfrac{4}{3}(6) + 7$

$$20 + (-5) \overset{?}{=} 8 + 7$$

$$15 = 15$$

The solution is 6.

129.

$$3x + (-2) + 5x = 7 + 2x + 21$$

$$8x + (-2) = 2x + 28$$

$$8x + (-2x) + (-2) = 2x + (-2x) + 28$$

$$6x + (-2) = 28$$

$$6x + (-2) + 2 = 28 + 2$$

$$6x = 30$$

$$\frac{6x}{6} = \frac{30}{6}$$

$$x = 5$$

Check: $3(5) + (-2) + 5(5) \overset{?}{=} 7 + 2(5) + 21$

$$15 + (-2) + 25 \overset{?}{=} 7 + 10 + 21$$

$$38 = 38$$

The solution is 5.

131. $5(3x+(-1))-6x = 3x+(-2)$

$$15x+(-5)-6x = 3x+(-2)$$

$$9x+(-5) = 3x+(-2)$$

$$9x+(-3x)+(-5) = 3x+(-3x)+(-2)$$

$$6x+(-5) = -2$$

$$6x+(-5)+5 = -2+5$$

$$6x = 3$$

$$\frac{6x}{6} = \frac{3}{6}$$

$$x = \frac{1}{2}$$

Check:

$$5\left(3\left(\frac{1}{2}\right)+(-1)\right)-6\left(\frac{1}{2}\right)\overset{?}{=}3\left(\frac{1}{2}\right)+(-2)$$

$$5\left(\frac{1}{2}\right)-3\overset{?}{=}\frac{3}{2}+(-2)$$

$$\frac{5}{2}-3\overset{?}{=}\frac{3}{2}+(-2)$$

$$-\frac{1}{2} = -\frac{1}{2}$$

The solution is $\dfrac{1}{2}$.

133. Let x be the number of hours the mechanic did the repair job.

His earning $= \$75(x)+\$225$

$$\$450 = \$75(x)+\$225$$

$$x = \frac{(450-225)}{75} = \frac{225}{75} = 3$$

The repair job took 3 hr.

135. Let x be the number.

$$4x+14 = 34$$

$$4x = 20$$

$$x = 5$$

The number is 5.

Chapter Test 4

1. $12 = 2\times2\times3$

$15 = \underline{3\times5}$

$2\times2\times3\times5$ Bring down the factors.

$2\times2\times3\times5 = 60$

So 60 is the LCD of 12 and 15.

3. Step 1: The LCD is 10.

Step 2: $\dfrac{2}{5} = \dfrac{4}{10}$

$\dfrac{4}{10} = \dfrac{4}{10}$

Step 3: $\dfrac{2}{5}+\dfrac{4}{10} = \dfrac{4}{10}+\dfrac{4}{10} = \dfrac{8}{10} = \dfrac{4}{5}$

5. $7-5\dfrac{7}{15} = 7-\dfrac{82}{15} = \dfrac{105}{15}-\dfrac{82}{15} = \dfrac{23}{15} = 1\dfrac{8}{15}$

7. Step 1: The LCD is 24.

Step 2: $\dfrac{3}{8} = \dfrac{9}{24}$

$\dfrac{5}{12} = \dfrac{10}{24}$

Step 3: $\dfrac{3}{8}+\dfrac{5}{12} = \dfrac{9}{24}+\dfrac{10}{24} = \dfrac{19}{24}$

9. $\dfrac{7}{9}-\dfrac{4}{9} = \dfrac{7-4}{9} = \dfrac{3}{9} = \dfrac{1}{3}$

11. $5\dfrac{3}{10}+2\dfrac{4}{10} = (5+3)+\dfrac{3}{10}+\dfrac{4}{10} = 7+\dfrac{7}{10}$

$= 7\dfrac{7}{10}$

13. $6\dfrac{3}{8}+5\dfrac{7}{10} = (6+5)+\dfrac{15}{40}+\dfrac{28}{40} = 11+\dfrac{43}{40}$

$= 11+1\dfrac{3}{43} = 12\dfrac{3}{40}$

15. $2\dfrac{2}{9}-3\dfrac{5}{6} = \dfrac{20}{9}-\dfrac{23}{6} = \dfrac{40}{18}-\dfrac{69}{18} = -\dfrac{29}{18}$

$= -1\dfrac{11}{18}$

17. $-\dfrac{1}{6}+\left(-\dfrac{3}{7}\right)=-\dfrac{7}{42}-\dfrac{18}{42}=-\dfrac{25}{42}$

19. $4\dfrac{2}{7}+3\dfrac{3}{7}+1\dfrac{3}{7}=(4+3+1)+\dfrac{2}{7}+\dfrac{3}{7}+\dfrac{3}{7}$

$\qquad\qquad =8+\dfrac{8}{7}=8+1\dfrac{1}{7}=9\dfrac{1}{7}$

21. $\dfrac{5}{12}+\dfrac{3}{12}=\dfrac{8}{12}=\dfrac{2}{3}$

23. Step 1: The LCD is 60.

Step 2: $\dfrac{11}{15}=\dfrac{44}{60}$

$\qquad\quad \dfrac{9}{20}=\dfrac{27}{60}$

Step 3: $\dfrac{11}{15}+\dfrac{9}{20}=\dfrac{44}{60}+\dfrac{27}{60}=\dfrac{71}{60}=1\dfrac{11}{60}$

25. $\underbrace{\left(1\dfrac{1}{3}+2\dfrac{1}{2}\right)}_{\text{parentheses}}\cdot 3\dfrac{1}{4}=\left(\dfrac{4}{3}+\dfrac{5}{2}\right)\cdot\dfrac{13}{4}$

$\qquad\qquad\qquad =\left(\dfrac{8}{6}+\dfrac{15}{6}\right)\cdot\dfrac{13}{4}$

$\qquad\qquad\qquad =\left(\dfrac{23}{6}\right)\cdot\dfrac{13}{4}=\dfrac{299}{24}=12\dfrac{11}{24}$

27. $\qquad 7x+(-5)=16$

$\qquad 7x+(-5)+5=16+5$

$\qquad\qquad\qquad 7x=21$

$\qquad\qquad\qquad \dfrac{7x}{7}=\dfrac{21}{7}$

$\qquad\qquad\qquad x=3$

Check: $7(3)+(-5)\overset{?}{=}16$

$\qquad\qquad 21+(-5)\overset{?}{=}16$

$\qquad\qquad\qquad 16=16$

The solution is 3.

29. $\qquad 7x+(-3)=4x+(-5)$

$7x+(-4x)+(-3)=4x+(-4x)+(-5)$

$\qquad\qquad 3x+(-3)=-5$

$\qquad\quad 3x+(-3)+3=-5+3$

$\qquad\qquad\qquad 3x=-2$

$\qquad\qquad\qquad x=-\dfrac{2}{3}$

Check: $7\left(-\dfrac{2}{3}\right)+(-3)\overset{?}{=}4\left(-\dfrac{2}{3}\right)+(-5)$

$\qquad -\dfrac{14}{3}+(-3)\overset{?}{=}-\dfrac{8}{3}+(-5)$

$\qquad\qquad -\dfrac{23}{3}=-\dfrac{23}{3}$

The solution is $-\dfrac{2}{3}$.

31. First, compute the total hours spent for the first and the second sections.

$\dfrac{1}{4}+\dfrac{1}{3}=\dfrac{3}{12}+\dfrac{4}{12}=\dfrac{7}{12}$

The time you have left to finish the last section is difference of the total hours allowed and the sum computed above.

$\dfrac{5}{6}-\dfrac{7}{12}=\dfrac{10}{12}-\dfrac{7}{12}=\dfrac{3}{12}=\dfrac{1}{4}$

You have $\dfrac{1}{4}$ hr to finish the last section of the test.

33. First, convert to an improper fraction,

$$51\frac{3}{4} \text{ years} = \frac{207}{4} \text{ years}$$

$$3\frac{1}{5} \text{cups} = \frac{16}{5} \text{cups}$$

Calculate how many cups of coffee a person will drink in a week:

$$\frac{16}{5}\frac{\text{cups}}{\text{day}} \times 5\frac{\text{days}}{\text{week}} = 16\frac{\text{cups}}{\text{week}}$$

Next, find the number of cups that an average person drinks in a work year.

$$16\frac{\text{cups}}{\text{week}} \times 50\frac{\text{weeks}}{\text{year}} = 800\frac{\text{cups}}{\text{year}}$$

In 52 years we have ,

$$\overset{200}{\cancel{800}}\frac{\text{cups}}{\text{year}} \times \frac{207}{\cancel{4}} \text{years} = 41,400 \text{ cups}.$$

The number of cups of coffee is equal to 41,400 cups.

Cumulative Review: Chapters 1–4

1.
$$\begin{array}{r} \overset{1\;\;1}{1,369} \\ +\,5,804 \\ \hline 7,173 \end{array}$$

3.
$$\begin{array}{r} \overset{1\;2}{357} \\ 28 \\ +\,2,346 \\ \hline 2,731 \end{array}$$

5.
$$\begin{array}{r} 289 \\ -\ 54 \\ \hline 235 \end{array}$$

7.
$$\begin{array}{r} \overset{4\;\;\;1}{\cancel{5}\,03} \\ -\ 74 \\ \hline \end{array}$$

$$\begin{array}{r} \overset{4\ \overset{9}{\cancel{10}}}{\cancel{5}\,\cancel{0}\,3} \\ -\ 74 \\ \hline 4\ 29 \end{array}$$

9.
$$\begin{array}{r} \overset{2}{58} \\ \times\ 3 \\ \hline 174 \end{array}$$

11.
$$\begin{array}{r} \overset{4}{\underset{5}{89}} \\ \times\ 56 \\ \hline 534 \\ 4450 \\ \hline 4,984 \end{array}$$

13.
$$\begin{array}{r} 24 \\ 281\overline{)6,935} \\ \underline{562} \\ 1315 \\ \underline{1124} \\ 191 \end{array}$$

We have $6,935 \div 281 = 24 \text{ r}191$.

15.
$$\begin{array}{r} 209 \\ 293\overline{)61,382} \\ \underline{586} \\ 278 \\ \underline{0} \\ 2782 \\ \underline{2637} \\ 145 \end{array}$$

We have $61,382 \div 293 = 209 \text{ r}145$.

17. $4 + 12 \div 4 = 4 + 3 = 7$

19. $28 \div 7 \times 4 = 4 \times 4 = 16$

21. $36 \div (3^2 + 3) = 36 \div (9 + 3) = 36 \div 12 = 3$

23. The opposite of 8 is -8.

25. $-(-12) = 12$

27. $-12 + (-6) = -18$

29. $(-6)(15) = -90$

31. $9\overline{)16}$ $\dfrac{16}{9} = 1\dfrac{7}{9}$

$\underline{9}$

7

33. $5\dfrac{3}{4} = \dfrac{(4 \times 5) + 3}{4} = \dfrac{23}{4}$

35. The cross products $8 \times 36 = 288$ and $32 \times 9 = 288$ are equal. The fractions are equivalent.

37. $\dfrac{7}{15} \times \dfrac{5}{21} = \dfrac{\overset{1}{\cancel{7}} \times \overset{1}{\cancel{5}}}{\underset{3}{\cancel{15}} \times \underset{3}{\cancel{21}}} = \dfrac{1 \times 1}{3 \times 3} = \dfrac{1}{9}$

39. $4 \times \dfrac{3}{8} = \dfrac{4}{1} \times \dfrac{3}{8} = \dfrac{\overset{1}{\cancel{4}} \times 3}{1 \times \underset{2}{\cancel{8}}} = \dfrac{1 \times 3}{1 \times 2} = \dfrac{3}{2} = 1\dfrac{1}{2}$

41. $5\dfrac{1}{3} \times 1\dfrac{4}{5} = \dfrac{16}{3} \times \dfrac{9}{5} = \dfrac{16 \times \overset{3}{\cancel{9}}}{\underset{1}{\cancel{3}} \times 5} = \dfrac{16 \times 3}{1 \times 5} = \dfrac{48}{5}$

$= 9\dfrac{3}{5}$

43. $3\dfrac{1}{5} \times \dfrac{7}{8} \times 2\dfrac{6}{7} = \dfrac{16}{5} \times \dfrac{7}{8} \times \dfrac{20}{7} = \dfrac{\overset{2}{\cancel{16}} \times \overset{1}{\cancel{7}} \times \overset{4}{\cancel{20}}}{\underset{1}{\cancel{5}} \times \underset{1}{\cancel{8}} \times \underset{1}{\cancel{7}}}$

$= \dfrac{2 \times 1 \times 4}{1 \times 1 \times 1} = \dfrac{8}{1} = 8$

45. $\dfrac{7}{15} \div \dfrac{14}{25} = \dfrac{7}{15} \times \dfrac{25}{14} = \dfrac{\overset{1}{\cancel{7}} \times \overset{5}{\cancel{25}}}{\underset{3}{\cancel{15}} \times \underset{2}{\cancel{14}}} = \dfrac{1 \times 5}{3 \times 2} = \dfrac{5}{6}$

47. $\dfrac{4}{15} + \dfrac{8}{15} = \dfrac{12}{15} = \dfrac{4}{5}$

49. Step 1: The LCD is 40.

Step 2: $\dfrac{2}{5} = \dfrac{16}{40}$

$\dfrac{3}{4} = \dfrac{30}{40}$

$\dfrac{5}{8} = \dfrac{25}{40}$

Step 3: $\dfrac{16}{40} + \dfrac{30}{40} + \dfrac{25}{40} = \dfrac{71}{40} = 1\dfrac{31}{40}$

51. Step 1: The LCD is 36.

Step 2: $\dfrac{5}{9} = \dfrac{20}{36}$

$\dfrac{5}{12} = \dfrac{15}{36}$

Step 3: $\dfrac{5}{9} - \dfrac{5}{12} = \dfrac{20}{36} - \dfrac{15}{36} = \dfrac{5}{36}$

53. $3\dfrac{5}{7} + 2\dfrac{4}{7} = (3 + 2) + \dfrac{5}{7} + \dfrac{4}{7} = 5 + \dfrac{9}{7} = 5 + 1\dfrac{2}{7}$

$= 6\dfrac{2}{7}$

55. $8\dfrac{1}{9} - 3\dfrac{5}{9} = (7 - 3) + \left(\dfrac{10}{9} - \dfrac{5}{9}\right) = 4 + \dfrac{5}{9} = 4\dfrac{5}{9}$

57. $9 - 5\dfrac{3}{8} = \dfrac{72}{8} - \dfrac{43}{8} = \dfrac{29}{8} = 3\dfrac{5}{8}$

59. $3\dfrac{5}{6} + 4\dfrac{3}{10} + 6\dfrac{1}{2} = (3 + 4 + 6) + \dfrac{25}{30} + \dfrac{9}{30} + \dfrac{15}{30}$

$= 13 + \dfrac{49}{30} = 13 + 1\dfrac{19}{30} = 14\dfrac{19}{30}$

Manuel worked $14\dfrac{19}{30}$ hr during the week.

61. $2x + 3 = 7x + 5$

$2x = 7x + 2$

$-5x = 2$

$x = -\dfrac{2}{5}$

Check: $2\left(-\dfrac{2}{5}\right) + 3 \overset{?}{=} 7\left(-\dfrac{2}{5}\right) + 5$

$-\dfrac{4}{5} + \dfrac{15}{5} \overset{?}{=} -\dfrac{14}{5} + \dfrac{25}{5}$

$\dfrac{11}{5} = \dfrac{11}{5}$

The solution is $-\dfrac{2}{5}$.

Chapter 5
Decimals

Prerequisite Check

1. Three and seven tenths

3. Seventeen and eighty-nine thousandths

5.
$$\begin{array}{r} \overset{1\ 10\ 13}{2\ 1\ 3} \\ -\ 49 \\ \hline 164 \end{array}$$

7.
$$\begin{array}{r} 79 \\ 10\overline{)792} \\ \underline{70} \\ 92 \\ \underline{90} \\ 2 \end{array}$$

We have $792 \div 10 = 79$ r2.

9. Area $=$ length $\times$ width
Area $= 17$ ft $\times 8$ ft
Area $= 136$ ft^2

11. The digit to the right of the thousands place, 4, is less than 5. Thus, the thousands digit remains the same and the digits to its right become zero. 23,456 is rounded to 23,000.

13. 4,913,457
7 ones, 5 tens, 4 hundreds, 3 thousands, 1 ten thousands, 9 hundred thousands, 4 millions
The place value of 1 is ten thousands.

Exercises 5.1

<Objective 1 >

1. $\dfrac{23}{100} = 0.23$

3. $\dfrac{209}{10,000} = 0.0209$

5. $-23\dfrac{56}{1,000} = -23.056$

7. $\dfrac{2}{10} = 0.2$

9. $-\dfrac{53}{10} = -5.3$

< Objective 2 >

11. The place value of 7 in 8.57932 is hundredths.

13. The place value of 3 in 8.57932 is ten-thousandths.

15. 32.06197
7 hundred-thousandths, 9 ten-thousandths, 1 thousandths, 6hundredths, 0 tenths, 2 ones, 3 tens.
The place value of 1 is thousandths.

17. 32.06197
7 hundred-thousandths, 9 ten-thousandths, 1 thousandths, 6 hundredths, 0 tenths, 2 ones, 3 tens.
The place value of 6 is hundredths.

19. "Fifty-one thousandths" is 0.051.

21. "Seven and three tenths" is 7.3.

23. "Negative eighteen and three tenths" is −18.7.

25. 0.23 is read as "twenty-three hundredths".

27. 0.071 is read as "seventy-one thousandths."

29. −12.07 is read as "negative twelve and seven hundredths".

< Objective 3 >

31. $0.\underbrace{65}_{\text{Two places}} = \dfrac{65}{\underbrace{100}_{\text{Two zeroes}}} = \dfrac{13}{20}$

(divide numerator and denominator by 5)

33. $5.\underbrace{231}_{\text{Three places}} = \dfrac{5231}{\underbrace{1,000}_{\text{Three zeroes}}} = 5\dfrac{231}{1,000}$

35. $-0.\underbrace{08}_{\text{Two places}} = -\dfrac{8}{\underbrace{100}_{\text{Two zeroes}}} = -\dfrac{2}{25}$

(divide numerator and denominator by 4)

37. $7.\underbrace{25}_{\text{Two places}} = \dfrac{725}{\underbrace{100}_{\text{Two zeroes}}} = \dfrac{29}{4} = 7\dfrac{1}{4}$

(divide numerator and denominator by 4)

< Objective 4 >

39. Write 0.69 as 0.690. Then we see that 0.69 is greater than 0.689, and we can write $0.69 > 0.689$.

41. Write 1.23 as 1.230. Then we see that 1.230 is equal to 1.23, and we can write $1.23 = 1.230$.

43. $-9.9 > -10$

45. Write 1.46 as 1.460. Then we see that 1.460 is greater than 1.459, and we can write $1.459 < 1.460$.

47.

Given	Decimal	Rank
4.0339	4.0339	1
4.034	4.0340	2
$4\dfrac{3}{10}$	4.3000	3
$\dfrac{432}{100}$	4.3200	4
4.33	4.3300	5

We see that the order from smallest to largest is 4.0339, 4.034, $4\dfrac{3}{10}$, $\dfrac{432}{100}$, 4.33.

49.

Given	Decimal	Rank
−1	−1.000	6
−1.01	−1.010	5
$-1\dfrac{1}{10}$	−1.100	3
−1.11	−1.110	2
$-1\dfrac{11}{1,000}$	−1.011	4

We see that the order from smallest to largest is $-1\dfrac{111}{1,000}$, -1.11, $-1\dfrac{1}{10}$, $-1\dfrac{11}{1,000}$, -1.01, -1.

51.

Given	Decimal	Rank
0.71	0.7100	9
0.072	0.0720	7
$\dfrac{7}{10}$	0.7000	8
0.007	0.0070	2
0.0069	0.0069	1
$\dfrac{7}{100}$	0.0700	4
0.0701	0.0701	5
0.0619	0.0619	3
0.0712	0.0712	6

We see that the order from smallest to largest is 0.0069, 0.007, 0.0619, $\dfrac{7}{100}$, 0.0701, 0.0712, 0.072, $\dfrac{7}{10}$, 0.71.

< Objective 5 >

53. The 3 is in the hundredths place. The next digit to the right, (4) is less than 5, so leave the hundredths digit as it is. Discard the remaining digits to the right. 21.534 is rounded to 21.53.

55. The 4 is in the hundredths place. The next digit to the right, (2) is less than 5, so leave the hundredths digit as it is. Discard the remaining digits to the right. 0.342 is rounded to 0.34.

57. The 8 is in the thousandths place. The next digit to the right, (2) is less than 5, so leave the thousandths digit as it is. Discard the remaining digits to the right. 2.71828 is rounded to 2.718.

59. The 9 is in the tenths place. The next to the right, (4) is less than 5, so leave the tenths digit as it is, Discard the remaining digits to the right. 2.942 is rounded to 2.9.

61. The 0 is in the tenths place. The next digit to the right, (4) is less than 5, so leave the tenths digit as it is. Discard the remaining digits to the right. 0.0475 is rounded to 0.0.

63. The 4 is in the ten-thousandths place. The next digit to the right, (4) is less than 5, so leave the ten-thousandths digit as it is. Discard the remaining digits to the right. −4.85344 is rounded to −4.8534.

65. The 9 is in the tenths place. The next digit to the right is (5), so we increase 9 by 1 and drop the remaining digits. Increasing 9 by 1 makes it 10, just like when rounding whole numbers. −2.95 is rounded to −3.0.

67. The 3 is in the hundredths place. The next digit to the right, (4) is less than 5, so leave the hundredths digit as it is. Discard the remaining digits to the right. 6.734 is rounded to 6.73.

69. The 3 is in the ten-thousandths place. The next digit to the right, (9) is 5 or more, so increase the digit you are rounding to by 1. Discard the remaining digits to the right. 6.58739 is rounded to 6.5874.

71. The 9 is in the thousandths place. The next digit to the right is (6), so we increase 9 by 1 and drop the remaining digits. Increasing 9 by 1 makes it 10, just like when rounding whole numbers. 0.59962 is rounded to 0.6 or 0.600.

73. The 4 is in the cents place. The next digit to the right, (5) is 5 or more, so increase the digit you are rounding to by 1. Discard the remaining digits to the right. $235.1457 is rounded to $235.15.

75. The 2 is in the dollars place. The next digit to the right, (5) is 5 or more, so increase the digit you are rounding to by 1. Discard the remaining digits to the right. $752.512 is rounded to $753.

77. The 9 is in the dollars place. The next digit to the right, (6) is 5 or more, so increase the digit you are rounding to by 1 and drop the remaining digits. Increasing 9 by 1 makes it 10, just like when rounding whole numbers. $49.605 is rounded to $50.

79.

81.

83. False

85. True

87. The 5 is in the tenths place. The next digit to the right, (3) is less than 5, so the digit in the tenths place stays the same. Discard the remaining digits to the right. 1.53 mg is rounded to 1.5 mg.

89. (a) 10.35 V
   (b) 0.00047 F
   (c) 0.0158 H

91.

Given	Decimal	Rank
0.308	0.308	3
0.297	0.297	1
0.31	0.310	4
0.3	0.300	2
0.311	0.311	5
0.32	0.320	6

Then we see that the order from smallest to largest is: 0.297, 0.3, 0.308, 0.31, 0.311, 0.32.

93. Above and Beyond

Exercises 5.2

< Objective 1 >

1.   0.28
      $\underline{+\ 0.79}$
        1.07

3.   13.580
        7.239
      $\underline{+\ 1.500}$
       22.319

5.   25.3582
        6.5000
        1.8980
      $\underline{+\ 0.6900}$
       34.4462

7.   0.86
      $\underline{+\ 5.91}$
        6.77

9.   5.0
      $\underline{+\ 0.7}$
        5.7

11.   4.743
       $\underline{+\ 12.000}$
        16.743

13.   0.430
         0.800
       $\underline{+\ 0.561}$
         1.791

15.   42.731
          1.058
       $\underline{+\ 103.240}$
        147.029

< Objective 3 >

17.   0.85
       $\underline{-\ 0.59}$
         0.26

19.   3.820
       $\underline{-\ 1.565}$
         2.255

21.   7.02
       $\underline{-\ 4.70}$
         2.32

23.   12.00
         $\underline{- \ 5.35}$
          6.65

25.   5.316
         $\underline{- \ 2.900}$
          2.416

27.   7.0
         $\underline{- \ 0.5}$
          6.5

29.   8.1
         $\underline{- \ 3.0}$
          5.1

31.   6.84
         $\underline{- \ 2.87}$
          3.97

33.   9.40
         $\underline{- \ 7.75}$
          1.65

35.   5.00
         $\underline{- \ 0.24}$
          4.76

37.   2.4
         $\underline{- \ 1.4}$
          1.0

39.   4.23
         $\underline{- \ 1.30}$
          2.93

< Objective 4 >

41. $15.2 - 22.8$
        22.8
        $\underline{-15.2}$
        7.6
        $15.2 - 22.8 = -7.6$

43. $-125.9 + 73.6$
        125.9
        $\underline{- \ 73.6}$
        52.3
        $-125.9 + 73.6 = -52.3$

45. $-8.4 + (-3.77) = -8.4 - 3.77$
        8.40
        $\underline{+3.77}$
        12.17
        $-8.4 + (-3.77) = -8.4 - 3.77 = -12.17$

47. $13.1 - (-18.55) = 13.1 + 18.55$
        13.10
        $\underline{+18.45}$
        31.55
        $13.1 - (-18.55) = 13.1 + 18.55 = 31.55$

49. $-81.63 - (-92.4) = -81.63 + 92.4$
        92.40
        $\underline{+81.63}$
        10.77
        $-81.63 - (-92.4) = -81.63 + 92.4 = 10.77$

< Objective 2>

51.   12.7
          15.9
         $\underline{+ \ 13.8}$
          42.4
        Dien bought 42.4 gal of gas on the trip.

53. **(a)** $5.38 + 3.2 + 4.79 = 13.37$
        13.37 cm of rain fell.
        (b)   5.38 cm
                  $\underline{- \ 4.79 \ cm}$
                   0.59 cm
        An extra 0.59 cm of rain fell.

55. $78.49
$129.45
$149.95
+ $8.80
$366.69
Nicole's expenses total was $366.69.

57. $8.16 + 8.16 + 12.68 + 12.68 = 41.68$
Lupe should purchase 41.68 yd of fence.

59. $2.321 + 2.887 + 2.417 + 2.007 + 1.903$
$= 11.535$
11.535 mi of fencing is needed for the property.

61.   2.8325 in.
− 2.7750 in.
0.0575 in.
The wall of the tubing is 0.0575 in. thick.

63. $37.25 + 8378 + 53.45 = 99.48$
Total charges on the credit card are $99.48.
99.48
− 73.50
25.98
You still owe $25.98.

65.

Beginning Balance	$456.00
Check #601	$199.29
Balance	$256.71
Service Charge	$18.00
Balance	$238.71
Check #602	$85.78
Balance	$152.93
Deposit	$250.45
Balance	$403.38
Check #603	$201.24
Ending Balance	$202.14

67.

Beginning Balance	$1,345.23
Check #821	$234.99
Balance	$1,110.24
Check #822	$555.77
Balance	$554.47
Deposit	$126.77
Balance	$681.24
Check #823	$53.89
Ending Balance	$627.35

69.   136.81 $\boxed{+}$ 713.566 $\boxed{=}$
Display: 850.376
The result is 850.376.

71.   56.8 $\boxed{-}$ 23.09 $\boxed{=}$
Display: 33.71
The result is 33.71.

73. $\boxed{-}$ 9.1 $\boxed{+}$ 8.32 $\boxed{=}$
Display: −0.78
The result is −0.78.

75. $-8 + (-4.67) = -8 + 4.67$
$\boxed{-}$ 8 $\boxed{+}$ 4.67 $\boxed{=}$
Display: −3.33
The result is −3.33.

77.   10345.2 $\boxed{+}$ 2308.35 $\boxed{+}$ 153.58 $\boxed{=}$
Display: 12807.13
The result is 12,807.13.

79.   532.89 $\boxed{-}$ 50 $\boxed{-}$ 27.54 $\boxed{-}$ 134.75
$\boxed{+}$ 50 $\boxed{=}$
Display: 370.6
Your ending balance is $370.60.

81.   Total amount = sum of amounts of sodium pertechnetate in each capsule.
$79.4 + 15.88 + 3.97 = 99.25$ mCi
The total amount administered to the patient was 99.25 mCi.

83.

Battery	Measured Voltage
1	12.20
2	13.84
3	11.42
4	13.00
5	12.45
6	12.82
7	11.93
8	11.01
9	12.77
10	12.03
Sum	**123.47**

85. The numbers are increasing by 0.25. Add 0.25 to the last number given in the sequence. $3.625 + 0.25 = 3.875$. The next number would be 3.875.

87. Adding the diagonal from the bottom left to top right gives $0.8 + 1 + 1.2 = 3$. So all rows, columns, and diagonals must have a sum of 3.

1.6	a	1.2
b	1	c
0.8	d	e

Since $1.6 + a + 1.2 = 3$, $a = 0.2$.
Since $1.6 + 1 + e = 3$, $e = 0.4$.
Since $1.2 + c + 0.4 = 3$, $c = 1.4$.
Since $b + 1 + 1.4 = 3$, $b = 0.6$.
Since $0.8 + d + 0.4 = 3$, $d = 1.8$.
As a check, note that all rows, columns, and diagonals add up to 3. The final square is

1.6	**0.2**	1.2
0.6	1	**1.4**
0.8	**1.8**	**0.4**

89. (a) The numbers are decreasing, so look at the differences.
$0.75 - 0.62 = 0.13$
$0.62 - 0.5 = 0.12$
$0.5 - 0.39 = 0.11$
The pattern shows that subtracting 0.10 from 0.39, and then 0.09 from this result, gives the next two numbers.
$0.39 - 0.10 = 0.29$
$0.29 - 0.09 = 0.20$
The next two numbers are 0.29, 0.20.
(b) Examining the sequence reveals every other number decreases by 0.1.
1.0, 0.9, 0.8, …
Similarly, the remaining numbers are increasing by 2.
1.5, 3.5, …
Subtract 0.1 from 0.8, and add 2 to 3.5 to get the next two numbers. Therefore, the next two numbers are 5.5, 0.7

Exercises 5.3

< Objective 1 >

1.
$$\begin{array}{r} 2.3 \\ \times 3.4 \\ \hline 92 \\ 69 \\ \hline 7.82 \end{array}$$

3.
$$\begin{array}{r} 8.4 \\ \times 5.2 \\ \hline 168 \\ 420 \\ \hline 43.68 \end{array}$$

5.
$$\begin{array}{r} 2.56 \\ \times\ 72 \\ \hline 512 \\ 1792 \\ \hline 184.32 \end{array}$$

7.
$$
\begin{array}{r}
0.78 \\
\times\ 2.3 \\
\hline
234 \\
156 \\
\hline
1.794
\end{array}
$$

9.
$$
\begin{array}{r}
15.7 \\
\times 2.35 \\
\hline
785 \\
471 \\
314 \\
\hline
36.895
\end{array}
$$

11.
$$
\begin{array}{r}
0.354 \\
\times\ 0.8 \\
\hline
0.2832
\end{array}
$$

13.
$$
\begin{array}{r}
3.28 \\
\times 5.07 \\
\hline
2296 \\
1640 \\
\hline
16.6296
\end{array}
$$

15.
$$
\begin{array}{r}
5.238 \\
\times\ 0.48 \\
\hline
41904 \\
20952 \\
\hline
2.51424
\end{array}
$$

17.
$$
\begin{array}{r}
1.053 \\
\times 0.552 \\
\hline
2106 \\
5265 \\
5265 \\
\hline
0.581256
\end{array}
$$

19.
$$
\begin{array}{r}
0.0056 \\
\times\ 0.082 \\
\hline
112 \\
448 \\
\hline
0.0004592
\end{array}
$$

21.
$$
\begin{array}{r}
2.376 \\
\times\ 0.8 \\
\hline
1.9008
\end{array}
$$

23.
$$
\begin{array}{r}
0.3085 \\
\times\ \ 4.5 \\
\hline
15425 \\
12340 \\
\hline
1.38825
\end{array}
$$

25.
$$
\begin{array}{r}
43.8 \\
\times 2.567 \\
\hline
3066 \\
26280 \\
219000 \\
876000 \\
\hline
112.4346
\end{array}
$$

27.
$$
\begin{array}{r}
2.5 \\
\times 2.5 \\
\hline
125 \\
500 \\
\hline
6.25
\end{array}
$$

29.
$$
\begin{array}{r}
3.28 \\
\times 3.28 \\
\hline
2624 \\
6560 \\
98400 \\
\hline
10.7584
\end{array}
$$

31.
$$
\begin{array}{r}
8.40 \\
\times 3.55 \\
\hline
4200 \\
42000 \\
252000 \\
\hline
29.8200
\end{array}
$$
$-8.4 \times 3.55 = -29.82$

33.
$$
\begin{array}{r}
0.06 \\
\times\,0.08 \\
\hline
48 \\
000 \\
\hline
0.0048
\end{array}
$$
$0.06 \times (-0.08) = -0.0048$

35.
$$
\begin{array}{r}
12.5 \\
\times\ 3.2 \\
\hline
250 \\
3750 \\
\hline
40.00
\end{array}
$$
$(-12.5) \times (-3.2) = 40$

< Objective 2 >

37.
$$
\begin{array}{r}
9.98 \\
\times\quad 4 \\
\hline
39.92
\end{array}
$$
The total cost of Kurt's purchase was $39.92.

39.
$$
\begin{array}{r}
8.34 \\
\times\ 2.5 \\
\hline
4170 \\
1668 \\
\hline
20.850
\end{array}
$$
2.5 gal weighs 20.85 lb.

41.
$$
\begin{array}{r}
0.095 \\
\times\,1500 \\
\hline
0000 \\
0000 \\
475 \\
95 \\
\hline
142.500
\end{array}
$$
The simple interest is $142.50.

43.
$$
\begin{array}{r}
43{,}640 \\
\times\,0.054 \\
\hline
174560 \\
2182000 \\
\hline
2{,}356.560
\end{array}
$$
Amount of his tax is $2356.56.

45. Area = Length × Width
$$
\begin{array}{r}
21.6 \\
\times\ 28 \\
\hline
1728 \\
432 \\
\hline
604.8
\end{array}
$$
Therefore, the area equals 604.8 cm^2.

< Objective 3 >

47. $5.89 \times 10 = 58.9$

49. $23.79 \times 100 = 2{,}379$

51. $10(0.045) = 10 \times 0.045 = 0.45$

53. $(0.431)100 = 0.431 \times 100 = 43.1$

55. $0.471 \times 100 = 47.1$

57. $1{,}000 \cdot 0.7125 = 1{,}000 \times 0.7125 = 712.5$

59. $4.25 \times 10^2 = 425$

61. $3.45 \times 10^4 = 34{,}500$

63. $1.38 \times 100 = 138$
The total cost is $138.

65. $2.2 \times 1{,}000 = 2{,}200$
There are 2,200 grams in 2.2 kg.

67. True

69. False

Exercises 5.4

< Objectives 1 and 3 >

1.
$$\begin{array}{r} 2.78 \\ 6{\overline{\smash{)}16.68}} \\ \underline{12} \\ 46 \\ \underline{42} \\ 48 \\ \underline{48} \\ 0 \end{array}$$

The quotient is 2.78.

3.
$$\begin{array}{r} 0.48 \\ 4{\overline{\smash{)}1.92}} \\ \underline{16} \\ 32 \\ \underline{32} \\ 0 \end{array}$$

The quotient is 0.48.

5.
$$\begin{array}{r} 0.685 \\ 8{\overline{\smash{)}5.480}} \\ \underline{48} \\ 68 \\ \underline{64} \\ 40 \\ \underline{40} \\ 0 \end{array}$$

The quotient is 0.685.

7.
$$\begin{array}{r} 2.315 \\ 6{\overline{\smash{)}13.890}} \\ \underline{12} \\ 18 \\ \underline{18} \\ 09 \\ \underline{6} \\ 30 \\ \underline{30} \\ 0 \end{array}$$

The quotient is 2.315.

9.
$$\begin{array}{r} 5.8 \\ 32{\overline{\smash{)}185.6}} \\ \underline{160} \\ 256 \\ \underline{256} \\ 0 \end{array}$$

The quotient is 5.8.

11.
$$\begin{array}{r} 8.05 \\ 32{\overline{\smash{)}257.60}} \\ \underline{254} \\ 16 \\ \underline{0} \\ 160 \\ \underline{160} \\ 0 \end{array}$$

The quotient is 8.05.

13.
$$\begin{array}{r} 0.265 \\ 52{\overline{\smash{)}13.780}} \\ \underline{104} \\ 338 \\ \underline{312} \\ 260 \\ \underline{260} \\ 0 \end{array}$$

The quotient is 0.265.

15. 0.3)6.24

$$
\begin{array}{r}
20.8 \\
3\overline{)62.4} \\
\underline{6} \\
02 \\
\underline{0} \\
24 \\
\underline{24} \\
0
\end{array}
$$

The quotient is 20.6.

17. 3.8)7.22

$$
\begin{array}{r}
1.9 \\
38\overline{)72.2} \\
\underline{38} \\
342 \\
\underline{342} \\
0
\end{array}
$$

The quotient is 1.9.

19. 3.4)1.717

$$
\begin{array}{r}
0.505 \\
34\overline{)17.170} \\
\underline{170} \\
17 \\
\underline{0} \\
170 \\
\underline{170} \\
0
\end{array}
$$

The quotient is 0.505.

21. 0.27)1.8495

$$
\begin{array}{r}
6.85 \\
27\overline{)184.95} \\
\underline{162} \\
229 \\
\underline{216} \\
135 \\
\underline{135} \\
0
\end{array}
$$

The quotient is 6.85.

23. 0.046)1.587

$$
\begin{array}{r}
34.5 \\
46\overline{)1587.0} \\
\underline{138} \\
207 \\
\underline{184} \\
230 \\
\underline{230} \\
0
\end{array}
$$

The quotient is 34.5.

25. 2.8)0.658

$$
\begin{array}{r}
0.235 \\
28\overline{)6.580} \\
\underline{56} \\
98 \\
\underline{84} \\
140 \\
\underline{140} \\
0
\end{array}
$$

The quotient is 0.235.

27.
$$
\begin{array}{r}
3.5 \\
2\overline{)7.0} \\
\underline{6} \\
10 \\
\underline{10} \\
0
\end{array}
$$

The quotient is 3.5.

29.

$$8\overline{)13.000}$$ quotient 1.625

$$
\begin{array}{r}
1.625 \\
8\overline{)13.000} \\
\underline{8} \\
50 \\
\underline{48} \\
20 \\
\underline{16} \\
40 \\
\underline{40} \\
0
\end{array}
$$

The quotient is 1.625.

31. The signs are opposite, so the quotient will be negative.

$$
\begin{array}{r}
1.2 \\
8\overline{)9.6} \\
\underline{8} \\
16 \\
\underline{16} \\
0
\end{array}
$$

$-9.6 \div 8 = -1.2$

33. The signs are the same, so the quotient will be positive.

$3.\underline{1}\overline{)2.\underline{1}08}$

$$
\begin{array}{r}
0.68 \\
31\overline{)21.08} \\
\underline{186} \\
248 \\
\underline{248} \\
0
\end{array}
$$

$-2.108 \div (-3.1) = 0.68$

< Objective 4 >

35. $5.8 \div 10 = 0\underline{5}.8 = 0.58$

37. $4.568 \div 100 = 0\underline{04}.568 = 0.04568$

39. $24.39 \div 1,000 = 0\underline{024}.39 = 0.02439$

41. $6.9 \div 1,000 = 0\underline{006}.9 = 0.0069$

43. $7.8 \div 10^2 = 0\underline{07}.8 = 0.078$

45. $45.2 \div 10^5 = 0\underline{00045}.2 = 0.000452$

47.

$$
\begin{array}{r}
2.64 \\
9\overline{)23.80} \\
\underline{18} \\
58 \\
\underline{54} \\
40 \\
\underline{36} \\
4
\end{array}
$$

Round 2.64 to 2.6. So $23.8 \div 9 = 2.6$ (rounded to the nearest tenths).

49.

$$
\begin{array}{r}
0.836 \\
46\overline{)38.480} \\
\underline{368} \\
168 \\
\underline{138} \\
300 \\
\underline{276} \\
24
\end{array}
$$

Round 0.836 to 0.84. So $38.48 \div 46 = 0.84$ (rounded to the nearest hundredths).

51.

$$
\begin{array}{r}
2.41 \\
52\overline{)125.40} \\
\underline{104} \\
214 \\
\underline{208} \\
60 \\
\underline{52} \\
8
\end{array}
$$

Round 2.41 to 2.4. So $125.4 \div 52 = 2.4$ (rounded to the nearest tenths).

53. $0.7\overline{)1.642}$

$$
\begin{array}{r}
2.345 \\
7\overline{)16.420} \\
\underline{14} \\
24 \\
\underline{21} \\
32 \\
\underline{28} \\
40 \\
\underline{35} \\
5
\end{array}
$$

Round 2.345 to 2.35. So $1.642 \div 0.7 = 2.35$
(rounded to the nearest hundredths).

55. $4.5\overline{)8.415}$

$$
\begin{array}{r}
1.87 \\
45\overline{)84.15} \\
\underline{45} \\
391 \\
\underline{360} \\
315 \\
\underline{315} \\
0
\end{array}
$$

Round 1.87 to 1.9. So $8.415 \div 4.5 = 1.9$
(rounded to the nearest tenths).

57. $3.4\overline{)27.44}$

$$
\begin{array}{r}
8.070 \\
34\overline{)274.400} \\
\underline{272} \\
240 \\
\underline{238} \\
20
\end{array}
$$

Round 8.070 to 8.07. So $27.44 \div 3.4 = 8.07$
(rounded to the nearest hundredths).

59.

$$
\begin{array}{r}
1.333 \\
3\overline{)4} \\
\underline{3} \\
10 \\
\underline{9} \\
10 \\
\underline{9} \\
1
\end{array}
$$

Round $1.\overline{3}$ to 1.333. So $4 \div 3 = 1.333$
(rounded to the nearest thousandths)

61.

$$
\begin{array}{r}
1.16 \\
6\overline{)7.00} \\
\underline{6} \\
10 \\
\underline{6} \\
40 \\
\underline{36} \\
4
\end{array}
$$

Round $1.1\overline{6}$ to 1.167. So $7 \div 6 = 1.167$
(rounded to the nearest thousandths)

63. The signs are opposite, so the quotient will be negative.

$0.7\overline{)4.23}$

$$
\begin{array}{r}
6.041 \\
7\overline{)42.3} \\
\underline{42} \\
30 \\
\underline{28} \\
20 \\
\underline{14} \\
6
\end{array}
$$

Round 6.041 to 6.04. So $-4.23 \div 0.7 = -6.04$
(rounded to the nearest hundredths)

65. The signs are the same, so the quotient will be positive.

$$4.\underline{5} \overline{)635.\underline{84}}$$

$$\begin{array}{r} 141.28 \\ 45\overline{)6358.4} \\ \underline{45} \\ 185 \\ \underline{180} \\ 58 \\ \underline{45} \\ 130 \\ \underline{90} \\ 400 \\ \underline{360} \\ 40 \end{array}$$

Round 141.28 to 141.3. So
$-635.84 \div (-4.5) = 141.3$ (rounded to the nearest tenths)

< Objective 2 >

67.

$$\begin{array}{r} 13.47 \\ 3\overline{)40.41} \\ \underline{3} \\ 10 \\ \underline{9} \\ 14 \\ \underline{12} \\ 21 \\ \underline{21} \\ 0 \end{array}$$

The cost per movie was $13.47.

69.

$$\begin{array}{r} 5.284 \\ 72\overline{)380.500} \\ \underline{360} \\ 205 \\ \underline{144} \\ 610 \\ \underline{576} \\ 340 \\ \underline{288} \\ 52 \end{array}$$

The average cost per book was $5.28.

71.

$$\begin{array}{r} 0.587 \\ 48\overline{)28.200} \\ \underline{240} \\ 420 \\ \underline{384} \\ 360 \\ \underline{336} \\ 24 \end{array}$$

The cost of an individual pen is $0.59 or 59¢.

73. First, subtract the down payment.
$736.12 - $100 = $636.12
Now, divide by the number of payments.

$$\begin{array}{r} 35.34 \\ 18\overline{)636.12} \\ \underline{54} \\ 96 \\ \underline{90} \\ 61 \\ \underline{54} \\ 72 \\ \underline{72} \\ 0 \end{array}$$

Al will be paying $35.34 per month.

75. $79 - 28.2 + 13.7 = 50.8 + 13.7 = 64.5$

77. $29.64 - (4.2 + 12.39) = 29.64 - 16.59$
$$= 13.05$$

79. $8.2 \div 0.25 \times 3.6 = 32.8 \times 3.6 = 118.08$

81. $\dfrac{7.8 + 4.2}{9.1 - 6.6} = \dfrac{12}{2.5} = 4.8$

83. $6.4 + 1.3^2 = 6.4 + 1.69 = 8.09$

85. $15.9 - 4.2 \times 3.5 = 15.9 - 14.7 = 1.2$

87. $6.1 + 7.3(5.9 - 8.08) = 6.1 + 7.3(-2.18)$
$$= 6.1 - 15.914$$
$$= -9.814$$

89. $5.2 - 10.5 \times 3.5 + (3.1 + 0.4)^2$
$$= 5.2 - 35.7 + (3.5)^2 = 5.2 - 35.7 + 12.25$$
$$= (5.2 + 12.25) - 35.7 = 17.45 - 36.06$$
$$= -18.25$$

91. $17.9 \times 1.1 - (2.3 \times 1.1)^2 + (13.4 - 2.1 \times 4.6)$
$$= 17.9 \times 1.1 - (2.53)^2 + (13.4 - 9.66)$$
$$= 17.9 \times 1.1 - (2.53)^2 + 3.74$$
$$= 17.9 \times 1.1 - 6.4009 + 3.74$$
$$= 19.69 - 6.4009 + 3.74 = 13.2891 + 3.74$$
$$= 17.0291$$

93. $127.85 \boxed{\times} 0.055 \boxed{\times} 15.84 \boxed{=}$
Display: 111.38292

95. $3.95 \boxed{y^x} 3 \boxed{=}$
Display: 61.629875

97. $2.546 \boxed{\div} 1.38 \boxed{=}$
Display: 1.844927536
The result is 1.84 (rounded to the nearest hundredths).

99. $0.5785 \boxed{\div} 1.236 \boxed{=}$
Display: 0.467799352
The result is 0.468 (rounded to the nearest thousandths).

101. $1.34 \boxed{\div} 2.63 \boxed{=}$
Display: 0.509505703
The result is 0.51 (rounded to two decimal places).

103. $A = b \times h$
$3.75 \boxed{\times} 2.35 \boxed{=}$
Display: 8.8125
The area is 8.8125 in.^{2}

105. Total cost of the fuel oil = cost per gallon
$\times$ total amount of fuel oil

$$
\begin{array}{r}
150.4 \\
\times\ 3.669 \\
\hline
13536 \\
9024 \\
9024 \\
4512 \\
\hline
551.8176
\end{array}
$$

Total cost of the fuel is $551.82.

107. (a) Current Plan:

Total cost for an year

= (lease rate of copy machine for a month
× 12 months) + (total number of copies
per year × cost per copy)

$= (\$325 \times 12) + (\$0.045 \times 100,000)$

$= 3900 + 4500 = \$8400$

3-year Plan:

Total cost for an year

= (lease rate of copy machine for a month
× 12 months) + (total number of copies
per year × cost per copy)

$= (\$375 \times 12) + (100,000 \times \$0.025)$

$= 4500 + 2500 = \$7000$

The 3-year Plan is better.

(b) Cost for making photocopies for 3 years
in the current plan

$= \$8400 \times 3 = \$25,200$

Cost for making photocopies for 3 years in
the 3-year plan $= \$7000 \times 3 = \$21,000$. The
better plan will save
$\$25,200 - \$21,000 = \$4,200$.

109. Cardiac index for a male patient $= \dfrac{4.8}{2.03}$

$\dfrac{4.8}{2.03} = 2.36453 \text{ L/(min} \cdot \text{m}^2)$

The cardiac index is $2.36 \text{ L/(min} \cdot \text{m}^2)$
rounded to the nearest hundredths.

111. Speed of her connection $= \dfrac{2.5}{1.7}$ MB/s

$$
\begin{array}{r}
1.470 \\
17\overline{)25.000} \\
\underline{17} \\
80 \\
\underline{68} \\
120 \\
\underline{119} \\
10
\end{array}
$$

The speed of her connection rounded to the
nearest hundredths is 1.47 MB/s.

113. Above and Beyond

115. $3.365\; \boxed{\times}\; 10\; \boxed{y^x}\; 3\; \boxed{=}$

The result will be 3,365.

117. $4.316\; \boxed{\times}\; 10\; \boxed{y^x}\; 5\; \boxed{=}$

The result will be 431,600.

119. $7.236\; \boxed{\times}\; 10\; \boxed{y^x}\; 8\; \boxed{=}$

The result will be 723,600,000.

121. Above and Beyond

Exercises 5.5

< Objective 1 >

1.
$$
\begin{array}{r}
0.75 \\
4\overline{)3.00} \\
\underline{28} \\
20 \\
\underline{20} \\
0
\end{array}
$$
$\dfrac{3}{4} = 0.75$

3.
$$
\begin{array}{r}
0.45 \\
20\overline{)9.00} \\
\underline{80} \\
100 \\
\underline{100} \\
0
\end{array}
$$
$\dfrac{9}{20} = 0.45$

5.
$$
\begin{array}{r}
0.2 \\
5\overline{)1.0} \\
\underline{10} \\
0
\end{array}
$$
$\dfrac{1}{5} = 0.2$

7.
$$16\overline{)5.0000} \quad \begin{array}{r} 0.3125 \end{array}$$

$$\begin{array}{r} \underline{48} \\ 20 \\ \underline{16} \\ 40 \\ \underline{32} \\ 80 \\ \underline{80} \\ 0 \end{array}$$

$\dfrac{5}{16} = 0.3125$

9.
$$10\overline{)7.0} \quad \begin{array}{r} 0.7 \end{array}$$

$$\begin{array}{r} \underline{70} \\ 0 \end{array}$$

$\dfrac{7}{10} = 0.7$

11.
$$40\overline{)27.000} \quad \begin{array}{r} 0.675 \end{array}$$

$$\begin{array}{r} \underline{240} \\ 300 \\ \underline{280} \\ 200 \\ \underline{200} \\ 0 \end{array}$$

$\dfrac{27}{40} = 0.675$

13.
$$6\overline{)5.0000} \quad \begin{array}{r} 0.8333 \end{array}$$

$$\begin{array}{r} \underline{48} \\ 20 \\ \underline{18} \\ 20 \\ \underline{18} \\ 20 \\ \underline{18} \\ 2 \end{array}$$

$\dfrac{5}{6} = 0.833$

(rounded to the nearest thousandths)

15.
$$15\overline{)4.0000} \quad \begin{array}{r} 0.2666 \end{array}$$

$$\begin{array}{r} \underline{30} \\ 100 \\ \underline{90} \\ 100 \\ \underline{90} \\ 100 \\ \underline{90} \\ 10 \end{array}$$

$\dfrac{4}{15} = 0.267$

(rounded to the nearest thousandths)

17.
$$9\overline{)15.00} \quad \begin{array}{r} 1.66 \end{array}$$

$$\begin{array}{r} \underline{9} \\ 60 \\ \underline{54} \\ 60 \\ \underline{54} \\ 6 \end{array}$$

$\dfrac{15}{9} = 1.\overline{6} = 1.67$

(rounded to two decimal places)

< Objective 2 >

19.
$$18\overline{)1.0000} \quad \begin{array}{r} 0.0555 \end{array}$$

$$\begin{array}{r} \underline{90} \\ 100 \\ \underline{90} \\ 100 \\ \underline{90} \\ 1 \end{array}$$

$\dfrac{1}{18} = 0.0\overline{5}$

21.

$$11\overline{)3.0000} 0.2727$$

$$\underline{22}$$
$$80$$
$$\underline{77}$$
$$30$$
$$\underline{22}$$
$$80$$
$$\underline{77}$$
$$3$$

$\dfrac{3}{11} = 0.\overline{27}$

23.

$$12\overline{)1.0000} 0.0833$$

$$\underline{96}$$
$$40$$
$$\underline{36}$$
$$40$$
$$\underline{36}$$
$$4$$

$\dfrac{1}{12} = 0.08\overline{3}$

25. $\dfrac{3}{5} = 0.6$

$5\dfrac{3}{5} = 5.6$

(Add 5 to the result.)

27. $\dfrac{7}{20} = 0.35$

$12\dfrac{7}{20} = 12.35$

(Add 12 to the result.)

29. $\dfrac{9}{4} = 2.25$

< Objective 3 >

31. $0.9 = \dfrac{9}{10}$

33. $0.8 = \dfrac{8}{10} = \dfrac{4}{5}$

35. $0.37 = \dfrac{37}{100}$

37. $0.587 = \dfrac{587}{1,000}$

39. $0.48 = \dfrac{48}{100} = \dfrac{12}{25}$

41. $18 = \dfrac{18}{1}$

43. $0.425 = \dfrac{425}{1,000} = \dfrac{17}{40}$

45. $0.375 = \dfrac{375}{1,000} = \dfrac{3}{8}$

47. $6.136 = 6\dfrac{136}{1,000} = 6\dfrac{17}{125}$

49. $0.059 = \dfrac{59}{1,000}$

51.

$$13\overline{)4.0000} 0.3076$$

$$\underline{39}$$
$$10$$
$$\underline{0}$$
$$100$$
$$\underline{91}$$
$$90$$
$$\underline{78}$$
$$12$$

$\dfrac{4}{13} = 0.3076$

Joel's hitting rounded to three decimal places is 0.308.

53. San Francisco wins = 10
Total games = $10 + 6 = 16$

$$
\begin{array}{r}
0.625 \\
16\overline{)10.000} \\
\underline{96} \\
40 \\
\underline{32} \\
80 \\
\underline{80} \\
0
\end{array}
$$

The fraction of wins for San Francisco, rounded to three decimal places, is 0.625.
St. Louis: wins = 7
Total games $= 7 + 9 = 16$

$$
\begin{array}{r}
0.4375 \\
16\overline{)7.0000} \\
\underline{64} \\
60 \\
\underline{48} \\
120 \\
\underline{112} \\
80 \\
\underline{80} \\
0
\end{array}
$$

The fraction of wins for St. Louis, rounded to three decimal places, is 0.438.
Seattle: wins = 7
Total games $= 7 + 9 = 16$

$$
\begin{array}{r}
0.4375 \\
16\overline{)7.0000} \\
\underline{64} \\
60 \\
\underline{48} \\
120 \\
\underline{112} \\
80 \\
\underline{80} \\
0
\end{array}
$$

The fraction of wins for Seattle, rounded to three decimal places, is 0.438.

Arizona: wins = 5
total games $= 5 + 11 = 16$

$$
\begin{array}{r}
0.3125 \\
16\overline{)5.0000} \\
\underline{48} \\
20 \\
\underline{16} \\
40 \\
\underline{32} \\
80 \\
\underline{80} \\
0
\end{array}
$$

The fraction of wins for Arizona, rounded to three decimal places, is 0.313.

55.
$$
\begin{array}{r}
0.090 \\
11\overline{)1.000} \\
\underline{99} \\
10
\end{array}
$$
$\dfrac{1}{11} = 0.\overline{09}$

57.
$$
\begin{array}{r}
0.00090 \\
1{,}111\overline{)1.00000} \\
\underline{9999} \\
10
\end{array}
$$
$\dfrac{1}{1{,}111} = 0.\overline{0009}$

59. The decimal representation of $\dfrac{31}{34}$ to the ten-thousandths is 0.9117. Comparing the digits in the ten-thousandths place, we see that 0.9118 is greater than $\dfrac{31}{34}$. Therefore, $\dfrac{31}{34} < 0.9118$.

61. The decimal representation of $\dfrac{13}{17}$ to the thousandths is 0.764. Comparing the digits in the thousandths place, we see that 0.7657 is greater than $\dfrac{13}{17}$. Therefore, $\dfrac{13}{17} < 0.7657$.

63. The decimal representation of $\dfrac{5}{16}$ to the thousandths is 0.3125. Comparing the digits in the thousandths place, we see 0.313 is greater than $\dfrac{5}{16}$. Therefore, $\dfrac{5}{16} < 0.313$.

65. $\dfrac{1}{2} + 0.385$

$\dfrac{1}{2} = 0.5000$

$\begin{array}{r} 0.5000 \\ + \ 0.3850 \\ \hline 0.8850 \end{array}$

The result is 0.885 (rounded to three decimal places).

67. $8.6245 + \dfrac{18}{11}$

$\dfrac{18}{11} = 1.6363$

$\begin{array}{r} 8.6245 \\ + \ 1.6363 \\ \hline 10.2608 \end{array}$

The result is 10.261 (rounded to three decimal places).

69. $3\dfrac{3}{5} + 5.608$

$3\dfrac{3}{5} = \dfrac{18}{5} = 3.6$

$\begin{array}{r} 3.6000 \\ + \ 5.6080 \\ \hline 9.2080 \end{array}$

The result is 9.208 (rounded to three decimal places).

71. 7 ÷ 8 =

Display: 0.875
The result is 0.875.

73. 5 ÷ 32 =

Display: 0.15625
The result is 0.156 (rounded to the nearest thousandth).

75. 3 ÷ 11 =

Display: 0.27272727
The result is $0.\overline{27}$.

77. 7 ÷ 8 + 3 =

Display: 3.875
The result is 3.875.

79. The internal diameter of an endotracheal tube for a child is $\dfrac{height}{20}$.

Height of the girl = 110 cm

Diameter $= \dfrac{110}{20} = 5.5$ cm or 55 mm.

81. $0.350 \text{ s} = \dfrac{350}{1,000} = \dfrac{7}{20} \text{ s}$

83. Above and Beyond

Exercises 5.6

< Objective 1 >

1. $3.2x = 12.8$

$\dfrac{3.2x}{3.2} = \dfrac{12.8}{3.2}$

$x = 4$

The solution is 4.

Check: $3.2(4) \overset{?}{=} 12.8$

$12.8 = 12.8$

3. $-4.5x = 13.5$

$\dfrac{-4.5x}{-4.5} = \dfrac{13.5}{-4.5}$

$x = -3$

The solution is –3.

Check: $-4.5(-3) \overset{?}{=} 13.5$

$13.5 = 13.5$

5. $1.3x + 2.8x = 12.3$

$4.1x = 12.3$

$\dfrac{4.1x}{4.1} = \dfrac{12.3}{4.1}$

$x = 3$

Check: $1.3(3) + 2.8(3) \overset{?}{=} 12.3$

$3.9 + 8.4 \overset{?}{=} 12.3$

$12.3 = 12.3$

The solution is 3.

7. $9.3x + (-6.2x) = 12.4$

$3.1x = 12.4$

$\dfrac{3.1x}{3.1} = \dfrac{12.4}{3.1}$

$x = 4$

Check: $9.3(4) + (-6.2(4)) \overset{?}{=} 12.4$

$37.2 - 24.8 \overset{?}{=} 12.4$

$12.4 = 12.4$

The solution is 4.

9. $5.3x + (-7) = 2.3x + 5$

$5.3x + (-2.3x) + (-7) = 2.3x + (-2.3x) + 5$

$3x + (-7) = 5$

$3x + (-7) + 7 = 5 + 7$

$3x = 12$

$\dfrac{3x}{3} = \dfrac{12}{3}$

$x = 4$

Check: $5.3(4) + (-7) \overset{?}{=} 2.3(4) + 5$

$21.2 + (-7) \overset{?}{=} 9.2 + 5$

$14.2 = 14.2$

The solution is 4.

11. $3x - 0.54 = 2(x - 0.15)$

$3x - 0.54 = 2x - 0.30$

$\begin{array}{rcr} -2x & & -2x \\ \hline x - 0.54 = & & -0.30 \\ +0.54 & & +0.54 \\ \hline x & = & 0.24 \end{array}$

Check: $3(0.24) - 0.54 \overset{?}{=} 2(0.24 - 0.15)$

$0.72 - 0.54 \overset{?}{=} 2(0.09)$

$0.18 = 0.18$

The solution is 0.24.

13. $6x + 3(x - 0.2789) = 4(2x + 0.3912)$

$6x + 3x - 0.8367 = 8x + 1.5648$

$9x - 0.8367 = 8x + 1.5648$

$\begin{array}{rcr} -8x & & -8x \\ \hline x - 0.8367 = & & 1.5648 \\ +0.8367 & & +0.8367 \\ \hline x & = & 2.4015 \end{array}$

Check:

$6(2.4015) + 3(2.4015 - 0.2789) \overset{?}{=} 4(2 \cdot 2.4015 + 0.3912)$

$14.409 + 3(2.1226) \overset{?}{=} 4(4.803 + 0.3912)$

$14.409 + 6.3678 \overset{?}{=} 4(5.1942)$

$20.7768 = 20.7768$

The solution is 2.4015.

15. $5x - (0.345 - x) = 5x + 0.8713$

$\quad 5x - 0.345 + x = 5x + 0.8713$

$\qquad 6x - 0.345 = 5x + 0.8713$

$\underline{\quad -5x \qquad\qquad -5x \qquad\qquad}$

$\qquad x - 0.345 = \qquad 0.8713$

$\underline{\qquad +0.345 \qquad\quad +0.345}$

$\qquad x \qquad = \qquad 1.2163$

Check:

$5(1.2163) - (0.345 - 1.2163) \overset{?}{=} 5(1.2163) + 0.8713$

$6.0815 - (-0.8713) \overset{?}{=} 6.0815 + 0.8713$

$6.0815 + 0.8713 \overset{?}{=} 6.8015 + 0.8713$

$6.9528 = 6.9258$

The solution is 1.2163.

17. $2.3x - 4.25 = 3.3x + 2.15$

$\quad -x - 4.25 = 2.15$

$\qquad -x = 6.4$

$\qquad x = -6.4$

19. $7.1x - 14 = 4.3x - 8.54$

$\quad 2.8x - 14 = -8.54$

$\qquad 2.8x = 5.46$

$\qquad x = 1.95$

21. $3.2x + 8.36 = 5x + 13.94$

$\quad -1.8x + 8.36 = 13.94$

$\qquad -1.8x = 5.58$

$\qquad x = -3.1$

23. $7.4(x - 1.2) = 8.4(x - 0.5)$

$\quad 7.4x - 8.88 = 8.4x - 4.2$

$\qquad -x = 4.68$

$\qquad x = -4.68$

25. $3.5(x - 1.4) = 1.3x - 3.14$

$\quad 3.5x - 4.9 = 1.3x - 3.14$

$\qquad 2.2x = 1.76$

$\qquad x = 0.8$

27. $5 - 1.6(x + 2) = 2(x - 1.3) - 0.1$

$\quad 5 - 1.6x - 3.2 = 2x - 2.6 - 0.1$

$\qquad -1.6x + 1.8 = 2x - 2.7$

$\qquad -3.6x = -4.5$

$\qquad x = 1.25$

29. $8x = 15$

$\quad \dfrac{8x}{8} = \dfrac{15}{8}$

$\quad x = 1.88$

31. $-3.9x = 12.25$

$\quad -\dfrac{3.9x}{3.9} = \dfrac{12.25}{3.9}$

$\qquad x = -3.14$

33. $\quad 3x + 10 = 9x - 12$

$\quad 3x + 10 - 3x = 9x - 3x - 12$

$\qquad 10 = 6x - 12$

$\qquad 10 + 12 = 6x - 12 + 12$

$\qquad 22 = 6x$

$\qquad x = 3.67$

35. $\qquad 7.2x - 4.65 = 1.9x - 17.5$

$\quad 7.2x - 1.9x - 4.65 = 1.9x - 1.9x - 17.5$

$\qquad 5.3x - 4.65 = -17.5$

$\quad 5.3x - 4.65 + 4.65 = -17.5 + 4.65$

$\qquad 5.3x = -12.85$

$\qquad x = -2.42$

37. $4.1x - (2.3x + 4.2) = 7.9x - 12.6$

$\quad 4.1x - 2.3x - 4.2 = 7.9x - 12.6$

$\qquad 1.8x - 4.2 = 7.9x - 12.6$

$\quad 1.8x - 1.8x - 4.2 = 7.9x - 1.8x - 12.6$

$\qquad -4.2 = 6.1x - 12.6$

$\quad -4.2 + 12.6 = 6.1x - 12.6 + 12.6$

$\qquad 8.4 = 6.1x$

$\qquad x = 1.38$

39. $12.9x + 2(3x - 5) = 3 - 2.1(x + 7.2)$

$12.9x + 6x - 10 = 3 - 2.1x - 15.12$

$13.5x - 10 = -2.1x - 12.12$

$13.5x + 2.1x - 10 = -2.1x + 2.1x - 12.12$

$15.6x - 10 = -12.12$

$15.6x = 2.12$

$x = 0.1$

< Objective 2 >

41. Julia's hourly wage multiplied by the number of hours she worked, gives her total earnings.
Let x = the number of hours Julia worked

$\$12.25(x) = \$1200.50$

$\dfrac{\$12.25(x)}{\$12.25} = \dfrac{\$1200.50}{\$12.25}$

$x = 98$

Julia worked 98 hours.

43. The expression for monthly long distance bill is $0.08t + 5.25$.
For $t = 173$ min

$0.08t + 5.25 = 0.08(173) + 5.25 = 19.09$

The monthly bill on the plan is $19.09.

45. **(a)** $5.5x + 4.5(400 - x) = 1,950$

(b) $5.5x + 4.5(400 - x) = 1,950$

$5.5x + 1,800 - 4.5x = 1,950$

$x = 1,950 - 1,800$

$= 150$

$400 - 150 = 250$

150 general tickets and 250 student tickets were sold.

47. The amount of dollars for each peso

$= \dfrac{\$34.20}{450 \text{ pesos}} = 0.076$.

The exchange rate she received is 0.076 dollar for each peso.

49. Let d be the diameter of the tree.

$110 = 3.14d$

$\dfrac{1}{3.14}(110) = \dfrac{1}{3.14}(3.14d)$

$35.03 = d$

The diameter of the tree is approximately 35 inches.

51. $x + yz = -2.34 + (-3.14) \cdot (4.12)$

Entering into the calculator:

$\boxed{(-)}\ 2.34\ \boxed{+}\ \boxed{(-)}\ 3.14\ \boxed{\times}\ 4.12\ \boxed{\text{ENTER}}$

Display: -15.2768

The result is -15.3 (rounded to the nearest tenth).

53. $x^2 - z^2 = (-2.34)^2 - (4.12)^2$

Entering into the calculator:

$\boxed{(}\ \boxed{(-)}\ 2.34\ \boxed{)}\ \boxed{y^x}\ 2 - \boxed{(}\ 4.12\ \boxed{)}\ \boxed{y^x}\ 2$

$\boxed{\text{ENTER}}$

Display: -11.4988

The result is -11.5 (rounded to the nearest tenth).

55. $\dfrac{xy}{z - x} = \dfrac{-2.34 \cdot (-3.14)}{4.12 - (-2.34)}$

Entering into the calculator:

$\boxed{(}\ \boxed{(-)}\ 2.34\ \boxed{\times}\ \boxed{(-)}\ 3.14\ \boxed{)}\ \boxed{\div}$

$\boxed{(}\ 4.12\ \boxed{-}\ \boxed{(-)}\ 2.34\ \boxed{)}\ \boxed{\text{ENTER}}$

Display: 1.137399381

The result is 1.1 (rounded to the nearest tenth).

57. $\dfrac{2x + y}{2x + z} = \dfrac{2 \cdot (-2.34) + (-3.14)}{2 \cdot (-2.34) + 4.12}$

Entering into the calculator:

$\boxed{(}\ 2\ \boxed{\times}\ \boxed{(-)}\ 2.34\ \boxed{+}\ \boxed{(-)}\ 3.14\ \boxed{)}\ \boxed{\div}$

$\boxed{(}\ 2\ \boxed{\times}\ \boxed{(-)}\ 2.34\ \boxed{+}\ 4.12\ \boxed{)}\ \boxed{\text{ENTER}}$

Display: 13.96428571

The result is 14.0 (rounded to the nearest tenth).

59. $230x = 157$

$$x = \frac{157}{230}$$

Entering into the calculator:

$\boxed{157}\ \boxed{\div}\ \boxed{230}\ \boxed{\text{ENTER}}$

Display: 0.6826
The result is 0.68 (rounded to the nearest hundredths).

61. $-29x = 432$

$$x = \frac{432}{-29}$$

Entering into the calculator:

$\boxed{432}\ \boxed{\div}\ \boxed{(-)}\ \boxed{29}\ \boxed{\text{ENTER}}$

Display: −14.896
The result is − 14.9 (rounded to the nearest hundredths).

63. $23.12x = 94.6$

$$x = \frac{94.6}{23.12}$$

Entering into the calculator:

$\boxed{94.6}\ \boxed{\div}\ \boxed{23.12}\ \boxed{\text{ENTER}}$

Display: 4.091
The result is 4.09 (rounded to the nearest hundredths).

65. $\dfrac{rT}{5,252}$ (For $r = 1,180$ and $T = 3$)

$$\frac{1,180 \times 3}{5,252} = \frac{3,540}{5,252} = 0.6740$$

The result is 0.674 (rounded to the nearest thousandth).

67. $\text{BMI} = \dfrac{703w}{h^2} = \dfrac{703 \times 190}{(69)^2} = \dfrac{133,570}{4,761}$

$\qquad = 28.055$
BMI is 28.1 (rounded to the nearest tenth).

69. **(c)**

71. **(a)**

Exercises 5.7

< Objective 1 >

1. $\sqrt{64} = 8$ Because $8 \times 8 = 64$

3. $\sqrt{169} = 13$ Because $13 \times 13 = 169$

< Objective 2 >

5. c is the hypotenuse, since it is the side opposite the right angle.

< Objective 3 >

7. $3^2 = 9$, $4^2 = 16$, $5^2 = 25$, and $9 + 16 = 25$, so $3^2 + 4^2 = 5^2$. Therefore, 3, 4, and 5 is a Pythagorean triple.

9. $7^2 = 49$, $12^2 = 144$, and $13^2 = 169$, but $49 + 144 \neq 169$ so $7^2 + 12^2 \neq 13^2$. 7, 12, and 13 is not a Pythagorean triple.

11. $8^2 = 64$, $15^2 = 225$, $17^2 = 289$, and $64 + 225 = 289$, so $8^2 + 15^2 = 17^2$. 8, 15, and 17 is a Pythagorean triple.

< Objective 4 >

13. $c^2 = a^2 + b^2 = (6)^2 + (8)^2 = 36 + 64 = 100$

$\qquad c = \sqrt{100} = 10$

15. $a^2 + b^2 = c^2$

$\qquad (8)^2 + b^2 = (17)^2$

$\qquad 64 + b^2 = 289$

$\qquad b^2 = 289 - 64 = 225$

$\qquad b = \sqrt{225} = 15$

< Objective 5 >

17. $\sqrt{16} = 4$ and $\sqrt{25} = 5$, so $\sqrt{23}$ must be between (b) 4 and 5.

19. $\sqrt{36} = 6$ and $\sqrt{49} = 7$, so $\sqrt{44}$ must be between (a) 6 and 7.

21. $a^2 + b^2 = c^2$

$a^2 + (6)^2 = (10)^2$

$a^2 + 36 = 100$

$a^2 = 100 - 36 = 64$

$a = \sqrt{64} = 8$

$P = 6 + 8 + 10 = 24$

The perimeter is 24.

23. $c^2 = a^2 + b^2 = (3)^2 + (4)^2 = 9 + 16 = 25$

$c = \sqrt{25} = 5$

$P = 3 + 4 + 5 = 12$

The perimeter is 12.

25. $(7)^2 + h^2 = (25)^2$

$49 + h^2 = 625$

$h^2 = 625 - 49 = 576$

$h = \sqrt{576} = 24$

The altitude of the triangle is 24.

27. Let d be the length of the diagonal.

$d^2 = (10)^2 + (24)^2 = 100 + 576 = 676$

$d = \sqrt{676} = 26$

The diagonal is 26 in. long.

29. False

31. always

33. With a scientific calculator:

64 $\boxed{\sqrt{}}$

With a graphing calculator:

$\boxed{\sqrt{}}$ 64 $\boxed{)}$ $\boxed{\text{ENTER}}$

Display: 8

35. With a scientific calculator:

289 $\boxed{\sqrt{}}$

With a graphing calculator:

$\boxed{\sqrt{}}$ 289 $\boxed{)}$ $\boxed{\text{ENTER}}$

Display: 17

37. With a scientific calculator:

1849 $\boxed{\sqrt{}}$

With a graphing calculator:

$\boxed{\sqrt{}}$ 1849 $\boxed{)}$ $\boxed{\text{ENTER}}$

Display: 43

39. With a scientific calculator:

8649 $\boxed{\sqrt{}}$

With a graphing calculator:

$\boxed{\sqrt{}}$ 8649 $\boxed{)}$ $\boxed{\text{ENTER}}$

Display: 93

41. With a scientific calculator:

23 $\boxed{\sqrt{}}$

With a graphing calculator:

$\boxed{\sqrt{}}$ 23 $\boxed{)}$ $\boxed{\text{ENTER}}$

Display: 4.795831523

$\sqrt{23} = 4.8$, rounded to the nearest tenth

43. With a scientific calculator:

51 $\boxed{\sqrt{}}$

With a graphing calculator:

$\boxed{\sqrt{}}$ 51 $\boxed{)}$ $\boxed{\text{ENTER}}$

Display: 7.141428429

$\sqrt{51} = 7.1$, rounded to the nearest tenth

45. With a scientific calculator:

134 $\boxed{\sqrt{}}$

With a graphing calculator:

$\boxed{\sqrt{}}$ 134 $\boxed{)}$ $\boxed{\text{ENTER}}$

Display: 11.5758369

$\sqrt{134} = 11.6$, rounded to the nearest tenth.

47. The ladder will represent the hypotenuse of a right triangle.
Let d be the length of the ladder.
$$d^2 = (7)^2 + (24)^2 = 49 + 576 = 625$$
$$d = \sqrt{625} = 25$$
The ladder must be at least 25 ft in length;
Yes, the 26-ft ladder is long enough.

49. $d = \sqrt{(0.86)^2 + (0.92)^2} = \sqrt{0.7396 + 0.8464}$
$$= \sqrt{1.596} = 1.26 \text{ in.}$$
(rounded to the hundredths)

Summary Exercises

1. The place value of 7 in 3.5742 is hundredths.

3. $\dfrac{37}{100} = 0.37$

5. $-\dfrac{4}{10} = -0.4$

7. 12.39 is read as "twelve and thirty-nine hundredths."

9. "Four and five tenths" is 4.5.

11. Write 0.79 as 0.790. Then we see that 0.790 is greater than 0.785, and we can write $0.79 > 0.785$.

13. Write 13 as 13.0. Then we see that 13.0 is greater than 12.8, and we can write $12.8 < 13$.

15. Write -0.4 as -0.400. Then we see that -0.400 is greater than -0.402, and we can write $-0.402 < -0.4$.

16. $-7.2 < -9.1$

17. The 3 is in the hundredths place. The next digit to the right (7) is 5 or more, so increase the digit you are rounding to by 1. Discard the remaining digits to the right. 5.837 is rounded to 5.84.

19. The 6 is in the thousandths place. The next digit to the right (2) is less than 5. Leave the thousandths digit as it is, and discard the remaining digits to the right. 4.87625 is rounded to 4.876.

21. $0.\underset{\substack{\uparrow \\ \text{Four places}}}{0067} = \dfrac{67}{\underset{\substack{\uparrow \\ \text{Four zeroes}}}{10,000}}$

23. $21.\underset{\substack{\uparrow \\ \text{Three places}}}{875} = \dfrac{21875}{\underset{\substack{\uparrow \\ \text{Three zeroes}}}{1,000}} = 21\dfrac{875}{1,000} = 21\dfrac{7}{8}$

25.
$$\begin{array}{r} 2.58 \\ + 0.89 \\ \hline 3.47 \end{array}$$

27.
$$\begin{array}{r} 0.324 \\ +4.691 \\ \hline 5.015 \end{array}$$

29.
$$\begin{array}{r} 5.91 \\ -3.70 \\ \hline 2.21 \end{array}$$
$$-3.70 + 5.91 = 2.21$$

31.
$$\begin{array}{r} 1.300 \\ 25.000 \\ 5.270 \\ + 6.158 \\ \hline 37.728 \end{array}$$

33.
$$\begin{array}{r} 29.21 \\ - 5.89 \\ \hline 23.32 \end{array}$$

35.
$$\begin{array}{r} 8.92 \\ -4.66 \\ \hline 4.26 \end{array}$$

37. $-7.12 - 5.6 = -(7.12 + 5.6)$

$$\begin{array}{r} 7.12 \\ +5.60 \\ \hline 12.72 \end{array}$$

39.
$$\begin{array}{r} 2.810 \\ -1.735 \\ \hline 1.075 \end{array}$$

41. $5.37 + 5.37 + 8.64 + 8.64 = 28.02$
The perimeter is 28.02 cm.

43. $1.85 + 3.10 + 1.20 = 6.15$
Dimension of a is 6.15 cm.

45.
$$\begin{array}{r} 22.8 \\ \times 0.72 \\ \hline 456 \\ 1596 \\ \hline 16.416 \end{array}$$

47.
$$\begin{array}{r} 1.24 \\ \times 56 \\ \hline 744 \\ 620 \\ \hline 69.44 \end{array}$$

49.
$$\begin{array}{r} 3.71 \\ \times 8.20 \\ \hline 000 \\ 7420 \\ 296800 \\ \hline 30.4220 \end{array}$$
$-3.71 \times 8.2 = -30.422$

51. $0.052 \times 1,000 = 52$

53.
$$\begin{array}{r} 37.4 \\ \times 7.25 \\ \hline 1870 \\ 748 \\ 2618 \\ \hline 271.150 \end{array}$$
Neal earned \$271.15.

55. The total amount paid is the monthly payment times the number of months.
$$\begin{array}{r} 27.15 \\ \times 24 \\ \hline 10860 \\ 5430 \\ \hline 651.60 \end{array}$$
The total amount paid is \$651.60.
Now, subtract the advertised price.
$$\begin{array}{r} 651.60 \\ -499.50 \\ \hline 152.10 \end{array}$$
You are paying \$152.10 extra.

57.
$$\begin{array}{r} 4.65 \\ 58\overline{)269.70} \\ \underline{232} \\ 377 \\ \underline{348} \\ 290 \\ \underline{290} \\ 0 \end{array}$$
The quotient is 4.65.

59. $4.\underline{5})\overline{85.2\underline{2}}$

$$
\begin{array}{r}
18.937 \\
45)\overline{852.2} \\
\underline{45} \\
402 \\
\underline{360} \\
422 \\
\underline{405} \\
170 \\
\underline{135} \\
350 \\
\underline{315} \\
35
\end{array}
$$

Round 18.937 to 18.94. So
$-85.22 \div 4.5 = -18.94$ (rounded to the nearest hundredth).

61. $0.\underline{7})\overline{1.\underline{8}65}$

$$
\begin{array}{r}
2.6642 \\
7)\overline{18.6500} \\
\underline{14} \\
46 \\
\underline{42} \\
45 \\
\underline{42} \\
30 \\
\underline{28} \\
20 \\
\underline{14} \\
6
\end{array}
$$

Round 2.6642 to 2.664. So
$1.865 \div 0.7 = 2.664$ (rounded to the nearest hundredths).

63. $5.\underline{3})\overline{6.\underline{7}48}$

$$
\begin{array}{r}
1.2732 \\
53)\overline{67.4800} \\
\underline{53} \\
144 \\
\underline{106} \\
388 \\
\underline{371} \\
170 \\
\underline{159} \\
110 \\
\underline{106} \\
4
\end{array}
$$

Round 1.2732 to 1.273. So
$6.748 \div 5.3 = 1.273$ (rounded to the nearest thousandths).

65. $3.\underline{8})\overline{12.\underline{5}6}$

$$
\begin{array}{r}
3.3052 \\
38)\overline{125.6} \\
\underline{114} \\
116 \\
\underline{114} \\
200 \\
\underline{190} \\
100 \\
\underline{76} \\
24
\end{array}
$$

Round 3.3052 to 3.305. So
$-12.56 \div (-3.8) = 3.305$ (rounded to the nearest thousandths)

67. $7.6 \div 10 = 0\underline{7}.6 = 0.76$

69. $457 \div 10^4 = 0\underline{0457}. = 0.0457$

71.

$$\begin{array}{r} 23.45 \\ 37\overline{)867.65} \\ \underline{74} \\ 127 \\ \underline{111} \\ 166 \\ \underline{148} \\ 185 \\ \underline{185} \\ 0 \end{array}$$

The donation per employee was \$23.45.

73.

$$\begin{array}{r} 18.5 \\ \underline{-\ 5.0} \\ 13.5 \end{array}$$

13.5 acres are available for lots.

$0.25\overline{)13.50}$

$$\begin{array}{r} 54 \\ 25\overline{)1350} \\ \underline{125} \\ 100 \\ \underline{100} \\ 0 \end{array}$$

54 lots are possible.

75.

$$\begin{array}{r} 0.4375 \\ 16\overline{)7.0000} \\ \underline{64} \\ 60 \\ \underline{48} \\ 120 \\ \underline{112} \\ 80 \\ \underline{80} \\ 0 \end{array}$$

$\dfrac{7}{16} = 0.4375$

77.

$$\begin{array}{r} 0.266 \\ 15\overline{)4.000} \\ \underline{30} \\ 100 \\ \underline{90} \\ 100 \\ \underline{90} \\ 10 \end{array}$$

$\dfrac{4}{15} = 0.2\overline{6}$

79.

$$\begin{array}{r} 0.1875 \\ 16\overline{)3.000} \\ \underline{16} \\ 140 \\ \underline{128} \\ 120 \\ \underline{112} \\ 80 \\ \underline{80} \\ 0 \end{array}$$

$-\dfrac{3}{16} = -0.1875$

81. $0.21 = \dfrac{21}{100}$

83. $2.03 = \dfrac{203}{100} = 2\dfrac{3}{100}$

85. $-0.44 = -\dfrac{44}{100} = -\dfrac{11}{25}$

87. $3.7x + 8 = 1.7x + 16$

$2x + 8 = 16$

$2x = 8$

$\dfrac{2x}{2} = \dfrac{8}{2}$

$x = 4$

Check: $3.7(4) + 8 \overset{?}{=} 1.7(4) + 16$

$14.8 + 8 \overset{?}{=} 6.8 + 16$

$22.8 = 22.8$

89. $2.9x = 4.9x - 3.3$

$-2x = -3.3$

$\dfrac{-2x}{-2} = \dfrac{-3.3}{-2}$

$x = 1.65$

Check: $2.9(1.65) \overset{?}{=} 4.9(1.65) - 3.3$

$4.785 \overset{?}{=} 8.085 - 3.3$

$4.785 = 4.785$

91. $2(x - 1.8) = 4.2(x + 0.9) + 3.18$

$2x - 3.6 = 4.2x + 3.78 + 3.18 = 4.2x + 6.96$

$-2.2x - 3.6 = 6.96$

$-2.2x = 10.56$

$x = -4.8$

Check:

$2(-4.8 - 1.8) \overset{?}{=} 4.2(-4.8 + 0.9) + 3.18$

$2(-6.6) \overset{?}{=} 4.2(-3.9) + 3.18$

$-13.2 \overset{?}{=} -16.38 + 3.18$

$-13.2 = -13.2$

93. $8.71 + 12.53 + 9.83 = 31.07$

$\$50.00$

$\underline{-\ \ 31.07}$

$\$18.93$

You have $18.93 left.

95. $\sqrt{324} = 18$ Because $18 \times 18 = 324$

97. $\sqrt{189} \approx 13.75$ Because $13.75 \times 13.75 \approx 189$

97. $c^2 = a^2 + b^2 = (33)^2 + (44)^2$

$= 1,089 + 1,936 = 3,025$

$c = \sqrt{3,025} = 55$

101. $a^2 + b^2 = c^2$

$a^2 + (8)^2 = (17)^2$

$a^2 + 64 = 289$

$a^2 = 289 - 64 = 225$

$a = \sqrt{225} = 15$

Chapter Test 5

1. The place value of 8 in 0.5248 is ten-thousandths.

3. "Twelve and seventeen thousandths" is 12.017.

5. The 7 is in the thousandths place. The next digit to the right (7) is 5 or more, so increase the digit you are rounding to by 1. Discard the remaining digits to the right. 0.5977 is rounded to 0.598.

7. The 6 is in the thousands place. The next digit to the right (1) is less than 5. Leave the thousands digit as it is, and discard the remaining digits to the right. 36,139.0023 is rounded to 36,000.

9.   3.45

     0.60

  $\underline{+\ 12.59}$

    16.64

11. $27\overline{)63.45}$ quotient 2.35

$\underline{54}$

94

$\underline{81}$

135

$\underline{135}$

0

The quotient is 2.35.

13.
$$
\begin{array}{r}
8.010 \\
-\,4.709 \\
\hline
3.301
\end{array}
$$
$-4.709 + 8.01 = 3.301$

15.
$$
\begin{array}{r}
8.70 \\
\times 5.65 \\
\hline
4350 \\
52200 \\
435000 \\
\hline
49.1550
\end{array}
$$
$(-8.7)(-5.65) = -49.155$

17. $4.983 \div 1,000 = 0\underline{004}.983 = 0.004983$

19. $0.735 \times 1,000 = 735$

21.
$$
\begin{array}{r}
5.630 \\
-\,1.742 \\
\hline
3.888
\end{array}
$$

23.
$$
\begin{array}{r}
0.049 \\
\times \; 0.57 \\
\hline
343 \\
245 \\
\hline
0.02793
\end{array}
$$

35.
$$
\begin{array}{r}
0.0513 \\
53\overline{)2.7200} \\
\underline{265} \\
70 \\
\underline{53} \\
170 \\
\underline{159} \\
11
\end{array}
$$
Round 0.0513 to 0.051. So $2.72 \div 53 = 0.051$ (rounded to the nearest thousandths).

27. Write 0.89 as 0.890. Then we see that 0.890 is greater than 0.889, and we can write $0.889 < 0.89$.

29. $\dfrac{3}{25} = 0.12$

Comparing the digits in hundredths place, we see 0.168 is greater than 0.12. Therefore, $0.168 > \dfrac{3}{25}$.

31.
$$
\begin{array}{r}
0.4375 \\
16\overline{)7.0000} \\
\underline{64} \\
60 \\
\underline{48} \\
120 \\
\underline{112} \\
80 \\
\underline{80} \\
0
\end{array}
$$
$\dfrac{7}{16} = 0.4375$

33.
$$
\begin{array}{r}
0.6363 \\
11\overline{)7.0000} \\
\underline{66} \\
40 \\
\underline{33} \\
70 \\
\underline{66} \\
40 \\
\underline{33} \\
7
\end{array}
$$
$\dfrac{7}{11} = 0.\overline{63}$

35. $0.072 = \dfrac{72}{1,000} = \dfrac{9}{125}$

37. $-0.08 = -\dfrac{8}{100} = \dfrac{2}{25}$

39.
$$
\begin{aligned}
2.3x + 18.6 &= 2.5 \\
2.3x + 18.8 - 18.6 &= 2.5 - 18.6 \\
2.3x &= -16.1 \\
\frac{2.3x}{2.3} &= -\frac{16.1}{2.3} \\
x &= -7
\end{aligned}
$$

41. $\sqrt{8} = 2.83$

(rounded to the nearest hundredth)

43. $c^2 = a^2 + b^2 = (12)^2 + (35)^2 = 144 + 1,225$

$= 1,369$

$c = \sqrt{1,369} = 37$

The length of the missing side is 37 cm.

45. $54.3¢ = \$0.543$

$0.543 \times 1,000 = 543$

The total cost is $543.

47. $14 - 2.8 = 11.2$

11.2 acres are available for lots.

$$
\begin{array}{r}
32 \\
0.35\overline{)11.20} \\
\underline{105} \\
70 \\
\underline{70} \\
0
\end{array}
$$

32 lots can be formed.

49. $13.99 + 18.75 + 9.20 + 5.00 = 46.94$

Total amount you pay for purchases is $46.94.

$$
\begin{array}{r}
50.00 \\
-\ 46.94 \\
\hline
3.06
\end{array}
$$

You have $3.06 left.

Cumulative Review: Chapters 1–5

1. $\underbrace{286}_{\text{thousands}}$, $\underbrace{543}_{\text{hundreds}}$

Two hundred eighty-six thousand, five hundred forty-three

3.
$$
\begin{array}{r}
\overset{1\ \ 1\ 1}{2,340} \\
685 \\
+\ 31,569 \\
\hline
34,594
\end{array}
$$

5.
$$
\begin{array}{r}
83 \\
\times\ 61 \\
\hline
83 \\
4980 \\
\hline
5,063
\end{array}
$$

7.
$$
\begin{array}{r}
17 \\
21\overline{)357} \\
\underline{21} \\
147 \\
\underline{147} \\
0
\end{array}
$$

We have $357 \div 21 = 17$.

9. $18 \div 2 + 4 \times 2^3 - (18 - 6)$

$= 18 \div 2 + 4 \times 2^3 - 12 = 18 \div 2 + 4 \times 8 - 12$

$= 9 + 4 \times 8 - 12 = 9 + 32 - 12 = 41 - 12 = 29$

11. $P = 7 \text{ ft} + 5 \text{ ft} + 7 \text{ ft} + 5 \text{ ft} = 24 \text{ ft}$

The perimeter is 24 ft.

$A = 7 \text{ ft} \times 5 \text{ ft} = 35 \text{ ft}^2$

The area is 35 ft^2.

13. $37 - (-43) = 37 + 43 = 80$

15. $\dfrac{-1,000}{-8} = 125$

17. $(3x^3 - 6x + 5) - (4x^2 - 5x - 3)$

$= 3x^3 - 6x + 5 - 4x^2 + 5x + 3$

$= 3x^3 - 4x^2 - x + 8$

19. $\dfrac{15}{51} = \dfrac{\cancel{3} \times 5}{\cancel{3} \times 17} = \dfrac{5}{17}$

21. $1\dfrac{2}{3} \times 1\dfrac{5}{7} = \dfrac{5}{3} \times \dfrac{12}{7} = \dfrac{5 \times \cancel{12}}{\cancel{3} \times 7} = \dfrac{5 \times 4}{1 \times 7} = \dfrac{20}{7}$

$= 2\dfrac{6}{7}$

23. $\dfrac{6}{7} - \dfrac{3}{7} + \dfrac{2}{7} = \dfrac{6-3+2}{7} = \dfrac{5}{7}$

25. $6\dfrac{3}{5} - 2\dfrac{7}{10} = \dfrac{33}{5} - \dfrac{27}{10} = \dfrac{66}{10} - \dfrac{27}{10} = \dfrac{39}{10} = 3\dfrac{9}{10}$

27. $4.6 + (-8.31) = -(8.31 - 4.6)$

$\begin{array}{r} 8.31 \\ -4.60 \\ \hline 3.71 \end{array}$

$4.6 + (-8.31) = -(8.31 - 4.6) = -3.71$

29. $\begin{array}{r} 1.2 \\ \times\, 5.4 \\ \hline 48 \\ 600 \\ \hline 6.48 \end{array}$

$(-1.2)(-5.4) = 6.48$

31. $523.8 \div 10^5 = 0\underline{00523}.8 = 0.005238$

33. $1.53 \times 10^4 = 15{,}300$

35. (a) $\begin{array}{r} 0.625 \\ 8\overline{)5.000} \\ \underline{48} \\ 20 \\ \underline{16} \\ 40 \\ \underline{40} \\ 0 \end{array}$ $\dfrac{5}{8} = 0.625$

(b) $\begin{array}{r} 0.391 \\ 23\overline{)9.000} \\ \underline{69} \\ 210 \\ \underline{207} \\ 30 \\ \underline{23} \\ 7 \end{array}$ $\dfrac{9}{23} = 0.39$

(rounded to the nearest hundredth)

37. $\begin{aligned} 18.4 - 3.16 \times 2.5 + 6.71 &= 18.4 - 7.9 + 6.71 \\ &= 10.5 + 6.71 = 17.21 \end{aligned}$

39. $\begin{aligned} 5x &= 75 \\ \dfrac{5x}{5} &= \dfrac{75}{5} \\ x &= 15 \end{aligned}$ Check: $5(15) \overset{?}{=} 75$
$ 75 = 75$

41. $\begin{aligned} 8x - 5 &= 19 \\ 8x - 5 + 5 &= 19 + 5 \\ 8x &= 24 \\ \dfrac{8x}{8} &= \dfrac{24}{8} \\ x &= 3 \end{aligned}$ Check: $8(3) - 5 \overset{?}{=} 19$
$ 19 = 19$

43. $\sqrt{64} = \sqrt{(8)^2} = 8$

45. Since all 4 sides of a square are equal, length of each side $= 19.2 \div 4 = 4.8$ cm.

47. The number of possible lots in the subdivision $= 80.5 \div 0.35 = 230$ lots.

Chapter 6
Ratios and Proportions

Prerequisite Check

1. $\dfrac{24 \div 8}{32 \div 8} = \dfrac{3}{4}$

3. $\dfrac{\frac{45}{2}}{30} = \dfrac{45}{2} \div 30 = \dfrac{45}{2} \times \dfrac{1}{30} = \dfrac{45 \div 15}{60 \div 15} = \dfrac{3}{4}$

5. $\dfrac{45}{2} \times \dfrac{1}{30} = \dfrac{45 \div 15}{60 \div 15} = \dfrac{3}{4}$

7. $280 = 2 \times 2 \times 2 \times 5 \times 7$

$525 = 3 \times 5 \times 5 \times 7$

$\text{GCF} = 5 \times 7 = 35$

9. $\dfrac{260}{36} = 36\overline{)260.000}^{\,7.222} \approx 7.22$

11. $\dfrac{7}{15} \overset{?}{=} \dfrac{84}{180}$

$7 \times 180 \overset{?}{=} 84 \times 15$

$1,260 = 1,260$

Therefore, the fractions are equivalent.

Exercises 6.1

< Objectives 1 and 2 >

1. The ratio of 9 to 13: $\dfrac{9}{13}$

3. The ratio of 9 to 4: $\dfrac{9}{4}$

5. The ratio of 10 to 15: $\dfrac{10}{15} = \dfrac{2}{3}$

7. The ratio of 18 to 15: $\dfrac{18}{15} = \dfrac{6}{5}$

9. The ratio of 3 to 21: $\dfrac{3}{21} = \dfrac{1}{7}$

11. The ratio of 24 to 6: $\dfrac{24}{6} = \dfrac{4}{1}$

13. The ratio of $3\frac{1}{2}$ to 14:

$\dfrac{3\frac{1}{2}}{14} = \dfrac{\frac{7}{2}}{14} = \dfrac{7}{2} \div \dfrac{14}{1} = \dfrac{7}{2} \times \dfrac{1}{14} = \dfrac{1}{4}$

15. The ratio of $1\frac{1}{4}$ to $\dfrac{3}{2}$: $\dfrac{1\frac{1}{4}}{\frac{3}{2}} = \dfrac{5}{4} \times \dfrac{2}{3} = \dfrac{5}{6}$

17. The ratio of 4.5 to 31.5:

$\dfrac{4.5}{31.5} = \dfrac{4.5}{31.5} \cdot \dfrac{10}{10} = \dfrac{45}{315} = \dfrac{1}{7}$

19. The ratio of 8.7 to 8.4:

$\dfrac{8.7}{8.4} = \dfrac{8.7}{8.4} \times \dfrac{10}{10} = \dfrac{87}{84} = \dfrac{29}{28}$

21. The ratio of 10.5 to 2.7:

$\dfrac{10.5}{2.7} = \dfrac{10.5 \times 10}{2.7 \times 10} = \dfrac{105}{27} = \dfrac{35}{9}$

23. The ratio of 12 mi to 18 mi: $\dfrac{12 \text{ mi}}{18 \text{ mi}} = \dfrac{12}{18} = \dfrac{2}{3}$

25. The ratio of 40 ft to 65 ft: $\dfrac{40 \text{ ft}}{65 \text{ ft}} = \dfrac{40}{65} = \dfrac{8}{13}$

27. The ratio of $48 to $42: $\dfrac{\$48}{\$42} = \dfrac{48}{42} = \dfrac{8}{7}$

29. The ratio of 75 s to 3 min:
$$\dfrac{75\text{ s}}{3\text{ min}} = \dfrac{75\text{ s}}{180\text{ s}} = \dfrac{5}{12}$$

31. The ratio of 4 nickels to 3 dimes:
$$\dfrac{4\text{ nickels}}{3\text{ dimes}} = \dfrac{20\text{ cents}}{30\text{ cents}} = \dfrac{2}{3}$$

33. The ratio of 2 days to 10 hr:
$$\dfrac{2\text{ days}}{10\text{ hr}} = \dfrac{48\text{ hr}}{10\text{ hr}} = \dfrac{24}{5}$$

35. The ratio of 5 gal to 12 qt: $\dfrac{5\text{ gal}}{12\text{ qt}} = \dfrac{20\text{ qt}}{12\text{ qt}} = \dfrac{5}{3}$

37. The ratio of men to women: $\dfrac{7}{13}$

The ratio of women to men: $\dfrac{13}{7}$

39. The ratio of yes votes to no votes: $\dfrac{4,500}{3,000} = \dfrac{3}{2}$

41. The ratio of $2\frac{2}{3}$ ft^3 to $5\frac{3}{4}$ ft^3:

$$\dfrac{2\frac{2}{3}\text{ ft}^3}{5\frac{3}{4}\text{ ft}^3} = \dfrac{2\frac{2}{3}}{5\frac{3}{4}} = \dfrac{\frac{8}{3}}{\frac{23}{4}} = \dfrac{8}{3} \div \dfrac{23}{4} = \dfrac{8}{3} \times \dfrac{4}{23} = \dfrac{32}{69}$$

43. The ratio of 18 men to 24 women: $\dfrac{18}{24} = \dfrac{3}{4}$

45. The ratio of rainfall in November to October:
$$\dfrac{2.4\text{ in.} + 0.4\text{ in.}}{2.4\text{ in.}} = \dfrac{2.8\text{ in.}}{2.4\text{ in.}} = \dfrac{2.8}{2.4} = \dfrac{7}{6}$$

47. (a) The ratio of high temperatures of June to September: $\dfrac{82.1^\circ\text{F}}{76.5^\circ\text{F}} = \dfrac{82.1}{76.5} = \dfrac{821}{765}$

(b) The high temperature ratio of August to July: $\dfrac{83.3^\circ\text{F}}{85.6^\circ\text{F}} = \dfrac{83.3}{85.6} = \dfrac{833}{856}$

49. True

51. always

53. $\dfrac{5}{4}$

55. $\dfrac{15}{24} = \dfrac{5}{8}$

57. $\dfrac{400\text{ lb}}{500\text{ lb}} = \dfrac{400}{500} = \dfrac{4}{5}$

59. Answers will vary.

61.
$$\dfrac{4.5}{1} = \dfrac{28}{N_s}$$
$$4.5\,N_s = 28$$
$$N_s = \dfrac{28}{4.5} = \dfrac{280}{45} = 6\frac{2}{9}\text{ V AC}$$

63. Above and Beyond

Exercises 6.2

< Objective 1 >

1. $\dfrac{300\text{ mi}}{4\text{ hr}} = \dfrac{300}{4} = 75\dfrac{\text{mi}}{\text{hr}}$

3. $\dfrac{\$10,000}{5\text{ years}} = \dfrac{10,000}{5} = 2,000\dfrac{\$}{\text{year}}$

5. $\dfrac{7,200\text{ revolutions}}{16\text{ mi}} = \dfrac{7,200}{16} = 450\dfrac{\text{rev}}{\text{mi}}$

7. $\dfrac{\$2,000,000}{4\text{ years}} = \dfrac{2,000,000}{4} = 500,000\dfrac{\$}{\text{year}}$

9. $\dfrac{240 \text{ lb of fertilizer}}{6 \text{ lawns}} = \dfrac{240}{6} = 40\dfrac{\text{lb}}{\text{lawn}}$

11. $\dfrac{323 \text{ mi}}{11 \text{ gal of fuel}} = \dfrac{323}{11} = 29.363\dfrac{\text{mi}}{\text{gal}} = 29.4\dfrac{\text{mi}}{\text{gal}}$

13. $\dfrac{210 \text{ ft}^2}{16 \text{ hr}} = \dfrac{210}{16} = 13.1\dfrac{\text{ft}^2}{\text{hr}}$

15. $\dfrac{141 \text{ pages}}{9 \text{ min}} = \dfrac{141}{9} = 15.7\dfrac{\text{pages}}{\text{min}}$

17. $\dfrac{447 \text{ lb}}{30 \text{ in.}^2} = \dfrac{447}{30} = 14.9\dfrac{\text{lb}}{\text{in.}^2}$

19. $\dfrac{189 \text{ points}}{42 \text{ games}} = \dfrac{63 \text{ points}}{14 \text{ game}} = \dfrac{9 \text{ points}}{2 \text{ game}}$
$= 4\dfrac{1}{2}\dfrac{\text{points}}{\text{game}}$

21. $\dfrac{72 \text{ lengths}}{40 \text{ min}} = \dfrac{9 \text{ lengths}}{5 \text{ min}} = 1\dfrac{4}{5}\dfrac{\text{lengths}}{\text{min}}$

23. $\dfrac{2 \text{ cups (c) of water}}{1\frac{1}{3}\text{ cups (c) of rice}} = \dfrac{2 \text{ c of water}}{\frac{4}{3}\text{ c of rice}}$
$= \dfrac{2 \times 3 \text{ c of water}}{4 \text{ c of rice}}$
$= 1\dfrac{1}{2}\dfrac{\text{c of water}}{\text{c of rice}}$

< Objective 2 >

25. **(a)** We write this rate as "Thirty-two per second".
(b) We write this rate as "One thirty-second second per foot".
(c) The rate $32\dfrac{\text{ft}}{\text{s}}$ describes how far you travel in one second. The rate $\dfrac{1}{32}\dfrac{\text{s}}{\text{ft}}$ describes how long it takes to travel one foot.

27. **(a)** We write this rate as "Eight and nine-tenths bushels per dollar".
(b) We write this rate as "Eleven-hundredths dollar per bushel (11¢ per bushel)".
(c) The rate $8.9\dfrac{\text{bushels}}{\text{dollar}}$ describes the number of bushels sold for one dollar. The rate $0.11\dfrac{\text{dollars}}{\text{bushel}}$ describes the cost of one bushel.

< Objective 3 >

29. $\dfrac{\$57.50}{5 \text{ shirts}} = \dfrac{57.50}{5} = 11.50\dfrac{\$}{\text{shirt}}$

31. $\dfrac{\$5.16}{1 \text{ dozen orange}} = \dfrac{5.16}{12} = 0.43\dfrac{\$}{\text{orange}}$

< Objective 4 >

33. Unit price of dishwashing liquid (a) is
$\dfrac{\$3.16}{12 \text{ fl oz}} = \dfrac{3.16}{12} = 0.263\dfrac{\$}{\text{fl oz}}$.
Unit price of dishwashing liquid (b) is
$\dfrac{\$5.16}{22 \text{ fl oz}} = \dfrac{5.16}{22} = 0.235\dfrac{\$}{\text{fl oz}}$.
Therefore, dishwashing liquid (b) is the best buy.

35. Unit price of syrup (a) is
$\dfrac{\$3.96}{12 \text{ fl oz}} = \dfrac{3.96}{12} = 0.330\dfrac{\$}{\text{fl oz}}$.
Unit price of syrup (b) is
$\dfrac{\$6.36}{24 \text{ fl oz}} = \dfrac{6.36}{24} = 0.265\dfrac{\$}{\text{fl oz}}$.
Unit price of syrup (c) is
$\dfrac{\$8.76}{36 \text{ fl oz}} = \dfrac{8.76}{36} = 0.243\dfrac{\$}{\text{fl oz}}$.
Therefore, syrup (c) is the best buy.

37. Unit price of salad oil (a) is

$$\frac{\$3.56}{18 \text{ fl oz}} = \frac{3.56}{18} = 0.198 \frac{\$}{\text{fl oz}}.$$

Unit price of salad oil (b) is

$$\frac{\$5.56}{1 \text{ qt}} = \frac{5.56}{32} = 0.174 \frac{\$}{\text{fl oz}}.$$

Unit price of salad oil (c) is

$$\frac{\$8.76}{1 \text{ qt } 16 \text{ fl oz}} = \frac{8.76}{48} = 0.183 \frac{\$}{\text{fl oz}}.$$

Therefore, salad oil (b) is the best buy.

39. Unit price of peanut butter (a) is

$$\frac{\$5.00}{12 \text{ oz}} = \frac{5.00}{12} = 0.416 \frac{\$}{\text{oz}}.$$

Unit price of peanut butter (b) is

$$\frac{\$6.88}{18 \text{ oz}} = \frac{6.88}{18} = 0.382 \frac{\$}{\text{oz}}.$$

Unit price of peanut butter (c) is

$$\frac{\$10.16}{1 \text{ lb } 12 \text{ oz}} = \frac{10.16}{28} = 0.363 \frac{\$}{\text{oz}}.$$

Unit price of peanut butter (d) is

$$\frac{\$15.04}{2 \text{ lb } 8 \text{ oz}} = \frac{15.04}{40} = 0.376 \frac{\$}{\text{oz}}.$$

Therefore, peanut butter (c) is the best buy.

41. $\dfrac{256 \text{ miles}}{8 \text{ gallons}} = 32 \dfrac{\text{mi}}{\text{gal}}$

43. $\dfrac{6{,}000 \text{ vehicles}}{2{,}400 \text{ spaces}} = \dfrac{6{,}000}{2{,}400} \dfrac{\text{vehicles}}{\text{space}}$

$$= 2.5 \frac{\text{vehicles}}{\text{space}}$$

45. $\dfrac{\$5{,}992}{214 \text{ shares}} = \dfrac{5{,}992}{214} \dfrac{\$}{\text{share}} = 28 \dfrac{\$}{\text{share}}$

47. $\dfrac{\$4.80}{12 \text{ oz}} = \dfrac{480¢}{12 \text{ oz}} = \dfrac{480}{12} \dfrac{¢}{\text{oz}} = 40 \dfrac{¢}{\text{oz}}$

49. Gerry: $\dfrac{634 \text{ bricks}}{35 \text{ min}} = 18.11 \dfrac{\text{bricks}}{\text{min}} \approx 18 \dfrac{\text{bricks}}{\text{min}}$

Matt: $\dfrac{515 \text{ bricks}}{27 \text{ min}} = 19.07 \dfrac{\text{bricks}}{\text{min}} \approx 19 \dfrac{\text{bricks}}{\text{min}}$

Matt is the faster bricklayer.

51. Adrian: $\dfrac{198 \text{ hits}}{585 \text{ bats}} = 0.338 \dfrac{\text{hits}}{\text{bat}}$

Miguel: $\dfrac{178 \text{ hits}}{533 \text{ bats}} = 0.334 \dfrac{\text{hits}}{\text{bat}}$

Adrian Gonzalez has the higher batting average.

53. False

55. never

57. $\dfrac{24 \text{ teeth}}{3 \text{ in.}} = 8 \dfrac{\text{teeth}}{\text{in.}}$

59. $\dfrac{\$99.99}{1{,}000 \text{ feet}} = 0.10 \dfrac{\$}{\text{ft}}$

(rounded to the nearest cent)

61. $\dfrac{\$1{,}780}{200 \text{ bushels}} = \$8.90/\text{bushel}$

63. Above and Beyond

65. Above and Beyond

Exercises 6.3

< Objective 1 >

1. $\dfrac{4}{9} = \dfrac{8}{18}$

3. $\dfrac{2}{9} = \dfrac{8}{36}$

5. $\dfrac{3}{5} = \dfrac{15}{25}$

7. $\dfrac{9}{13} = \dfrac{27}{39}$

< Objective 2 >

9. $4 \times 9 = 36$

$3 \times 12 = 36$

Because the products are equal, the fractions are proportional.

11. $4 \times 15 = 60$

$3 \times 20 = 60$

Because the products are equal, the fractions are proportional.

13. $15 \times 9 = 135$

$11 \times 13 = 143$

Because the products are not equal, the fractions are not proportional.

15. $3 \times 24 = 72$

$8 \times 9 = 72$

Because the products are equal, the fractions are proportional.

17. $17 \times 9 = 153$

$6 \times 11 = 66$

Because the products are not equal, the fractions are not proportional.

19. $16 \times 21 = 336$

$7 \times 48 = 336$

Because the products are equal, the fractions are proportional.

21. $3 \times 150 = 450$

$10 \times 50 = 500$

Because the products are not equal, the fractions are not proportional.

23. $7 \times 18 = 126$

$3 \times 42 = 126$

Because the products are equal, the fractions are proportional.

25. $7 \times 180 = 1260$

$15 \times 84 = 1260$

Because the products are equal, the fractions are proportional.

27. $60 \times 15 = 900$

$36 \times 25 = 900$

Because the products are equal, the fractions are proportional.

29. $\dfrac{1}{5} \times 30 = 6$

$3 \times 6 = 18$

Because the products are not equal, the fractions are not proportional.

31. $12 \times 1 = 12$

$\dfrac{3}{4} \times 16 = 12$

Because the products are equal, the fractions are proportional.

33. $60 \times 0.3 = 18$

$3 \times 6 = 18$

Because the products are equal, the fractions are proportional.

35. $15 \times 2 = 30$

$0.6 \times 75 = 45$

Because the products are not equal, the fractions are not proportional.

< Objective 3 >

37. $4 \times 4 = 16$

$7 \times 3 = 21$

The rates are not proportional.

39. $15 \times 55 = 825$

$22 \times 35 = 770$

The rates are not proportional.

41. $57 \times 6 = 342$

$9 \times 38 = 342$

The rates are proportional.

43. $18 \times 200 = 3,600$

$300 \times 12 = 3,600$

The rates are proportional.

45. $1.4 \times 36 = 50.4$

$12 \times 7 = 84$

The rates are not proportional.

47. $\dfrac{15 \text{ lb}}{\$20} = \dfrac{45 \text{ lb}}{\$60}$

49. $\dfrac{3 \text{ credits}}{\$216} = \dfrac{12 \text{ credits}}{\$864}$

51. $\dfrac{180 \text{ mi}}{3 \text{ hr}} = \dfrac{300 \text{ mi}}{5 \text{ hr}}$

53. True

55. sometimes

57. Does $\dfrac{80 \text{ mg}}{1 \text{ mL}} = \dfrac{300 \text{ mg}}{3.75 \text{ mL}}$?

$80 \times 3.75 = 300$

$1 \times 300 = 300$

Because the products are equal, these rates are proportional.

59. Does $\dfrac{5 \text{ in.}}{20 \text{ teeth}} = \dfrac{18 \text{ in.}}{68 \text{ teeth}}$?

$5 \times 68 = 340$

$20 \times 18 = 360$

Because the products are not equal, the fractions are not proportional and hence the two gears will not mesh.

Exercises 6.4

< Objective 1 >

1. $\dfrac{x}{3} = \dfrac{6}{9}$

$9x = 18$

$\dfrac{9x}{9} = \dfrac{18}{9}$

$x = 2$

3. $\dfrac{10}{n} = \dfrac{15}{6}$

$15n = 60$

$\dfrac{15n}{15} = \dfrac{60}{15}$

$n = 4$

5. $\dfrac{4}{7} = \dfrac{y}{14}$

$7y = 56$

$\dfrac{7y}{7} = \dfrac{56}{7}$

$y = 8$

7. $\dfrac{5}{7} = \dfrac{x}{35}$

$7x = 175$

$\dfrac{7x}{7} = \dfrac{175}{7}$

$x = 25$

9. $\dfrac{11}{a} = \dfrac{2}{44}$

$2a = 484$

$\dfrac{2a}{2} = \dfrac{484}{2}$

$a = 242$

11. $\dfrac{x}{8} = \dfrac{15}{24}$

$24x = 120$

$\dfrac{24x}{24} = \dfrac{120}{24}$

$x = 5$

13. $\dfrac{x}{8} = \dfrac{7}{16}$

$16x = 56$

$\dfrac{16x}{16} = \dfrac{56}{16}$

$x = \dfrac{7}{2} = 3\dfrac{1}{2}$

15. $\dfrac{4}{y} = \dfrac{6}{1}$

$4 = 6y$

$\dfrac{4}{6} = \dfrac{6y}{6}$

$y = \dfrac{2}{3}$

17. $\dfrac{a}{4} = \dfrac{9}{14}$

$14a = 36$

$\dfrac{14a}{14} = \dfrac{36}{14}$

$a = 2\dfrac{4}{7}$

19. $\dfrac{5}{m} = \dfrac{12}{5}$

$25 = 12m$

$\dfrac{25}{12} = \dfrac{12m}{12}$

$m = 2\dfrac{1}{12} = 2.08$

21. $\dfrac{x}{10} = \dfrac{9}{7}$

$7x = 90$

$\dfrac{7x}{7} = \dfrac{90}{7}$

$x = 12\dfrac{6}{7} = 12.86$

23. $\dfrac{x}{8} = \dfrac{1}{12}$

$12x = 8$

$\dfrac{12x}{12} = \dfrac{8}{12}$

$x = 0.67$

25. $\dfrac{\frac{1}{2}}{2} = \dfrac{3}{a}$

$\dfrac{1}{2} \times \dfrac{1}{2} = \dfrac{3}{a}$

$\dfrac{1}{4} = \dfrac{3}{a}$

$a = 12$

27. $\dfrac{0.2}{2} = \dfrac{1.2}{a}$

$0.2a = 2.4$

$\dfrac{0.2a}{0.2} = \dfrac{2.4}{0.2}$

$a = 12$

29.
$$\frac{\frac{2}{5}}{8} = \frac{1.2}{n}$$

$$\frac{2}{5} \times \frac{1}{8} = \frac{1.2}{n}$$

$$\frac{2}{40} = \frac{1.2}{n}$$

$$n = 24$$

31.
$$\frac{\frac{1}{4}}{12} = \frac{m}{40}$$

$$\frac{1}{4} \times \frac{1}{12} = \frac{m}{40}$$

$$\frac{1}{48} = \frac{m}{40}$$

$$m = \frac{5}{6}$$

33.
$$\frac{12}{\frac{1}{3}} = \frac{80}{y}$$

$$\frac{12}{1} \times \frac{3}{1} = \frac{80}{y}$$

$$36y = 80$$

$$y = 2\frac{2}{9}$$

35.
$$\frac{x}{3.3} = \frac{1.1}{6.6}$$

$$6.6x = 3.63$$

$$\frac{6.6x}{6.6} = \frac{3.63}{6.6}$$

$$x = 0.6$$

37.
$$\frac{4}{a} = \frac{\frac{1}{4}}{0.8}$$

$$\frac{4}{a} = \frac{1}{4} \div \frac{0.8}{1}$$

$$\frac{4}{a} = \frac{1}{4} \times \frac{1}{0.8} = \frac{1}{3.2}$$

$$a = 12.8$$

< Objective 2 >

39.
$$\frac{12}{80} = \frac{18}{x}$$

$$12x = 1440$$

$$\frac{12x}{12} = \frac{1440}{12}$$

$$x = 120$$

You will pay $120 for 18 books.

41.
$$\frac{18}{4.89} = \frac{48}{x}$$

$$x = \frac{48 \times 4.89}{18} = \frac{234.72}{18} = 13.04$$

48 tea bags cost $13.04.

43.
$$\frac{3 \text{ yes votes}}{2 \text{ no votes}} = \frac{2{,}880 \text{ yes votes}}{x \text{ no votes}}$$

$$3x = 5{,}760$$

$$x = 1{,}920$$

1,920 no votes were cast.

45.
$$\frac{5 \text{ in. wide}}{6 \text{ in. high}} = \frac{15 \text{ in. wide}}{x \text{ in. high}}$$

$$5x = 90$$

$$x = 18$$

The height of enlargement will be 18 in.

47.
$$\frac{120,000}{2,100} = \frac{150,000}{x}$$
$$120,000x = 315,000,000$$
$$\frac{120,000x}{120,000} = \frac{315,000,000}{120,000}$$
$$x = 2,625$$
You will pay $2,625 taxes on a $150,000 home.

49. The scale on the map given is $\frac{1}{2}$ in. $= 40$ mi.

The measured distance from Harrisburg to Philadelphia is $1\frac{3}{8}$ in. $= \frac{11}{8}$ in. Use the fact that the ratio of inches (on the map) to miles remains the same.

$$\frac{\frac{1}{2} \text{ in.}}{40 \text{ mi}} = \frac{\frac{11}{8} \text{ in.}}{x \text{ mi}}$$

$$\frac{1}{2}x = 55$$

$$x = 110$$

The distance from Harrisburg to Philadelphia is 110 mi.

51. The scale on the map given is $\frac{1}{2}$ in. $= 40$ mi.

The measured distance from Gettysburg to Meadville is $2\frac{11}{16}$ in. $= \frac{43}{16}$ in. Use the fact that the ratio of inches (on the map) to miles remains the same.

$$\frac{\frac{1}{2} \text{ in.}}{40 \text{ mi.}} = \frac{\frac{43}{16} \text{ in.}}{x \text{ mi}}$$

$$\frac{1}{2}x = \frac{1,720}{12}$$

$$x = 215$$

The distance from Gettysburg to Meadville is 215 mi.

53. Let x represent the missing side.
$$\frac{x}{2} = \frac{6}{4}$$

$$4x = 12$$

$$x = 3$$
The length of the missing side is 3 cm.

55. Let x represent the missing side.
$$\frac{12}{8} = \frac{x}{5}$$

$$x = \frac{3 \times 5}{2}$$

$$x = 7\frac{1}{2}$$

The length of the missing side is $7\frac{1}{2}$ in.

57. Since the triangles are similar, their corresponding sides have the same ratio.
$$\frac{5}{3} = \frac{x}{12}$$
$$3 \cdot x = 5 \cdot 12$$
$$3x = 60$$
$$x = 20 \text{ in.}$$

59. Since the triangles are similar, their corresponding sides have the same ratio.
$$\frac{24}{28} = \frac{x}{14}$$
$$28 \cdot x = 24 \cdot 14$$
$$28x = 336$$
$$x = 12 \text{ in.}$$

61. Since the triangles are similar, their corresponding sides have the same ratio.
$$\frac{30.1}{38.7} = \frac{x}{50.4}$$
$$38.7 \cdot x = 30.1 \cdot 50.4$$
$$38.7x = 1517.04$$
$$x = 39.2 \text{ mm}$$

63. Let h be the height of the tree.

$$\frac{9}{15} = \frac{h}{40}$$

$$15h = 360$$

$$h = 24$$

The height of the tree is 24 ft.

65. The yardstick and its shadow form a similar triangle to the light pole and its shadow. Because of this, we can use the common ratio to find the height of the pole. Note that 1 yd = 36 in.

$$\frac{36 \text{ in.}}{9 \text{ in.}} = \frac{x \text{ ft}}{4 \text{ ft}}$$

$$\frac{36}{9} = \frac{x}{4}$$

$$9x = 144$$

$$x = 16$$

The light pole is 16 ft tall.

67. $\dfrac{500 \text{ shipment}}{27 \text{ defective parts}} = \dfrac{1{,}200 \text{ shipment}}{x \text{ defective parts}}$

$$x = \frac{1{,}200 \times 27}{500} = 64.8 \approx 65$$

65 defective parts should be expected in a shipment of 1,200.

69. $\dfrac{9 \text{ ft}}{5 \text{ ft}} = \dfrac{15 \text{ ft}}{x}$

$$x = \frac{15 \times 5}{9} = 8.33$$

A 8.3-ft pole casts a 15-ft shadow.

71. $\dfrac{5 \text{ in.}}{7 \text{ ft}} = \dfrac{10 \text{ in.}}{x \text{ ft}}$

$$5x = 70$$

$$x = 14$$

The actual length of the bedroom will be 14 ft.

73. $\dfrac{12°\text{F}}{\frac{1}{4} \text{ in.}} = \dfrac{44°\text{F}}{x \text{ in.}}$

$$x = \frac{44}{12 \times 4} = \frac{11}{12}$$

The metal bar will expand $\dfrac{11}{12}$ in., if the temperature rises 44°F.

75. $\dfrac{7}{10} = \dfrac{x}{115{,}000}$

$$10x = 805{,}000$$

$$x = 80{,}500$$

80,500 cars will have one person in them.

77. False

79. always

81. $a = \dfrac{(1365)(15)}{630}$

$1365 \boxed{\times} 15 \boxed{\div} 630 \boxed{=} 32.5\,5$

$a = 32.5$

83. $x = \dfrac{(11.8)(4.7)}{16.9}$

$11.8 \boxed{\times} 4.7 \boxed{\div} 16.9 \boxed{=} 3.3$

$x = 3.3$

85. $n = \dfrac{(3.8)(5.9)}{2.7}$

$3.8 \boxed{\times} 5.9 \boxed{\div} 2.7 \boxed{=} 8.3$

$n = 8.3$

87. $\dfrac{\$496.80}{34.5 \text{ hr}} = \dfrac{\$x}{31.75 \text{ hr}}$

$$x = \frac{496.8 \times 31.75}{34.5} = 457.2$$

Bill will receive \$457.20.

89.
$$\dfrac{88\,\frac{\text{ft}}{\text{s}}}{60\,\frac{\text{mi}}{\text{hr}}} = \dfrac{x\,\frac{\text{ft}}{\text{s}}}{750\,\frac{\text{mi}}{\text{hr}}}$$

$$x = \dfrac{88 \times 750}{60}$$

$$88 \boxed{\times} 750 \boxed{\div} 60 \boxed{=} 1100$$

The speed of sound is 1,100 ft/s.

91.
$$\dfrac{5\text{ Web pages}}{2\text{ s}} = \dfrac{x}{1\text{ min}}$$

$$\dfrac{5}{2} = \dfrac{x}{60}$$

$$300 = 2x$$

$$\dfrac{300}{2} = \dfrac{2x}{2}$$

$$x = 150$$

The computer can transmit 150 Web pages in 1 min.

93.
$$\dfrac{7\text{ holes}}{0.322\text{ lb}} = \dfrac{43\text{ holes}}{x}$$

$$7x = 13.846$$

$$\dfrac{7x}{7} = \dfrac{13.846}{7}$$

$$x = 1.978$$

1.978 lb will be removed.

95. Above and Beyond

Summary Exercises

1. The ratio of 4 to 17: $\dfrac{4}{17}$

3. The ratio of wins to games played: $\dfrac{10}{16} = \dfrac{5}{8}$

5. The ratio of $2\frac{1}{3}$ to $5\frac{1}{4}$:

$$\dfrac{2\frac{1}{3}}{5\frac{1}{4}} = \dfrac{\frac{7}{3}}{\frac{21}{4}} = \dfrac{7}{3} \div \dfrac{21}{4} = \dfrac{7}{3} \cdot \dfrac{4}{21} = \dfrac{4}{9}$$

7. The ratio of 7 in. to 3 ft: $\dfrac{7\text{ in.}}{3\text{ ft}} = \dfrac{7\text{ in.}}{36\text{ in.}} = \dfrac{7}{36}$

9. $\dfrac{600\text{ mi}}{6\text{ hr}} = \dfrac{600}{6}\dfrac{\text{mi}}{\text{hr}} = 100\dfrac{\text{mi}}{\text{hr}}$

11. $\dfrac{350\text{ cal}}{7\text{oz}} = \dfrac{350}{7}\dfrac{\text{cal}}{\text{oz}} = 50\dfrac{\text{cal}}{\text{oz}}$

13. $\dfrac{5{,}000\text{ feet}}{30\text{ seconds}} = \dfrac{500}{3}\dfrac{\text{ft}}{\text{s}} = 166\dfrac{2}{3}\dfrac{\text{ft}}{\text{s}}$

15. $\dfrac{117\text{ hits}}{18\text{ games}} = 6\dfrac{1}{2}\dfrac{\text{hits}}{\text{games}}$

17. Taniko:
$$\dfrac{246\text{ points}}{20\text{ games}} = \dfrac{246}{20}\dfrac{\text{points}}{\text{games}} = 12.3\dfrac{\text{points}}{\text{game}}$$
Marisa:
$$\dfrac{216\text{ points}}{16\text{ games}} = \dfrac{216}{16}\dfrac{\text{points}}{\text{games}} = 13.5\dfrac{\text{points}}{\text{game}}$$
Marisa has the highest points per game rate.

19. $\dfrac{\$2.88}{32\text{ oz}} = \dfrac{288¢}{32\text{ oz}} = \dfrac{288}{32}\dfrac{¢}{\text{oz}} = 9\dfrac{¢}{\text{oz}}$ or $0.09\dfrac{\$}{\text{oz}}$

21. $\dfrac{\$2.28}{24\text{ oz}} = \dfrac{228¢}{24\text{ oz}} = \dfrac{228}{24}\dfrac{¢}{\text{oz}} = 9.5\dfrac{¢}{\text{oz}}$ or $9.5\dfrac{\$}{\text{oz}}$

23. $\dfrac{\$44.85}{3\text{ CDs}} = \dfrac{44.85}{3}\dfrac{\$}{\text{CDs}} = 14.95\dfrac{\$}{\text{CD}}$

25. $\dfrac{4}{9} = \dfrac{20}{45}$

27. $\dfrac{110\text{ mi}}{2\text{ hr}} = \dfrac{385\text{ mi}}{7\text{ hr}}$

29. $13 \times 7 = 91$

$4 \times 22 = 88$

Because the products are not equal, the fractions are not proportional.

31. $24 \times 12 = 288$

$9 \times 32 = 288$

Because the products are equal, the fractions are proportional.

33. $5 \times 4 = 20$

$\dfrac{1}{6} \times 120 = 20$

Because the products are equal, the fractions are proportional.

35. $40 \times 75.25 = 3,010$

$35 \times 86 = 3,010$

Because the products are equal, the fractions are proportional.

37. $\dfrac{16}{24} = \dfrac{m}{3}$

$24m = 48$

$\dfrac{24m}{24} = \dfrac{48}{24}$

$m = 2$

39. $\dfrac{14}{35} = \dfrac{t}{10}$

$35t = 140$

$\dfrac{35t}{35} = \dfrac{140}{35}$

$t = 4$

41. $\dfrac{\frac{1}{2}}{18} = \dfrac{5}{w}$

$\dfrac{1}{2}w = 90$

$\dfrac{\frac{1}{2}w}{\frac{1}{2}} = \dfrac{90}{\frac{1}{2}}$

$w = 180$

43. $\dfrac{5}{x} = \dfrac{0.6}{12}$

$0.6x = 60$

$\dfrac{0.6x}{0.6} = \dfrac{60}{0.6}$

$x = 100$

45. $\dfrac{4 \text{ tickets}}{\$90} = \dfrac{6 \text{ tickets}}{x}$

$4x = 540$

$\dfrac{4x}{4} = \dfrac{540}{4}$

$x = 135$

Therefore, the price of 6 tickets is $135.

47. $\dfrac{5 \text{ in. wide}}{7 \text{ in. tall}} = \dfrac{x \text{ in. wide}}{21 \text{ in. tall}}$

$7x = 105$

$x = 15$

The width of the enlargement will be 15 in.

49. $\dfrac{14 \text{ defective parts}}{400 \text{ total parts}} = \dfrac{x \text{ defective parts}}{800 \text{ total parts}}$

$400x = 11,200$

$x = 28$

28 defective parts can be expected in a shipment of 800 parts.

51. $\dfrac{16.5 \text{ cm}}{55 \text{ g}} = \dfrac{42 \text{ cm}}{x \text{ g}}$

$16.5x = 2,310$

$x = 140$

The piece of tubing weighs 140 g.

53. Let x represent the missing length.

$\dfrac{3}{4} = \dfrac{6}{x}$

$3x = 24$

$x = 8$

The missing length is 8 in.

55. Let x represent the missing length.

$$\frac{8}{10} = \frac{12}{x}$$

$$8x = 120$$

$$x = 15$$

The missing length is 15 cm.

57. Let x represent the missing length.

$$\frac{4}{10} = \frac{6}{x}$$

$$4x = 60$$

$$x = 15$$

The missing length is 24 mi.

Chapter Test 6

1. The ratio of 7 to 19: $\dfrac{7}{19}$

3. The ratio of 8 ft to 4 yd:

$$\frac{8 \text{ ft}}{4 \text{yd}} = \frac{8 \text{ ft}}{12 \text{ ft}} = \frac{8}{12} = \frac{2}{3}$$

5. $\dfrac{840 \text{ mi}}{175 \text{ gal}} = \dfrac{840}{175} \dfrac{\text{mi}}{\text{gal}} = 4.8 \dfrac{\text{mi}}{\text{gal}}$

7. $45x = 900$

$$\frac{45x}{45} = \frac{900}{45}$$

$$x = 20$$

9. $\dfrac{\frac{1}{2}}{p} = \dfrac{5}{30}$

$$5p = 15$$

$$\frac{15p}{15} = \frac{15}{5}$$

$$p = 3$$

11. $9 \times 27 = 243$

$3 \times 81 = 243$

Because the products are equal, the fractions are proportional.

13. $10 \times 27 = 270$

$9 \times 30 = 270$

Because the products are equal, the fractions are proportional.

15. $\dfrac{\$28.16}{11 \text{ gal}} = \dfrac{28.16}{11} \dfrac{\$}{\text{gal}} = 2.56 \dfrac{\$}{\text{gal}}$

17. $\dfrac{95¢}{5 \text{ pens}} = \dfrac{x ¢}{12 \text{ pens}}$

$$5x = 11.40$$

$$\frac{5x}{5} = \frac{11.40}{5}$$

$$x = 228$$

A dozen pens will cost 228¢ or \$2.28.

19. $\dfrac{5 \text{ mufflers}}{4 \text{ min}} = \dfrac{x \text{ mufflers}}{8 \text{ hr}} = \dfrac{x \text{ mufflers}}{480 \text{ min}}$

$$4x = 2,400$$

$$x = 600$$

600 mufflers can be installed in an 8-hr shift.

Cumulative Review: Chapters 1–6

1. $\underset{\text{thousands}}{\underline{45}}$, $\underset{\text{hundreds}}{\underline{789}}$

Forty-five thousand, seven hundred eighty-nine

3.
$$\begin{array}{r} \overset{2\ 2}{2,790} \\ 831 \\ + 22,683 \\ \hline 26,304 \end{array}$$

5.

$$\overset{\overset{3}{4}}{76}$$
$$\underline{\times 58}$$
$$608$$
$$\underline{3800}$$
$$4{,}408$$

7.

Amount owed	815
Minus first payment	-125
	690
Minus second payment	$-\ 80$
	610
Minus third payment	$-\ 90$
	520
Plus interest charges	$+\ 48$
	568

Luis still owes $568 on the account.

9. Perimeter $= 6$ ft $+ 2$ ft $+ 6$ ft $+ 2$ ft $= 16$ ft

Area $= 6$ ft $\times 2$ ft $= 12$ ft^2

11. $-|-14| = -14$

13. $(-5)(-17) = 85$

15. $x + 2$

17.

$$7 - 5x = 4 - 2(2x + 1) = 4 - 4x - 2$$
$$7 - 5x + 4x = 2 - 4x + 4x$$
$$7 - x = 2$$
$$x = 5$$

Check: $7 - 5(5) \overset{?}{=} 4 - 2[2(5) + 1]$

$$7 - 25 \overset{?}{=} 4 - 2[10 + 1]$$
$$-18 \overset{?}{=} 4 - 22$$
$$-18 = -18$$

19. $2\overline{)924}\overset{462}{}\searrow 2\overline{)462}\overset{231}{}\searrow 3\overline{)231}\overset{77}{}\searrow 7\overline{)77}\overset{11}{}$

$924 = 2 \times 2 \times 3 \times 7 \times 11$

21. $\dfrac{42}{168} = \dfrac{\cancel{2} \times \cancel{3} \times \cancel{7}}{\cancel{2} \times 2 \times 2 \times \cancel{3} \times \cancel{7}} = \dfrac{1}{4}$

23. $2\dfrac{2}{3} \times 3\dfrac{3}{4} = \dfrac{8}{3} \times \dfrac{15}{4} = \dfrac{\overset{2}{\cancel{8}} \times \overset{5}{\cancel{15}}}{\cancel{3} \times \cancel{4}} = \dfrac{2 \times 5}{1 \times 1} = \dfrac{10}{1} = 10$

25. $5\dfrac{1}{2} \div 3\dfrac{1}{4} = \dfrac{11}{2} \div \dfrac{13}{4} = \dfrac{11 \times \overset{2}{\cancel{4}}}{\cancel{2} \times 13} = \dfrac{11 \times 2}{1 \times 13} = \dfrac{22}{13}$

$$= 1\dfrac{9}{13}$$

27. $\left(-1\dfrac{7}{8}\right)\left(-5\dfrac{3}{20}\right) = \left(-\dfrac{15}{8}\right)\left(-\dfrac{113}{20}\right)$

$$= \dfrac{\overset{3}{\cancel{15}} \times 113}{8 \times \underset{4}{\cancel{20}}} = \dfrac{339}{32} = 9\dfrac{21}{32}$$

29. Step 1: The LCD is 44.

Step 2: $\dfrac{9}{11} = \dfrac{36}{44}$

$$\dfrac{3}{4} = \dfrac{33}{44}$$

$$\dfrac{1}{2} = \dfrac{22}{44}$$

Step 3: $\dfrac{36}{44} - \dfrac{33}{44} + \dfrac{22}{44} = \dfrac{25}{44}$

31. $8\dfrac{2}{7} - 3\dfrac{11}{14} = \dfrac{58}{7} - \dfrac{53}{14} = \dfrac{116}{14} - \dfrac{53}{14} = \dfrac{63}{14} = 4\dfrac{7}{14}$

$$= 4\dfrac{1}{2}$$

33. $36 = 2 \times 2 \times 3 \times 3$

$\underline{60 = 2 \times 2 \times 3 \times\quad\ 5}$

$\quad\ 2 \times 2 \times 3 \times 3 \times 5$ Bring down the factors.

So 180 is the LCM.

35. $132 \text{ mi} \div 2\frac{3}{4}\text{ hr} = \frac{132}{1}\text{ mi} \div \frac{11}{4}\text{ hr}$

$$= \frac{132}{1}\text{ mi} \times \frac{4}{11}\frac{1}{\text{hr}}$$

$$= \frac{12 \times 4}{1 \times 1}\frac{\text{mi}}{\text{hr}} = 48\frac{\text{mi}}{\text{hr}}$$

37. $2.56\overline{)3.1488}$

$$
\begin{array}{r}
1.23 \\
256\overline{)314.88} \\
\underline{256} \\
588 \\
\underline{512} \\
768 \\
\underline{768} \\
0
\end{array}
$$

39. $0.36 = \frac{36}{100} = \frac{9}{25}$

41. $4.23 \text{ m} + 4.23 \text{ m} + 2.8 \text{ m} + 2.8 \text{ m} = 14.06 \text{ m}$
The perimeter is 14.06 m.

43.
$$2.8x + 8.2 = 23.88$$
$$2.8x + 8.2 - 8.2 = 23.88 - 8.2$$
$$2.8x = 15.68$$
$$\frac{2.8x}{2.8} = \frac{15.68}{2.8}$$
$$x = 5.6$$

45.
$$c^2 = a^2 + b^2$$
$$(36)^2 = (27)^2 + b^2$$
$$b^2 = (36)h - (27)^2 = 1296 - 729 = 567$$
$$b = 23.81 \text{ in.}$$
(rounded to the nearest hundredth)

47. $\frac{12}{26} = \frac{6}{13}$

49. 6 dimes = \$0.60
3 quarters = \$0.75

$$\frac{\$0.60}{\$0.75} = \frac{0.60}{0.75} = \frac{0.60 \times 100}{0.75 \times 100} = \frac{60}{75} = \frac{4}{5}$$

51. $7 \times 9 = 63$
$3 \times 22 = 66$
Because the products are not equal, the fractions are not proportional.

53. $\frac{5}{x} = \frac{4}{12}$

$$4x = 60$$

$$\frac{4x}{4} = \frac{60}{4}$$

$$x = 15$$

55. $\frac{3 \text{ cm}}{250 \text{ km}} = \frac{7.2 \text{ cm}}{x \text{ km}}$

$$3x = 1,800$$

$$x = 600$$
The two cities are 600 km apart.

Chapter 7
Percents

Prerequisite Check

1. $\dfrac{44}{100} = \dfrac{44 \div 4}{100 \div 4} = \dfrac{11}{25}$

3. $\dfrac{3}{8} = 8\overline{)3.000}^{\,0.375}$

5. $0.08 = \dfrac{8}{100} = \dfrac{8 \div 4}{100 \div 4} = \dfrac{2}{25}$

7. $0.04 \times 1{,}040 = 41.6$

9. $1\dfrac{1}{3} \times 62.1 = \dfrac{4}{3} \times 62.1 = \dfrac{4 \times 62.1}{3} = 4 \times 20.7$
$\phantom{1\dfrac{1}{3} \times 62.1} = 82.8$

11. $\dfrac{5{,}250}{100} = 100\overline{)5{,}250}^{\,52.5}$

13. $\dfrac{4}{5} = \dfrac{x}{60}$
$5x = 240$
$x = 48$

15. $\dfrac{21}{100} = \dfrac{x}{18}$
$100x = 378$
$x = \dfrac{378}{100} = 3.78$

17. The store's profit margin on the electric range.

Exercise 7.1

< Objective 1 >

1. 35 of 100 squares are shaded. As a fraction we write this as $\dfrac{35}{100}$.
$\dfrac{35}{100} = 35\left(\dfrac{1}{100}\right) = 35\%$
35 percent of the drawing is shaded.

3. 3 of 4 parts are shaded. As a fraction we write this as $\dfrac{3}{4}$.
$\dfrac{3}{4} = \dfrac{75}{100} = 75\left(\dfrac{1}{100}\right) = 75\%$
75 percent of the drawing is shaded.

5. $\dfrac{53}{100} = 53\left(\dfrac{1}{100}\right) = 53\%$

7. $\dfrac{74}{100} = 74\left(\dfrac{1}{100}\right) = 74\%$

9. $\dfrac{3}{10} = \dfrac{30}{100} = 30\left(\dfrac{1}{100}\right) = 30\%$

11. $\dfrac{27}{50} = \dfrac{54}{100} = 54\left(\dfrac{1}{100}\right) = 54\%$

13. $\dfrac{23}{50} = \dfrac{46}{100} = 46\left(\dfrac{1}{100}\right) = 46\%$

15. $\dfrac{5}{20} = \dfrac{25}{100} = 25\left(\dfrac{1}{100}\right) = 25\%$

< Objective 2 >

17. $6\% = 6\left(\dfrac{1}{100}\right) = \dfrac{6}{100} = \dfrac{3}{50}$

19. $75\% = 75\left(\dfrac{1}{100}\right) = \dfrac{75}{100} = \dfrac{3}{4}$

21. $65\% = 65\left(\dfrac{1}{100}\right) = \dfrac{65}{100} = \dfrac{13}{20}$

23. $50\% = 50\left(\dfrac{1}{100}\right) = \dfrac{50}{100} = \dfrac{1}{2}$

25. $46\% = 46\left(\dfrac{1}{100}\right) = \dfrac{46}{100} = \dfrac{23}{50}$

27. $66\% = 66\left(\dfrac{1}{100}\right) = \dfrac{66}{100} = \dfrac{33}{50}$

29. $150\% = 150\left(\dfrac{1}{100}\right) = \dfrac{150}{100} = 1\dfrac{50}{100} = 1\dfrac{1}{2}$

31. $225\% = 225\left(\dfrac{1}{100}\right) = \dfrac{225}{100} = 2\dfrac{25}{100} = 2\dfrac{1}{4}$

33. $166\dfrac{2}{3}\% = \dfrac{3\times166+2}{3}\% = \dfrac{500}{3}\%$

$\qquad = \dfrac{500}{3}\left(\dfrac{1}{100}\right) = \dfrac{5}{3} = 1\dfrac{2}{3}$

35. $212\dfrac{1}{2}\% = \dfrac{425}{2}\% = \dfrac{425}{2}\left(\dfrac{1}{100}\right) = \dfrac{425}{200} = \dfrac{17}{8}$

$\qquad = 2\dfrac{1}{8}$

< Objective 3 >

37. $20\% = 20\left(\dfrac{1}{100}\right) = 0.20$

39. $35\% = 35\left(\dfrac{1}{100}\right) = 0.35$

41. $39\% = 39\left(\dfrac{1}{100}\right) = 0.39$

43. $5\% = 5\left(\dfrac{1}{100}\right) = 0.05$

45. $135\% = 135\left(\dfrac{1}{100}\right) = 1.35$

47. $240\% = 240\left(\dfrac{1}{100}\right) = 2.40$

49. $23.6\% = 23.6\left(\dfrac{1}{100}\right) = 0.236$

51. $6.4\% = 6.4\left(\dfrac{1}{100}\right) = 0.064$

53. $0.2\% = 0.2\left(\dfrac{1}{100}\right) = 0.002$

55. $1.05\% = 1.05\left(\dfrac{1}{100}\right) = 0.0105$

57. $7\dfrac{1}{2}\% = 7.5\% = 7.5\left(\dfrac{1}{100}\right) = 0.075$

59. $87\dfrac{1}{2}\% = 87.5\% = 87.5\left(\dfrac{1}{100}\right) = 0.875$

61. $128\dfrac{3}{4}\% = 128.75\% = 128.75\left(\dfrac{1}{100}\right)$

$\qquad = 1.2875$

63. $\dfrac{1}{2}\% = 0.5\% = 0.5\left(\dfrac{1}{100}\right) = 0.005$

65. $50\% = 50\left(\dfrac{1}{100}\right) = \dfrac{50}{100} = \dfrac{1}{2}$

$\qquad \dfrac{1}{2} = 1 \div 2 = 0.5$

67. $43\dfrac{1}{2}\% = 43.5\left(\dfrac{1}{100}\right) = \dfrac{435}{1000} = \dfrac{87}{200}$

$\dfrac{87}{200} = 87 \div 200 = 0.435$

69. $33\dfrac{1}{3}\% = \dfrac{100}{3}\% = \dfrac{100}{3}\left(\dfrac{1}{100}\right) = \dfrac{1}{3}$

$\dfrac{1}{3} = 1 \div 3 = 0.\overline{3}$

71. $24\dfrac{1}{6}\% = \dfrac{145}{6}\% = \dfrac{145}{6}\left(\dfrac{1}{100}\right) = \dfrac{29}{120}$

$\dfrac{29}{120} = 29 \div 120 = 0.241\overline{6}$

73. $85\% = 85\left(\dfrac{1}{100}\right) = \dfrac{85}{100} = \dfrac{17}{20}$

75. $15\% = 15\left(\dfrac{1}{100}\right) = \dfrac{15}{100} = 0.15$

$\dfrac{15}{100} = \dfrac{3}{20}$

77. **(a)** $30\% = 30\left(\dfrac{1}{100}\right) = \dfrac{30}{100} = 0.3$

$30\% = \dfrac{30}{100} = \dfrac{3}{10}$

(b) $10\% = 10\left(\dfrac{1}{100}\right) = \dfrac{10}{100} = 0.1$

$10\% = \dfrac{10}{100} = \dfrac{1}{10}$

(c) $12\% = 12\left(\dfrac{1}{100}\right) = \dfrac{12}{100} = 0.12$

$12\% = \dfrac{12}{100} = \dfrac{3}{25}$

(d) $8\% = 8\left(\dfrac{1}{100}\right) = \dfrac{8}{100} = 0.08$

$8\% = \dfrac{8}{100} = \dfrac{2}{25}$

(e) $7\% = 7\left(\dfrac{1}{100}\right) = \dfrac{7}{100} = 0.07$

$7\% = \dfrac{7}{100}$

(f) $2\% = 2\left(\dfrac{1}{100}\right) = \dfrac{2}{100} = 0.02$

$2\% = \dfrac{2}{100} = \dfrac{1}{50}$

79. Expenditure is 10.2% of the total budget, hence in fraction it can be expressed as

$10.2\left(\dfrac{1}{100}\right) = \dfrac{10.2}{100} = \dfrac{102}{1,000} = \dfrac{51}{500}$

81. $12\% = 12\left(\dfrac{1}{100}\right) = \dfrac{12}{100} = 0.12$

$\dfrac{12}{100} = \dfrac{3}{25}$

83. Since the metal shrinks 3.125%, the fraction of the lost size can be expressed as

$3.125\left(\dfrac{1}{100}\right) = \dfrac{3.125}{100} = \dfrac{3125}{100,000} = \dfrac{1}{32}.$

85. (a) (4)
   (b) (3)
   (c) (6)
   (d) (5)
   (e) (1)
   (f) (2)

Exercise 7.2

< Objectives 1 and 2 >

1. $0.08 = \dfrac{8}{100} = 8\left(\dfrac{1}{100}\right) = 8\%$

3. $0.05 = \dfrac{5}{100} = 5\left(\dfrac{1}{100}\right) = 5\%$

5. $0.18 = \dfrac{18}{100} = 18\left(\dfrac{1}{100}\right) = 18\%$

7. $0.86 = \dfrac{86}{100} = 86\left(\dfrac{1}{100}\right) = 86\%$

9. $0.4 = \dfrac{4}{10} = 40\left(\dfrac{1}{100}\right) = 40\%$

11. $0.7 = \dfrac{7}{10} = 70\left(\dfrac{1}{100}\right) = 70\%$

13. $1.10 = \dfrac{110}{100} = 110\left(\dfrac{1}{100}\right) = 110\%$

15. $4.40 = \dfrac{440}{100} = 440\left(\dfrac{1}{100}\right) = 440\%$

17. $0.065 = \dfrac{65}{1,000} = 6.5\left(\dfrac{1}{100}\right) = 6.5\%$ or $6\dfrac{1}{2}\%$

19. $0.025 = \dfrac{25}{1,000} = 2.5\left(\dfrac{1}{100}\right) = 2.5\%$ or $2\dfrac{1}{2}\%$

21. $\dfrac{1}{4} = 0.25 = 25\%$

23. $\dfrac{2}{5} = 0.4 = 40\%$

25. $\dfrac{1}{5} = 0.2 = 20\%$

27. $\dfrac{5}{8} = 0.625 = 62.5\%$ or $62\dfrac{1}{2}\%$

29. $\dfrac{5}{16} = 0.3125 = 31.25\%$ or $31\dfrac{1}{4}\%$

31. $3\dfrac{1}{2} = 3.5 = 350\%$

33. $0.002 = \dfrac{2}{1,000} = 0.2\left(\dfrac{1}{100}\right) = 0.2\%$

35. $0.004 = \dfrac{4}{1,000} = 0.4\left(\dfrac{1}{100}\right) = 0.4\%$

37. $\dfrac{1}{6} = \dfrac{100}{6}\left(\dfrac{1}{100}\right) = \dfrac{50}{3}\left(\dfrac{1}{100}\right) = 16\dfrac{2}{3}\left(\dfrac{1}{100}\right)$
$= 16\dfrac{2}{3}\%$

Exact value of $\dfrac{1}{6}$ as a percent is $16\dfrac{2}{3}\%$.

39. $\dfrac{7}{9} = \dfrac{700}{9}\left(\dfrac{1}{100}\right) = 77.7\overline{7}\left(\dfrac{1}{100}\right) = 77.8\%$

$\dfrac{7}{9}$ to the nearest tenth of a percent is 77.8%.

41. $\dfrac{7}{9} = \dfrac{700}{9}\left(\dfrac{1}{100}\right) = 77\dfrac{7}{9}\left(\dfrac{1}{100}\right) = 77\dfrac{7}{9}\%$

Exact value of $\dfrac{7}{9}$ as percent is $77\dfrac{7}{9}\%$.

43. $5\dfrac{1}{4} = 5.25 = 525\left(\dfrac{1}{100}\right) = 525\%$

45. $4\dfrac{1}{3} = 4.33\overline{3} = 433.3\%$

$4\dfrac{1}{3}$ to the nearest tenth of a percent is 433.3%.

47. $4\dfrac{1}{3} = \dfrac{5}{3} = \dfrac{500}{3}\left(\dfrac{1}{100}\right) = 433\dfrac{1}{3}\%$

Exact value of $4\dfrac{1}{3}$ as percent is $433\dfrac{1}{3}\%$.

49. $\dfrac{11}{6} = 1.833\overline{3} = 183.3\%$

$\dfrac{11}{6}$ to the nearest tenth of a percent is 183.3%.

51. $\dfrac{11}{6} = \dfrac{1100}{6}\left(\dfrac{1}{100}\right) = \dfrac{550}{3}\left(\dfrac{1}{100}\right) = 183\dfrac{1}{3}\%$

Exact value of $\dfrac{11}{6}$ as percent is $183\dfrac{1}{3}\%$.

53. $\dfrac{10}{7} = 1.\overline{428571} = 142.9\%$

$\dfrac{10}{7}$ to the nearest tenth of a percent is 142.9%.

55. $\dfrac{10}{7} = \dfrac{1000}{7}\left(\dfrac{1}{100}\right) = 142\dfrac{6}{7}\%$

Exact value of $\dfrac{10}{7}$ as percent is $142\dfrac{6}{7}\%$.

57. 25 of 100 squares are shaded.

$\dfrac{25}{100} = 0.25$ or $\dfrac{1}{4}$ or 25%

59. 47 of 100 squares are shaded.

$\dfrac{47}{100} = 0.47$ or $\dfrac{47}{100}$ or 47%

61. 77 of 100 squares are shaded.

$\dfrac{77}{100} = 0.77$ or $\dfrac{77}{100}$ or 77%

63. 4 of 100 squares are shaded.

$\dfrac{4}{100} = 0.04$ or $\dfrac{4}{100} = \dfrac{1}{25}$ or 4%

65. $\dfrac{32.9 - 26.9}{26.9} = \dfrac{6}{26.9} = 0.223 = 22.3\%$

(rounded to the nearest tenth of a percent)

67. Percent spent on car expenses is

$\dfrac{200}{1,000} = 0.2 = 20\%$.

69. Least amount of money spent was $100 in miscellaneous. Miscellaneous percent is

$\dfrac{100}{1,000} = 0.1 = 10\%$

71. True

73. always

75. $\dfrac{13}{16} = 0.8125 = 81.25\%$

77. $\dfrac{165}{172} = 0.96 = 96\%$

(rounded to the nearest whole percent)

79. As per the definition given, the efficiency of motor is $\dfrac{400}{435} = \dfrac{80}{87} = 0.9195 = 91.95\% = 92\%$.

(rounded to the nearest whole percent)

81. Percent of products to be returned is

$\dfrac{2}{27} = 0.\overline{074} = 7.4\%$.

(rounded to the nearest tenth of a percent)

83. $\dfrac{7}{12} = 0.5833 = 58.33\%$

$0.0\underline{8} = 8\% = 8\left(\dfrac{1}{100}\right) = \dfrac{8}{100} = \dfrac{2}{25}$

$35\% = 35\left(\dfrac{1}{100}\right) = \dfrac{35}{100} = \dfrac{7}{20} = 0.35$

$\dfrac{11}{18} = 0.6111 = 61.11\%$

$0.2\underline{6}5 = 26.5\%; \quad 0.265 = \dfrac{265}{1{,}000} = \dfrac{53}{200}$

$4\dfrac{3}{8}\% = 4.375\% = 4.375\left(\dfrac{1}{100}\right) = 0.04375$

$\qquad = \dfrac{4{,}375}{100{,}000} = \dfrac{7}{160}$

Fraction	Decimal	Percent
$\dfrac{7}{12}$	0.5833	58.3%
$\dfrac{2}{25}$	0.08	8%
$\dfrac{7}{20}$	0.35	35%
$\dfrac{11}{18}$	0.6111	61.11%
$\dfrac{53}{200}$	0.265	26.5%
$\dfrac{7}{160}$	0.0438	$4\dfrac{3}{8}\%$

85. Above and Beyond

Exercise 7.3

< Objective 1 >

1. 23% of 400 is 92.
    R   B   A

3. 40% of 600 is 240.
    R   B   A

5. What is 7% of 325?
    A   R   B

7. 16% of what number is 56?
    R   B   A

9. 480 is 60% of what number?
    A   R   B

11. What percent of 120 is 40?
    R   B   A

13. To find the rate, look for the percent symbol. 5% is the rate. The base is the whole, or the amount of merchandise that Jan sells. $40,000 is the base. Her commission, the unknown, is the amount.

15. The rate is the unknown percent in this problem. "What percent" asks for the unknown rate. The base is the whole, or the number of students in the chemistry class. 30 is the base. The amount is the portion of the base, here the number of students who received a grade of A. 5 is the amount.

17. To find the rate, look for the percent symbol. 5.5% is the rate. The selling price, the unknown, is the base or the whole in the problem. The amount is the portion of the base, here the tax. $3.30 is the amount.

19. To find the rate, look for the percent symbol. 6% is the rate. The base is the whole, or the number of students at the start of the school year. 9,000 is the base. The number of additional students, the unknown, is the amount or the portion of the base.

< Objectives 2 and 3 >

21. $\dfrac{A}{600} = \dfrac{35}{100}$

$100A = 21,000$

$A = \dfrac{21,000}{100} = 210$

210 is 35% of 600

23. $\dfrac{A}{200} = \dfrac{45}{100}$

$100A = 9,000$

$A = \dfrac{9,000}{100} = 90$

45% of 200 is 90.

25. $\dfrac{A}{2,500} = \dfrac{40}{100}$

$100A = 1,00,000$

$A = \dfrac{1,00,000}{100} = 1,000$

40% of 2,500 is 1,000.

27. $\dfrac{4}{50} = \dfrac{R}{100}$

$50R = 400$

$R = \dfrac{400}{50} = 8$

8% of 50 is 4.

29. $\dfrac{45}{500} = \dfrac{R}{100}$

$500R = 4,500$

$R = \dfrac{4,500}{500} = 9$

9% of 500 is 45.

31. $\dfrac{340}{200} = \dfrac{R}{100}$

$200R = 34,000$

$R = \dfrac{34,000}{200} = 170$

170% of 200 is 340.

33. $\dfrac{46}{B} = \dfrac{8}{100}$

$8B = 4,600$

$B = \dfrac{4,600}{8} = 575$

46 is 8% of 575.

35. $\dfrac{55}{B} = \dfrac{11}{100}$

$11B = 5,500$

$B = \dfrac{5,500}{11} = 500$

The base is 500.

37. $\dfrac{58.5}{B} = \dfrac{13}{100}$

$13B = 5,850$

$B = \dfrac{5,850}{13} = 450$

58.5 is 13% of 450.

39. $\dfrac{A}{800} = \dfrac{110}{100}$

$100A = 88,000$

$A = \dfrac{88,000}{100} = 880$

110% of 800 is 880.

41. $\dfrac{A}{4,000} = \dfrac{108}{100}$

$100A = 432,000$

$A = \dfrac{432,000}{100} = 4,320$

108% of 400 is 4,320.

43. $\dfrac{210}{120} = \dfrac{R}{100}$

$120R = 21,000$

$R = \dfrac{21,000}{120} = 175$

210 is 175% of 120.

45. $\dfrac{360}{90} = \dfrac{R}{100}$

$90R = 36,000$

$R = \dfrac{36,000}{90} = 400$

360 is 400% of 90.

47. $\dfrac{625}{B} = \dfrac{125}{100}$

$125B = 62,500$

$B = \dfrac{62,500}{125} = 500$

625 is 125% of 500.

49. $\dfrac{935}{B} = \dfrac{110}{100}$

$110B = 93,500$

$B = \dfrac{93,500}{110} = 850$

The base is 850.

51. $\dfrac{A}{300} = \dfrac{8.5}{100}$

$100A = 2,550$

$A = \dfrac{2,550}{100} = 25.5$

8.5% of 300 is 25.5.

53. $\dfrac{A}{6,000} = \dfrac{11\frac{3}{4}}{100} = \dfrac{11.75}{100}$

$100A = 70,500$

$A = \dfrac{70,500}{100} = 705$

$11\frac{3}{4}\%$ of 6,000 is 705.

55. $\dfrac{A}{3,000} = \dfrac{5.25}{100}$

$100A = 15,750$

$A = \dfrac{15,750}{100} = 157.5$

5.25% of 3,000 is 157.5.

57. $\dfrac{60}{800} = \dfrac{R}{100}$

$800R = 6,000$

$R = \dfrac{6,000}{800} = 7.5$

60 is 7.5% of 800.

59. $\dfrac{120}{180} = \dfrac{R}{100}$

$180R = 12,000$

$R = \dfrac{12,000}{180} = 66\frac{2}{3}$

$66\frac{2}{3}\%$ of 180 is 120.

61. $\dfrac{750}{1,200} = \dfrac{R}{100}$

$1,200R = 75,000$

$R = \dfrac{75,000}{1,200} = 62.5$

62.5% of 1,200 is 750.

63. $\dfrac{A}{112} = \dfrac{87}{100}$

$100A = 9,744$

$A = \dfrac{9,744}{100} = 97.44$

87% of 112 is 97.44.

65. $\dfrac{17}{B} = \dfrac{30}{100}$

$30B = 1,700$

$B = \dfrac{1,700}{30} = \dfrac{170}{3} = 56\dfrac{2}{3}$

17 is 30% of $\dfrac{170}{3}$ or $56\dfrac{2}{3}$.

67. $\dfrac{A}{180} = \dfrac{66\dfrac{2}{3}}{100} = \dfrac{200}{3}\left(\dfrac{1}{100}\right)$

$300A = 36,000$

$A = \dfrac{36,000}{300} = 120$

$66\dfrac{2}{3}$% of 180 is 120.

69. $\dfrac{75}{B} = \dfrac{33\dfrac{1}{3}\%}{100} = \dfrac{100}{3}\left(\dfrac{1}{100}\right)$

$100B = 22,500$

$B = \dfrac{22,500}{100} = 225$

75 is $33\dfrac{1}{3}$% of 225.

71. $\dfrac{A}{120} = \dfrac{55\dfrac{5}{9}}{100} = \dfrac{500}{9}\left(\dfrac{1}{100}\right)$

$900A = 60,000$

$A = \dfrac{60,000}{900} = \dfrac{200}{3} = 66\dfrac{2}{3}$

$55\dfrac{5}{9}$% of 120 is $\dfrac{170}{3}$ or $56\dfrac{2}{3}$.

73. $\dfrac{420}{B} = \dfrac{10.5}{100}$

$10.5B = 42,000$

$B = \dfrac{42,000}{10.5} = 4,000$

10.5% of 4,000 is 420.

75. $\dfrac{58.5}{B} = \dfrac{13}{100}$

$13B = 5,850$

$B = \dfrac{5,800}{13} = 450$

58.5 is 13% of 450.

77. $\dfrac{195}{B} = \dfrac{7.5}{100}$

$7.5B = 19,500$

$B = \dfrac{19,500}{7.5} = 2,600$

195 is 7.5% of 2,600.

79. False

81. To find the rate, look for the percent symbol. 25% is the rate. The base is the whole, or the unknown. The 225 mL of ethyl alcohol, is the amount.

83. To find the rate, look for the percent symbol. 3.5% is the rate. The base is the whole or the total milk quantity which is 938 g. The amount of butterfat or the unknown which is a part of the base is the amount.

Exercise 7.4

< Objectives 1–4 >

1. The base is $3,400, and the rate is 12%. The amount is the interest that you will pay.

$$\frac{A}{3,400} = \frac{12}{100}$$

$$100A = 40,800$$

$$A = 408$$

The interest that you will pay is $408.

3. The amount is $140, and the base is $2,800. To find the rate, we have

$$\frac{140}{2,800} = \frac{R}{100}$$

$$2,800R = 14,000$$

$$R = 5$$

The commission rate is 5%.

5. The rate is 80%, and the amount is 20 problems. The base is the number of questions on the test.

$$\frac{20}{B} = \frac{80}{100}$$

$$80B = 2,000$$

$$B = 25$$

25 questions were on the test.

7. The rate is 6.4%, and the base is $260. The amount is the tax that one must pay.

$$\frac{A}{260} = \frac{6.4}{100}$$

$$100A = 1,664$$

$$A = 16.64$$

The tax that one must pay is $16.64.

9. The amount is 102 people, and the base is 1,200 people. To find the rate, we have

$$\frac{102}{1,200} = \frac{R}{100}$$

$$1,200R = 10,200$$

$$R = 8.5$$

The town's employment rate is 8.5%.

11. The rate is 65%, and the amount is 780 people. The base is the number of people who responded to the survey.

$$\frac{780}{B} = \frac{65}{100}$$

$$65B = 78,000$$

$$B = 1,200$$

1,200 people responded to the survey.

13. The rate is 22%, and the base is $1,200. The amount is the markup.

$$\frac{A}{1,200} = \frac{22}{100}$$

$$100A = 26,400$$

$$A = 264$$

The markup is $264, making the selling price $1,200 + $264 = $1,464.

15. The amount is \$2,030, and the rate is 14%. The base is the price of the van before the increase.

$$\frac{2,030}{B} = \frac{14}{100}$$

$$14B = 203,000$$

$$B = 14,500$$

The price of the van before the increase was \$14,500.

17. The amount is \$162.50, and the rate is 6.5%. The base is Carlotta's salary before the raise.

$$\frac{162.50}{B} = \frac{6.5}{100}$$

$$6.5B = 16,250$$

$$B = 2,500$$

Her salary before the raise was \$2,500.

19. The base is 6,000, and the rate is 14%. The amount is the increase in the population.

$$\frac{A}{6,000} = \frac{14}{100}$$

$$100A = 84,000$$

$$A = 840$$

The population after the increase is $6,000 + 840 = 6,840$.

21. The rate is 40%, and the amount is 300 s. The base is the time that it should take to check all the files.

$$\frac{300}{B} = \frac{40}{100}$$

$$40B = 30,000$$

$$B = 750$$

It should take 750 s to check all the files.

23. The base is 254.2 million, which is the total registered vehicles. The amount is 194 million, which is the number of passenger cars registered. The rate is

$$\frac{194}{254.2} = \frac{R}{100}$$

$$R = \frac{19,400}{254.2} = 76.3$$

(rounded to the nearest tenth percent)
76.3% of passenger cars are registered.

25. The total green house emissions in 1990 is the base B. The increase in the emission in 2008 is the amount A. The rate is 13.3%

$$\frac{A}{B} = \frac{13.3}{100}$$

$$A = \frac{13.3B}{100} = 0.133B$$

Now, total emission in 2008 is 6,924.56 mmt which can be represented as:

$$A + B = 6,924.56$$

$$0.133B + B = 6,924.56$$

$$1.133B = 6,924.56$$

$$B = \frac{6,924.56}{1.133} = 6,111.70$$

(using $A = 0.133B$)
Hence greenhouse gas emissions in 1990 were 6,111.70 mmt (rounded to the nearest hundredth mmt).

27. Simple interest $(I) = PRT$

$$I = (6,000)(0.03)(2) = \$360$$

29. Principal $(P) = \dfrac{I}{r \bullet t}$

$$P = \frac{(150)}{(0.04)(2)} = 1,875$$

The principal was \$1,875.

31. 26% of Robert's pay

$$= \frac{26}{100} \times \$850 = 0.26 \times 850 = 221$$

The amount withheld is \$221.

33. First, find the number of people who dropped out,

$$60 - 45 = 15$$
$$A = RB$$
$$15 = 60R$$
$$\frac{15}{60} = \frac{60R}{60}$$
$$0.25 = R$$

Therefore, the drop-out rate is 25%.

35. First, find the amount of water,

$$900 - 117 = 783$$
$$A = RB$$
$$783 = 900R$$
$$\frac{783}{900} = \frac{900R}{900}$$
$$0.87 = R$$

Therefore, 87% of the solution is water.

37. First, we must find the increase in exports from 2006 to 2010

$$136,473 - 133,722 = 2,751$$

To find the percent

$$\frac{2,751}{133,722} = \frac{R}{100}$$
$$133,722R = 275,100$$
$$R \approx 22$$

(rounded to the nearest whole percent)
Rate of increase of exports from 2006 to 2010 is 22%.

39. The difference between imports and exports in 2006 is

$$198,253 - 133,722 = 64,531$$

to find the percent

$$\frac{64,531}{133,722} = \frac{R}{100}$$
$$133,722R = 6,453,100$$
$$R \approx 48$$

(rounded to the nearest whole percent)
Hence, imports exceeded exports by 48% in 2006.

41. For year 1, the interest is 6% of $4,000, or $0.06 \times \$4,000 = \$240$. At the end of year 1, the amount in the account will be $\$4,000 + \$240 = \$4,240$.
For year 2, the interest is 6% of $4,240, or $0.06 \times \$4,240 = \$254.40$. At the end of year 2, the amount in the account will be $\$4,240 + \$254.40 = \$4,494.40$.

43. For year 1, the interest is 5% of $4,000, or $0.05 \times \$4,000 = \$200$. At the end of year 1, the amount in the account will be $\$4,000 + \$200 = \$4,200$.
For year 2, the interest is 5% of $4,200, or $0.05 \times \$4,200 = \$210$. At the end of year 2, the amount in the account will be $4,200 + 210 = \$4,410$.
For year 3, the interest is 5% of $4,410, or $0.05 \times \$4,410 = \$220.50$. At the end of year 2, the amount in the account will be $\$4,410 + \$220.50 = \$4,630.50$.

45. Length AC, 4 units, is the amount, and length AB, 16 units, is the base. To find the percent, we have

$$\frac{4}{16} = \frac{R}{100}$$
$$16R = 400$$
$$R = 25$$

Length AC is 25% of length AB.

47. Length AE, 6 units, is the amount, and length AB, 16 units, is the base. To find the percent, we have

$$\frac{6}{16} = \frac{R}{100}$$
$$16R = 600$$
$$R = 37.5$$

Length AE is 37.5% of length AB.

49. The population after the increase is
$100 + 4.2 = 104.2\%$ of the original population.
The base is 19,500, and the rate is 104.2%.

$$\frac{A}{19,500} = \frac{104.2}{100}$$

$$A = \frac{19,500 \times 104.2}{100}$$

Entering into the calculator:

$$19500 \;\boxed{\times}\; 104.2 \;\boxed{\div}\; 100 \;\boxed{=}$$

We get 20,319. The increased population is
20,319.

51. The sale price is $100 - 12.5 = 87.5\%$ of the
original price. The base is $98.50 and the rate
is 87.5%.

$$\frac{A}{98.50} = \frac{87.5}{100}$$

$$A = \frac{98.50 \times 87.5}{100}$$

Entering into the calculator:

$$98.50 \;\boxed{\times}\; 87.5 \;\boxed{\div}\; 100 \;\boxed{=}$$

We get 86.1875. The sale price is $86.19
(rounding to the nearest hundredth).

53. Carolyn's take home pay is
$100 - 24.6 = 75.4\%$ of her monthly salary.
The rate is 75.4%, and the base is $5,220. The
amount is her take home pay.

$$\frac{A}{5,220} = \frac{75.4}{100}$$

$$A = \frac{5,220 \times 75.4}{100}$$

Entering into the calculator:

$$5220 \;\boxed{\times}\; 75.4 \;\boxed{\div}\; 100 \;\boxed{=}$$

We get 3,935.88. Her take home pay is
$3,935.88 per month.

55. The rate is 25%. The portion of the solution
that is alcohol is 225 mL. This is the amount.
The total quantity of the solution is the
unknown base. To solve for the base, we have

$$\frac{225}{B} = \frac{25}{100}$$

$$25B = 22,500$$

$$B = 900$$

900 mL of alcohol solution can be prepared.

57. The rate is 3.5%. The total quantity of milk is
938 g. This is the base. The portion of the
milk that is butterfat is the unknown amount.
To solve for the unknown amount, we have

$$\frac{A}{938} = \frac{3.5}{100}$$

$$100A = 3,283$$

$$A = 32.83$$

32.83 grams of butterfat are in 1 liter of 3.5%
milk.

59.
$$A = RB$$
$$300 = 500R$$
$$\frac{300}{500} = \frac{500R}{500}$$
$$0.6 = R$$

Therefore, 60% of the hard drive is full.

61.
$$A = RB$$
$$6 = 0.08B$$
$$\frac{6}{0.08} = \frac{0.08B}{0.08}$$
$$75 = B$$

The store sold 75 alternators.

63. 2.414 gal of water needs to be added to the
mixture for it to be 13% water.

65. Above and Beyond

67. Above and Beyond

Summary Exercises

1. 3 of 4 parts are shaded. As a fraction we write this as $\dfrac{3}{4}$.

$$\dfrac{3}{4} = \dfrac{300}{4}\left(\dfrac{1}{100}\right) = 75\%$$

75% of the diagram is shaded.

3. $2\% = 2\left(\dfrac{1}{100}\right) = \dfrac{2}{100} = \dfrac{1}{50}$

5. $37.5\% = 37.5\left(\dfrac{1}{100}\right) = \dfrac{375}{1000} = \dfrac{3}{8}$

7. $233\dfrac{1}{3}\% = 233\dfrac{1}{3}\left(\dfrac{1}{100}\right) = \dfrac{700}{3}\left(\dfrac{1}{100}\right) = \dfrac{7}{3}$

$$= 2\dfrac{1}{3}$$

9. $75\% = 75\left(\dfrac{1}{100}\right) = 0.75$

11. $6.25\% = 6.25\left(\dfrac{1}{100}\right) = 0.0625$

13. $0.6\% = 0.6\left(\dfrac{1}{100}\right) = 0.006$

15. $0.06 = \dfrac{6}{100} = 6\left(\dfrac{1}{100}\right) = 6\%$

17. $2.4 = \dfrac{24}{10} = 240\left(\dfrac{1}{100}\right) = 240\%$

19. $0.035 = \dfrac{35}{1,000} = 3.5\left(\dfrac{1}{100}\right) = 3.5\%$

21. $\dfrac{43}{100} = 0.43 = 43\%$

23. $\dfrac{2}{5} = 0.40 = 40\%$

25. $2\dfrac{2}{3} = 2.66\overline{6} = 266\dfrac{2}{3}\%$ (exact value)

27. $\dfrac{80}{B} = \dfrac{4}{100}$

$$4B = 8,000$$

$$B = \dfrac{8,000}{4} = 2,000$$

80 is 4% of 2,000.

29. $\dfrac{A}{3,000} = \dfrac{11}{100}$

$$100A = 33,000$$

$$A = \dfrac{33,000}{100} = 330$$

11% of 3,000 is 330.

31. $\dfrac{625}{B} = \dfrac{12.5}{100}$

$$12.5B = 62,500$$

$$B = \dfrac{62,500}{12.5} = 5,000$$

The base is 5,000.

33. $\dfrac{A}{700} = \dfrac{9.5}{100}$

$$100A = 6,650$$

$$A = \dfrac{6,650}{100} = 66.5$$

66.5 is 9.5% of 700.

35. $\dfrac{780}{B} = \dfrac{130}{100}$

$$130B = 78,000$$

$$B = \dfrac{78,000}{130} = 600$$

The base is 600.

37. $\dfrac{28.8}{960} = \dfrac{R}{100}$

$$960R = 2,880$$

$$R = \dfrac{2,880}{960} = 3$$

28.8 is 3% of 960.

39. The rate is 4%, and the base is $45,000. The amount is Joan's commission

$$\frac{A}{45,000} = \frac{4}{100}$$

$$100A = 180,000$$

$$A = 1,800$$

Joan's commission was $1,800.

41. The base is 400 mL, and the amount is 30 mL. To find the percent, we have

$$\frac{30}{400} = \frac{R}{100}$$

$$400R = 3,000$$

$$R = 7.5$$

The solution contains 7.5% of acid.

43. The rate is 25%, and the base is $136. The amount is the markdown.

$$\frac{A}{136} = \frac{25}{100}$$

$$100A = 3,400$$

$$A = 34$$

To find the sale price, we subtract the discount price from the original price $136 - \$34 = \$102$. The sale price is $102.

45. The rate is 35%, and the amount is 252. The base is the total number of science students.

$$\frac{252}{B} = \frac{35}{100}$$

$$35B = 25,200$$

$$B = 720$$

There are 720 science students altogether.

47. The rate is 5.25%, and the base is $3,000. The amount is the interest that you have earned.

$$\frac{A}{3,000} = \frac{5.25}{100}$$

$$100A = 15,750$$

$$A = 157.50$$

The interest you have earned is $157.50, making $\$3,000 + \$157.50 = \$3,157.50$ the total at the end of the year.

49. The rate is 30%, and the amount is 150 s. The base is the time that it should take to check all the files.

$$\frac{150}{B} = \frac{30}{100}$$

$$30 = 15,000$$

$$B = 500$$

It should take 500 s to check all the files, or 8 min 20 s.

Chapter Test 7

1. 4 out of 5 squares are shaded.

$$\frac{4}{5} = \frac{80}{100} = 80\left(\frac{1}{100}\right) = 80\%$$

80% of the drawing is shaded.

3. $0.042 = \dfrac{42}{1000} = 4.2\left(\dfrac{1}{100}\right) = 4.2\%$

5. $\dfrac{5}{8} = 0.625 = 62.5\%$

7. $6\% = 6\left(\dfrac{1}{100}\right) = 0.06$

9. $7\% = 7\left(\dfrac{1}{100}\right) = \dfrac{7}{100}$

11. The rate is 25%; the base is 200; the amount is 50.

13. We want to find the amount.

$$\frac{A}{250} = \frac{4.5}{100}$$

$$100A = 1,125$$

$$A = 11.25$$

4.5% of 250 is 11.25.

15. We want to find the amount.

$$\frac{A}{1,500} = \frac{33\frac{1}{3}}{100} = \frac{100}{3}\left(\frac{1}{100}\right)$$

$$A = \frac{150,000}{300} = 500$$

$33\frac{1}{3}\%$ of 1,500 is 500.

17. We want to find the rate.

$$\frac{875}{500} = \frac{R}{100}$$

$$R = \frac{87,500}{500} = 175$$

875 is 175% of 500.

19. We want to find the base.

$$\frac{25.5}{B} = \frac{8.5}{100}$$

$$B = \frac{2,500}{8.5}$$

$$A = 300$$

8.5% of 300 is 25.5.

21. The rate (R) is 6%; the base (B) is amount of purchase; the amount (A) is $30.

23. The rate is 75%, and the base is 80. To find the amount

$$\frac{A}{80} = \frac{75}{100}$$

$$A = \frac{6,000}{100} = 60$$

60 questions were correct.

25. The amount is $300, and the base is $2,500. To find the rate, we have

$$\frac{300}{2,500} = \frac{R}{100}$$

$$2,500R = 30,000$$

$$R = 12$$

The interest rate for the loan is 12%.

27. The amount is $540, and the rate is 3%. The base is the total sales.

$$\frac{540}{B} = \frac{3}{100}$$

$$3B = 54,000$$

$$B = 18,000$$

Sarah's total sales were $18,000.

29. The rate is 12%, and the amount is $2,220. The base is the money that Shawn borrowed to finance the car.

$$\frac{2,220}{B} = \frac{12}{100}$$

$$12B = 222,000$$

$$B = 18,500$$

He borrowed $18,500 to finance the car.

Cumulative Review: Chapters 1–7

1. 234,768
8 ones, 6 tens, 7 hundreds, 4 thousands,
   3 ten thousands, 2 hundred thousands
The place value of 4 is thousands.

3.

$$\begin{array}{r} 89 \\ 34\overline{)3,026} \\ \underline{272} \\ 306 \\ \underline{306} \\ 0 \end{array}$$

We have $3,026 \div 34 = 89$.

5. $15 - 3 \times 2 = 15 - 6 = 9$

7. $14 + (-9) = 5$

9. $-8 - 5 = -13$

11. $3x^2 + 10x - 4 = 3(-2)^2 - 10(-2) - 4$
$$= 12 + 20 - 4 = 28$$

13. $2x - 3(4x - 1) = 2x - 12x + 3 = -10x + 3$

15. Since each has exactly two factors, 1 and itself, the prime numbers between 50 and 70 are 53, 59, 61, and 67.

17. The factors of each of the two numbers are,
84: 1, 2, 3, 4, 6, 7, 12, 14, 21, 28, 42, 84
140: 1, 2, 4, 5, 7, 10, 14, 20, 28, 35, 70, 140
The GCF is 28.

19. $3\dfrac{2}{5} \times 2\dfrac{1}{2} = \dfrac{17}{5} \times \dfrac{5}{2} = \dfrac{17 \times \cancel{5}^{1}}{\cancel{5}_{1} \times 2} = \dfrac{17}{2} = 8\dfrac{1}{2}$

21. $4\dfrac{3}{4} + 3\dfrac{5}{6} = \dfrac{19}{4} + \dfrac{23}{6} = \dfrac{57}{12} + \dfrac{46}{12} = \dfrac{103}{12} = 8\dfrac{7}{12}$

23. $\left(5\dfrac{1}{2}\,\text{yd} \times 3\dfrac{1}{4}\,\text{yd}\right) \times \dfrac{\$16}{\text{yd}^2}$

$= \left(\dfrac{11}{2}\,\text{yd} \times \dfrac{13}{4}\,\text{yd}\right) \times \dfrac{\$16}{\text{yd}^2} = \dfrac{143}{8}\,\cancel{\text{yd}^2} \times \dfrac{\$16}{\cancel{\text{yd}^2}}$

$= \dfrac{143 \times \$\cancel{16}^{2}}{\cancel{8}_{1}} = \$286$

It will cost \$286 to cover the floor.

25. First, convert the length of the board to inches
$8\cancel{\text{ft}} \times \dfrac{12\ \text{in.}}{1\cancel{\text{ft}}} = 96\ \text{in}$
Then, subtract the length of the bookshelf.
$96 - 54\dfrac{5}{8} = 96 - \dfrac{437}{8} = \dfrac{768}{8} - \dfrac{437}{8} = \dfrac{331}{8}$
Then, subtract the amount wasted in the cut.
$\dfrac{331}{8} - \dfrac{1}{8} = \dfrac{330}{8} = 41\dfrac{2}{8} = 41\dfrac{1}{4}$
The length of the board that remains is
$41\dfrac{1}{4}$ in.

27.
$$3x + 5 = 2x - 11$$
$$3x + (-2x) + 5 = 2x + (-2x) - 11$$
$$x + 5 = -11$$
$$x = -16$$

29. 8 is in the hundredths place.

31. 6.28 is 6 and 28 hundredths.
6.30 is 6 and 30 hundredths.
$6.28 < 6.30$

33.
$$\begin{array}{r} 4.03 \\ \times\ 2.8 \\ \hline 3224 \\ 806 \\ \hline 11.284 \end{array}$$

35. Calculate the amount you will pay by multiplying the monthly payment by the number of months: $\$29.50 \times 24 = \$708$. You will pay \$708. Now subtract the advertised price from this amount: $\$708 - \$599.95 = \$108.05$. You are paying \$108.05 extra on the investment plan.

37. Volume = length × width × height
Volume $= 5.1 \times 3.6 \times 2.4 = 44.064\ \text{ft}^3$

39. $\sqrt{20} = 4.5$
(rounded to the nearest tenth)

41. $0.36 = \dfrac{36}{100} = \dfrac{9}{25}$

43. $\dfrac{8\frac{1}{2}}{12\frac{3}{4}} = \dfrac{\frac{17}{2}}{\frac{51}{4}} = \dfrac{17}{2} \div \dfrac{51}{4} = \dfrac{17}{2} \times \dfrac{4}{51} = \dfrac{2}{3}$

45. $\dfrac{3}{7} = \dfrac{8}{x}$

$3x = 56$

$\dfrac{3x}{3} = \dfrac{56}{3}$

$x = 18\dfrac{2}{3}$

47. $\dfrac{\frac{1}{4}\text{ in.}}{25\text{ mi}} = \dfrac{3\frac{1}{2}\text{ in.}}{x\text{ mi}}$

$\dfrac{1}{4}x = 25 \cdot 3\dfrac{1}{2}$

$4\left(\dfrac{1}{4}\right)x = 4 \cdot 25 \cdot \dfrac{7}{2}$

$1 \cdot x = 4 \cdot 25 \cdot \dfrac{7}{2}$

$x = 350$

The two towns are 350 mi apart.

49. $34\% = 34\left(\dfrac{1}{100}\right) = 0.34$ (as decimal)

$34\% = 34\left(\dfrac{1}{100}\right) = \dfrac{34}{100} = \dfrac{17}{50}$ (as fraction)

51. $\dfrac{A}{250} = \dfrac{18}{100}$

$100A = 4{,}500$

$A = \dfrac{4{,}500}{100} = 45$

45 is 18% of 250.

53. $\dfrac{10}{B} = \dfrac{8}{100}$

$8B = 1{,}000$

$B = \dfrac{1{,}000}{8} = 125$

The company had 125 employees before the reduction.

Chapter 8
Measurement and Geometry

Prerequisite Check

1. $42.84 \times 10^4 = 428,400$

3. $5^2 = 5 \times 5 = 25$

5. $\frac{1}{2}(8)(6+3) = \frac{1}{2}(8)(9) = 36$

7. $\frac{1}{2}(9)(5) = \frac{45}{2}$

9. $\frac{144}{108} = \frac{4}{3}$

11. $15 \text{ to } 6 = \frac{15}{6} = \frac{5}{2}$

13. $\frac{\$344}{32 \text{ hr}} = \$10.75/\text{hr}$

15. $\frac{\$3.29}{64 \text{ fl oz}} = \$0.0514/\text{fl oz} = 5.14 \text{ cents/fl oz}$

17. $P = 16.5 + 10.5 + 16.5 + 10.5 = 54 \text{ in.}$

19. $A = (\text{side})^2 = (12 \text{ mm})^2 = 144 \text{ mm}^2$

Exercises 8.1

< Objective 1 >

1. $8 \text{ ft} = 8 \cancel{\text{ft}}\left(\dfrac{12 \text{ in.}}{1 \cancel{\text{ft}}}\right) = 96 \text{ in.}$

3. $3 \text{ lb} = 3 \cancel{\text{lb}}\left(\dfrac{16 \text{ oz}}{1 \cancel{\text{lb}}}\right) = 48 \text{ oz}$

5. $360 \text{ min} = 360 \cancel{\text{min}}\left(\dfrac{1 \text{ hr}}{60 \cancel{\text{min}}}\right) = 6 \text{ hr}$

7. $4 \text{ days} = 4 \cancel{\text{days}}\left(\dfrac{24 \text{ hr}}{1 \cancel{\text{day}}}\right) = 96 \text{ hr}$

9. $16 \text{ qt} = 16 \cancel{\text{qt}}\left(\dfrac{1 \text{ gal}}{4 \cancel{\text{qt}}}\right) = 4 \text{ gal}$

11. $10,000 \text{ lb} = 10,000 \cancel{\text{lb}}\left(\dfrac{1 \text{ ton}}{2000 \cancel{\text{lb}}}\right) = 5 \text{ tons}$

13. $30 \text{ pt} = 30 \cancel{\text{pt}}\left(\dfrac{1 \text{ qt}}{2 \cancel{\text{pt}}}\right) = 15 \text{ qt}$

15. $64 \text{ oz} = 64 \cancel{\text{oz}}\left(\dfrac{1 \text{ lb}}{16 \cancel{\text{oz}}}\right) = 4 \text{ lb}$

17. $7 \text{ yd} = 7 \cancel{\text{yd}}\left(\dfrac{3 \text{ ft}}{1 \cancel{\text{yd}}}\right) = 21 \text{ ft}$

19. $39 \text{ ft} = 39 \cancel{\text{ft}}\left(\dfrac{1 \text{ yd}}{3 \cancel{\text{ft}}}\right) = 13 \text{ yd}$

21. $8 \text{ min} = 8 \cancel{\text{min}}\left(\dfrac{60 \text{ s}}{1 \cancel{\text{min}}}\right) = 480 \text{ s}$

23. $192 \text{ hr} = 192 \cancel{\text{hr}}\left(\dfrac{1 \text{ day}}{24 \cancel{\text{hr}}}\right) = 8 \text{ days}$

25. $16 \text{ qt} = 16 \cancel{\text{qt}}\left(\dfrac{2 \text{ pt}}{1 \cancel{\text{qt}}}\right) = 32 \text{ pt}$

27. $7\dfrac{1}{4}\,\text{hr} = \dfrac{29}{4}\,\cancel{\text{hr}}\left(\dfrac{60\,\text{min}}{1\,\cancel{\text{hr}}}\right) = 435\,\text{min}$

29. $56\,\text{oz} = 56\,\cancel{\text{oz}}\left(\dfrac{1\,\text{lb}}{16\,\cancel{\text{oz}}}\right) = 3.5\,\text{lb}$

31. $225\,\text{s} = 225\,\cancel{\text{s}}\left(\dfrac{1\,\text{min}}{60\,\cancel{\text{s}}}\right) = 3.75\,\text{min}$

33. $1.55\,\text{lb} = 1.55\,\cancel{\text{lb}}\left(\dfrac{16\,\text{oz}}{1\,\cancel{\text{lb}}}\right) = 24.8\,\text{oz}$

35. $40\,\text{in.} = 40\,\cancel{\text{in.}}\left(\dfrac{1\,\cancel{\text{ft}}}{12\,\cancel{\text{in.}}}\right)\left(\dfrac{1\,\text{yd}}{3\,\cancel{\text{ft}}}\right) = \dfrac{10}{9}\,\text{yd}$

37. 16 gal

$= 16\,\cancel{\text{gal}}\left(\dfrac{4\,\cancel{\text{qt}}}{1\,\cancel{\text{gal}}}\right)\left(\dfrac{2\,\cancel{\text{pt}}}{1\,\cancel{\text{qt}}}\right)\left(\dfrac{2\,\cancel{\text{c}}}{1\,\cancel{\text{pt}}}\right)\left(\dfrac{8\,\text{fl oz}}{1\,\cancel{\text{c}}}\right)$

$= 2,048\,\text{fl oz}$

39. $1\,\text{ton} = 1\,\cancel{\text{ton}}\left(\dfrac{2,000\,\cancel{\text{lb}}}{1\,\cancel{\text{ton}}}\right)\left(\dfrac{16\,\text{oz}}{1\,\cancel{\text{lb}}}\right)$

$= 32,000\,\text{oz}$

41. 6 weeks

$= 6\,\cancel{\text{weeks}}\left(\dfrac{7\,\cancel{\text{day}}}{1\,\cancel{\text{week}}}\right)\left(\dfrac{24\,\cancel{\text{hr}}}{1\,\cancel{\text{day}}}\right)\left(\dfrac{60\,\text{min}}{1\,\cancel{\text{hr}}}\right)$

$= 60,480\,\text{min}$

< Objective 2 >

43. $4\,\text{ft}\,18\,\text{in.} = 4\,\text{ft} + 1\,\text{ft} + 6\,\text{in.} = 5\,\text{ft}\,6\,\text{in.}$

45. $7\,\text{qt} + 5\,\text{pt} = 7\,\text{qt} + 2\,\text{qt} + 1\,\text{pt} = 9\,\text{qt}\,1\,\text{pt}$

47. $5\,\text{gal}\,9\,\text{qt} = 5\,\text{gal} + 2\,\text{gal} + 1\,\text{qt} = 7\,\text{gal}\,1\,\text{qt}$

49. $9\,\text{min}\,75\,\text{s} = 9\,\text{min} + 1\,\text{min} + 15\,\text{s}$
$= 10\,\text{min}\,15\,\text{s}$

51.
$$\begin{array}{r} 8\text{ lb }\ \,7\text{ oz} \\ +\ 6\text{ lb }15\text{ oz} \\ \hline 14\text{ lb }22\text{ oz} \end{array}$$
Since $22\,\text{oz} = 1\,\text{lb}\,6\,\text{oz}$, the final result is
15 lb 6 oz.

53.
$$\begin{array}{r} 3\text{ hr }20\text{ min} \\ 4\text{ hr }25\text{ min} \\ +\ 5\text{ hr }35\text{ min} \\ \hline 12\text{ hr }80\text{ min} \end{array}$$
Since $80\,\text{min} = 1\,\text{hr}\,20\,\text{min}$, the final result is
13 hr 20 min.

55.
$$\begin{array}{r} 4\text{ lb }\ \,7\text{ oz} \\ 3\text{ lb }11\text{ oz} \\ +\ 5\text{ lb }\ \,8\text{ oz} \\ \hline 12\text{ lb }26\text{ oz} \end{array}$$
Since $26\,\text{oz} = 1\,\text{lb}\,10\,\text{oz}$, the final result is
13 lb 10 oz.

57.
$$\begin{array}{r} 9\text{ lb }15\text{ oz} \\ -\ 5\text{ lb }\ \,8\text{ oz} \\ \hline 4\text{ lb }\ \,7\text{ oz} \end{array}$$

59. To complete the subtraction, we borrow
1 hr.

$\begin{array}{r} 6\text{ hr }30\text{ min} \\ -\ 3\text{ hr }50\text{ min} \\ \hline \end{array}$ $\rightarrow$ $\begin{array}{r} 5\text{ hr }90\text{ min} \\ -\ 3\text{ hr }50\text{ min} \\ \hline 2\text{ hr }40\text{ min} \end{array}$

61. To complete the subtraction, we borrow
1 yd.

$\begin{array}{r} 5\text{ yd }1\text{ ft} \\ -\ 2\text{ yd }2\text{ ft} \\ \hline \end{array}$ $\rightarrow$ $\begin{array}{r} 4\text{ yd }4\text{ ft} \\ -\ 2\text{ yd }2\text{ ft} \\ \hline 2\text{ yd }2\text{ ft} \end{array}$

63. $4 \times 13\,\text{oz} = 52\,\text{oz}$ or 3 lb 4 oz

65.
$$\begin{array}{r} 4\text{ ft }\ \,5\text{ in.} \\ \times\qquad\quad 3 \\ \hline 12\text{ ft }15\text{ in.} = 13\text{ ft }3\text{ in.} \end{array}$$

67. $\dfrac{4\text{ ft }6\text{ in.}}{2} = 2\,\text{ft}\,3\,\text{in.}$

69. $\dfrac{16 \text{ min } 28 \text{ s}}{4} = 4 \text{ min } 7 \text{ s}$

71.   2 gal 3 qt 1 pt
          + 3 gal 2 qt 1 pt
          5 gal 5 qt 2 pt = 5 gal 6 qt = 6 gal 2 qt

73.   13 yd 15 ft 10 in.
          − 9 yd 16 ft 15 in.
          We borrow 1 yd.

   13 yd 15 ft 10 in.   12 yd 18 ft 10 in.
   − 9 yd 16 ft 15 in.   →   − 9 yd 16 ft 15 in.

   We borrow 1 ft.
   12 yd 18 ft 10 in.   12 yd 17 ft 22 in.
   − 9 yd 16 ft 15 in.   →   − 9 yd 16 ft 15 in.
                                      3 yd   1 ft   7 in.

75.   2 weeks   7 days 18 hr 40 min
          ×   2
          4 weeks 14 days 36 hr 80 min
        = 4 weeks 14 days 37 hr 20 min
        = 4 weeks 15 days 13 hr 20 min
        = 6 weeks   1 day 13 hr 20 min

< Objective 3 >

77. 9,000,000 tons

   $= 9{,}000{,}000 \text{ tons} \left(\dfrac{2{,}000 \text{ lb}}{1 \text{ ton}} \right)$

   $= 18{,}000{,}000{,}000 \text{ lb}$

   18 billion lb of particulates were emitted in that year.

79.   4 ft   8 in.
           11 ft   7 in.
          + 9 ft   3 in.
           24 ft 18 in. = 25 ft 6 in.
   The total length of material needed is 25 ft 6 in.

81. $2 \text{ yd} = 2 \text{ yd} \left(\dfrac{3 \text{ ft}}{1 \text{ yd}} \right) = 6 \text{ ft}$

   We borrow 1 ft.
      6 ft   5 ft 12 in.
   − 2 ft 10 in.   →   − 2 ft 10 in.
                              3 ft   2 in.
   3 ft 2 in. of the fabric remains.

83.
      2 ft   6 in.
   ×   2
      4 ft 12 in. = 5 ft

      1 ft   8 in.
   ×   2
      2 ft 16 in. = 3 ft 4 in.

      5 ft
   + 3 ft 4 in.
      8 ft 4 in.
   Yes, 8 inches will remain.

85.   1 pt   9 fl oz
          + 2 pt 10 fl oz
           3 pt 19 fl oz = 4 pt 3 fl oz

   $3 \text{ qt} = 3 \text{ qt} \left(\dfrac{2 \text{ pt}}{1 \text{ qt}} \right) = 6 \text{ pt}$

   We borrow 1 pt.
      6 pt   5 pt 16 fl oz
   − 4 pt 3 fl oz   →   − 2 pt   3 fl oz
                              1 pt 13 fl oz
   1 pt 13 fl oz of the developer remains.

87.   2 lb   9 oz
          ×   6
           12 lb 54 oz = 15 lb 6 oz
   The total weight is 15 lb 6 oz.

89. $3 \times 12 \text{ oz} = 36 \text{ oz}$ or 2 lb 4 oz
   The 2-lb-8-oz can is the better buy.

91.

$$30 \text{ mi} = 30 \text{ mi}\left(\frac{5,280 \text{ ft}}{1 \text{ mi}}\right)\left(\frac{1 \text{ yd}}{3 \text{ ft}}\right)$$

$$= 52,800 \text{ yards}$$

$$1 \text{ gal} = 1 \text{ gal}\left(\frac{4 \text{ qt}}{1 \text{ gal}}\right)\left(\frac{2 \text{ pt}}{1 \text{ qt}}\right)\left(\frac{16 \text{ fl oz}}{1 \text{ pt}}\right)$$

$$= 128 \text{ fl oz}$$

$$\frac{52,800 \text{ yds}}{128 \text{ fl oz}} = 412.5 \text{ yd/fl oz}$$

Colette can travel 412.5 yd on 1 fl oz of fuel.

93. $0.5 \text{ gal} = 0.5 \text{ gal}\left(\frac{4 \text{ qt}}{1 \text{ gal}}\right)\left(\frac{2 \text{ pt}}{1 \text{ qt}}\right)\left(\frac{16 \text{ fl oz}}{1 \text{ pt}}\right)$

$$= 64 \text{ fl oz}$$

$$\frac{\$3.29}{64 \text{ fl oz}} = \$0.05/\text{fl oz} = 5¢/\text{fl oz}$$

The cost per fluid ounce is 5¢/fl oz.

95. $55 \text{ mi} = 55 \text{ mi}\left(\frac{5,280 \text{ ft}}{1 \text{ mi}}\right) = 290,400 \text{ ft}$

$$1 \text{ hr} = 1 \text{ hr}\left(\frac{60 \text{ min}}{1 \text{ hr}}\right)\left(\frac{60 \text{ s}}{1 \text{ min}}\right) = 3,600 \text{ s}$$

$$\frac{3,600 \text{ s}}{290,400 \text{ ft}} = 0.00124 \text{ s/ft}$$

$$0.00124 \times 1,000 = 12.4 \text{ s}$$

(rounded to the nearest tenth of a second)
The driver takes 12.4 s to travel 1,000 ft.

97. True

99. Always

101. 13 ft 8 in. is equivalent to 164 in.
164 in. ÷ 4 = 41 in. = 3 ft 5 in.
Each piece will be 3 ft 5 in.

103.   218 lb 12 oz
              36 lb   3 oz
          + 36 lb   3 oz
           ‾‾‾‾‾‾‾‾‾‾‾‾
            290 lb 18 oz = 291 lb 2 oz
The total weight of the block with two heads
is 291 lb 2 oz.

105. 4 ft 9 in is equivalent to 57 in.
57 in. × 13 = 741 in. = 61 ft 9 in.
The total length required for the studs is
61 ft 9 in.

107. (a) $\frac{72 \text{ beats}}{\text{min}}\left(\frac{60 \text{ min}}{\text{hr}}\right)\left(\frac{24 \text{ hr}}{\text{day}}\right)$

$$= \frac{103,680 \text{ beats}}{\text{day}}$$

$$1 \times 10^9 \text{ beats}\left(\frac{1 \text{ day}}{103,680 \text{ beats}}\right)$$

$$\approx 9,645 \text{ days}$$

(rounded to the nearest whole day)

(b) $9,645 \text{ days}\left(\frac{1 \text{ yr}}{365 \text{ days}}\right) \approx 26 \text{ yr}$

(rounded to the nearest whole year)

109. Above and Beyond

111. Above and Beyond

113. Above and Beyond

Exercises 8.2

< Objective 1 >

1. The height of a ceiling: **(b)** 2.5 m.

3. The height of a kitchen counter: **(c)** 90 cm.

5. The height of a two-story building: **(a)** 7 m.

7. The width of a roll of cellophane tape:
(b) 12.7 mm.

9. The thickness of window glass: **(a)** 5 mm.

11. The length of a ballpoint pen: **(c)** 16 cm.

13. A playing card is 6 cm wide.

15. A doorway is 2 m high.

17. A basketball court is 28 m long.

19. The width of a nail file is 12 mm.

21. A recreation room is 6 m long.

23. A long-distance run is 35 km.

< Objective 2 >

25. $3,000 \text{ mm} = 3,000 \text{ mm} \left(\dfrac{1 \text{ m}}{1000 \text{ mm}} \right) = 3 \text{ m}$

27. $8 \text{ m} = 8 \text{ m} \left(\dfrac{100 \text{ cm}}{1 \text{ m}} \right) = 800 \text{ cm}$

29. $250 \text{ km} = 250 \text{ km} \left(\dfrac{100,000 \text{ cm}}{1 \text{ km}} \right)$
$= 25,000,000 \text{ cm}$

31. $25 \text{ cm} = 25 \text{ cm} \left(\dfrac{10 \text{ mm}}{1 \text{ cm}} \right) = 250 \text{ mm}$

33. $7,000 \text{ m} = 7,000 \text{ m} \left(\dfrac{1 \text{ km}}{1,000 \text{ m}} \right) = 7 \text{ km}$

35. $8 \text{ cm} = 8 \text{ cm} \left(\dfrac{10 \text{ mm}}{1 \text{ cm}} \right) = 80 \text{ mm}$

37. $5 \text{ km} = 5 \text{ km} \left(\dfrac{1,000 \text{ m}}{1 \text{ km}} \right) = 5,000 \text{ m}$

39. $5 \text{ m} = 5 \text{ m} \left(\dfrac{1,000 \text{ mm}}{1 \text{ m}} \right) = 5,000 \text{ mm}$

41. $250 \text{ cm} = 250 \text{ cm} \left(\dfrac{1 \text{ m}}{100 \text{ cm}} \right) = 2.5 \text{ m}$

43. $12 \text{ mm} = 12 \text{ mm} \left(\dfrac{1 \text{ cm}}{10 \text{ mm}} \right) = 1.2 \text{ cm}$

45. $3.4 \text{ m} = 3.4 \text{ m} \left(\dfrac{1,000 \text{ mm}}{1 \text{ m}} \right) = 3,400 \text{ mm}$

47. $132 \text{ km} = 132 \text{ km} \left(\dfrac{1,000 \text{ m}}{1 \text{ km}} \right) = 132,000 \text{ m}$

49. $90 \text{ km} = 90 \text{ km} \left(\dfrac{1,000 \text{ m}}{1 \text{ km}} \right) = 90,000 \text{ m}$

$1 \text{ hr} = 1 \text{ hr} \left(\dfrac{60 \text{ min}}{1 \text{ hr}} \right) \left(\dfrac{60 \text{ s}}{1 \text{ min}} \right) = 3,600 \text{ s}$

$90 \dfrac{\text{km}}{\text{hr}} = \dfrac{90,000 \text{ m}}{3,600 \text{ s}} = 25 \dfrac{\text{m}}{\text{s}}$

51. $800 \text{ km} = 800 \text{ km} \left(\dfrac{1,000 \text{ m}}{1 \text{ km}} \right) = 800,000 \text{ m}$

$1 \text{ day} = 1 \text{ day} \left(\dfrac{24 \text{ hr}}{1 \text{ day}} \right) = 24 \text{ hr}$

$800 \dfrac{\text{km}}{\text{day}} = \dfrac{800,000 \text{ m}}{24 \text{ hr}} = 33,333 \dfrac{1}{3} \dfrac{\text{m}}{\text{hr}}$

< Objective 3 >

53. $250 \text{ km} = 250 \text{ km} \left(\dfrac{0.62 \text{ mi}}{1 \text{ km}} \right) = 155 \text{ mi}$

55. $150 \text{ mi} = 150 \text{ mi} \left(\dfrac{1.61 \text{ km}}{1 \text{ mi}} \right) = 241.5 \text{ km}$

57. $2.6 \text{ m} = 2.6 \text{ m} \left(\dfrac{39.37 \text{ in.}}{1 \text{ m}} \right) = 102.36 \text{ in.}$

59. $3 \text{ ft} = 3 \text{ ft} \left(\dfrac{12 \text{ in.}}{1 \text{ ft}} \right) \left(\dfrac{2.5 \text{ cm}}{1 \text{ in.}} \right) \left(\dfrac{10 \text{ mm}}{1 \text{ cm}} \right)$
$= 900 \text{ mm}$

61. 19.4 mm
$= 19.4 \text{ mm} \left(\dfrac{1 \text{ m}}{1,000 \text{ mm}} \right) \left(\dfrac{1 \text{ yd}}{0.91 \text{ m}} \right)$
$= 0.02 \text{ yd}$

63. $75 \text{ km} = 75 \cancel{\text{km}} \left(\dfrac{0.62 \cancel{\text{mi}}}{1 \cancel{\text{km}}} \right) \left(\dfrac{5,280 \text{ ft}}{1 \cancel{\text{mi}}} \right)$

$= 245,520 \text{ ft}$

$1 \text{ hr} = 1 \cancel{\text{hr}} \left(\dfrac{60 \cancel{\text{min}}}{1 \cancel{\text{hr}}} \right) \left(\dfrac{60 \text{ s}}{1 \cancel{\text{min}}} \right) = 3,600 \text{ s}$

$75 \dfrac{\text{km}}{\text{hr}} = \dfrac{245,520 \text{ ft}}{3,600 \text{ s}} = 68.2 \dfrac{\text{ft}}{\text{s}}$

< Objective 4 >

65. Evaluate $C = \dfrac{5(F-32)}{9}$ with $F = 52$.

$C = \dfrac{5(52-32)}{9} = \dfrac{100}{9} = 11.11$

$52°F = 11.11°C$

67. Evaluate $F = \dfrac{9C}{5} + 32$ with $C = 24$.

$F = \dfrac{9C}{5} + 32 = \dfrac{9 \times 24}{5} + 32 = 43.2 + 32$

$= 75.2$

$24°C = 75.2°F$

69. Evaluate $C = \dfrac{5(F-32)}{9}$ with $F = 8$.

$C = \dfrac{5(86-32)}{9} = \dfrac{270}{9} = 30$

$86°F = 30°C$

71. Evaluate $F = \dfrac{9C}{5} + 32$ with $C = 20$.

$F = \dfrac{9C}{5} + 32 = \dfrac{9 \times 20}{5} + 32 = 36 + 32$

$= 68$

$20°C = 68°F$

73. Evaluate $C = \dfrac{5(F-32)}{9}$ with $F = 100$.

$C = \dfrac{5(100-32)}{9} = \dfrac{340}{9} = 37.78$

$100°F = 37.78°C$

75. Evaluate $F = \dfrac{9C}{5} + 32$ with $C = 37$.

$F = \dfrac{9C}{5} + 32 = \dfrac{9 \times 37}{5} + 32 = 130.6 + 32$

$= 98.6$

$37°C = 98.6°F$

77. $55 \text{ mi} = 55 \cancel{\text{mi}} \left(\dfrac{1.61 \cancel{\text{km}}}{1 \cancel{\text{mi}}} \right) = 88.55 \text{ km}$

$55 \dfrac{\text{mi}}{\text{hr}} = \dfrac{88.5 \text{ km}}{1 \text{ hr}} = 88.55 \dfrac{\text{km}}{\text{hr}}$

Samantha can drive at 88.55 km/hr.

79. A Boeing 747 can travel 8,336 mi on one 57,285-gal tank of airplane fuel. Therefore, fuel efficiency

$= \dfrac{8,336}{57,285} = 0.1455 = 0.146 \text{ mi/gal}$.

81. Boeing 747 travels 0.146 mi with one gallon of fuel. A fuel tank has a capacity of 57,285 gallon, therefore a Boeing 747 can travel $= 57,285 \times 0.146 \text{ mi} = 8,363.61 \text{ mi}$.

Converting miles to kilometres $= 8,363.61 \times 1.61 = 13,456.4121$

$= 13,456 \text{ km}$.

83. A Boeing 747 can travel 6,838 mi on one 45,536-gal tank of airplane fuel. Therefore, the amount of fuel consumed to travel each

mile $= \dfrac{45,536}{6,838} = 6.659 = 6.7 \text{ gal/mi}$.

85. $0.8 \text{ cm} + 1.5 \text{ cm} + 0.8 \text{ cm} + 1.5 \text{ cm} = 4.6 \text{ cm}$
The perimeter of the parallelogram is 4.6 cm.

87. $2 \text{ cm} + 1.5 \text{ cm} + 2 \text{ cm} + 1.5 \text{ cm} = 7 \text{ cm}$
The perimeter of the rectangle is 7 cm.

89. $15 \text{ mm} + 15 \text{ mm} + 15 \text{ mm} + 15 \text{ mm} = 60 \text{ mm}$
The perimeter of the square is 60 mm.

91. True

93. False

95. always

97. Start by converting all units into the smallest units needed, cm.

$$3 \text{ m} = 3 \text{ m}\left(\frac{100 \text{ cm}}{1 \text{ m}}\right) = 300 \text{ cm}$$

$$9.3 \text{ dm} = 9.3 \text{ dm}\left(\frac{10 \text{ cm}}{1 \text{ dm}}\right) = 93 \text{ cm}$$

We start with 300 cm and need to subtract 86 cm, 93 cm, and 29 cm.

$$300 \text{ cm} - (86 + 93 + 29)\text{ cm}$$

$$= 300 \text{ cm} - 208 \text{ cm} = 92 \text{ cm}$$

The remaining piece is 92 cm long.

99. First, convert 81.2 m into cm.

$$81.2 \text{ m} = 81.2 \text{ m}\left(\frac{100 \text{ cm}}{1 \text{ m}}\right) = 8,120 \text{ cm}$$

Now, divide by 40.

$$8,120 \text{ cm} \div 40 = 203 \text{ cm}$$

101. $300,000 \text{ km/s} = \dfrac{300,000 \text{ km}}{1 \text{ s}}\left(\dfrac{1,000 \text{ m}}{1 \text{ km}}\right)$

$$= 300,000,000 \frac{\text{m}}{\text{s}}$$

103. Evaluate $F = \dfrac{9C}{5} + 32$ with $C = 10$.

$$F = \frac{9C}{5} + 32 = \frac{9 \times 10}{5} + 32 = 18 + 32 = 50$$

$10°C = 50°F$

The corresponding temperature is 50°F.

105. Evaluate $F = \dfrac{9C}{5} + 32$ with $C = 12.5$.

$$F = \frac{9C}{5} + 32 = \frac{9 \times 12.5}{5} + 32 = 22.5 + 32$$

$$= 54.5$$

$12.5°C = 54.5°F$

The corresponding temperature is 54.5°F.

107. $51 \text{ cm}\left(\dfrac{0.394 \text{ in.}}{1 \text{ cm}}\right) = 20.094 \approx 20 \text{ in.}$

(rounded to the nearest inch)

109. Above and Beyond

111. Above and Beyond

113. $156 \times 8,336 = 1,300,416$ p-mi

The Boeing 747 has travelled 1,300,416 passenger-miles.

115. Above and Beyond

117. $\dfrac{957,320 \text{ p-mi}}{45,536 \text{ gal}} = 21$ p-mi/gal

(rounded to one decimal place)
The passenger-miles per gallon that the Boeing 777 gets is 21 p-mi/gal.

119. Above and Beyond

Exercises 8.3

< Objective 1 >

1.

The figure has two endpoints, A and B, making it a line segment.

3.

The figure continues forever in both directions, through points A and C, making it a line.

5. The figure continues forever in both directions. This is a line.

7. The figure has two endpoints, C and D. This is a line segment.

9. The figure has two endpoints, A and B. This is a line segment.

11. The figure continues forever in both directions. This is a line.

< Objective 2 >

13. The lines are parallel, since these lines never intersect.

< Objective 3 >

15. The vertex of the angle is O, and the angle begins at P and ends at Q, so we would name the angle as $\angle POQ$ or $\angle QOP$.

17. The vertex of the angle is N, and the angle begins at M and ends at L, so we would name the angle as $\angle MNL$ or $\angle LNM$.

19. The vertex of the angle is E, and the angle begins at F and ends at G, so we would name the angle as $\angle FEG$ or $\angle GEF$.

21. The vertex of the angle is V, and the angle begins at S and ends at T, so we would name the angle $\angle SVT$ or $\angle TVS$.

< Objective 4 >

23. Using the protractor, we find $m\angle AOB = 135°$. $\angle AOB$ is an obtuse angle, since its measure is between 90° and 180°.

25. Using the protractor, we find $m\angle DOC = 90°$. $\angle DOC$ is a right angle, since its measure is 90°.

27. Using the protractor, we find $m\angle FOG = 30°$. $\angle FOG$ is an acute angle, since its measure is between 0° and 90°.

< Objective 5 >

29. Since $m\angle x + 29° = 90°$ (complementary angles), we must have $m\angle x = 90° - 29° = 61°$. The measure of $\angle x$ is 61°.

31. Since $m\angle y + 53° = 180°$ (supplementary angles), we must have $m\angle y = 180° - 53° = 127°$. The measure of $\angle y$ is 127°.

33. Since $m\angle x + 39° = 90°$ (complementary angles), we must have $m\angle x = 90° - 39° = 51°$. The measure of $\angle x$ is 51°.

35. Since the measure of the given angle is 102°, the measure of $\angle x = 102°$ (vertical angles). $m\angle y + m\angle x = 180°$ (supplementary angles) so $m\angle y = 180° - m\angle x = 180° - 102° = 78°$. $m\angle x = 102°$; $m\angle y = 78°$.

37. Since $m\angle w + 77° = 180°$ (supplementary angles), we must have $m\angle w = 180° - 77° = 103°$. The measure of $\angle w$ is 103°.

39. Since the measure of the given angle is 62°, $m\angle a = 62°$ (corresponding angles). $m\angle b = m\angle a$ (vertical angles), so $m\angle b = 62°$. Since $m\angle c + m\angle a = 180°$ (supplementary angles), we must have $m\angle c = 180° - 62° = 118°$.

41. Since the measure of the given angle is 132°, $m\angle c = 132°$ (vertical angles). The supplement of $132° = 180° - 132° = 48°$. $m\angle a = 48°$ (alternate interior angles). $m\angle b = m\angle a$ (vertical angles), so $m\angle b = 48°$.

43. $\dfrac{1}{6} \times 360 = 60$

The measure of $\angle A$ is 60°.

45. $\dfrac{7}{12} \times 360 = 210$

The measure of $\angle C$ is 210°.

< Objective 6 >

47. The triangle has three acute angles. It is an acute triangle.

49. The triangle has three acute angles. It is an acute triangle.

51. Begin by finding the missing angle. To find the missing angle, add the two given measurements and then subtract that from $180°$.

$$60° + 60° = 120°$$

$$180° - 120° = 60°$$

The triangle has three angles with the same measure. It is an equilateral triangle.

53. Begin by finding the missing angle. To find the missing angle, add the two given measurements and then subtract that from $180°$.

$$70° + 40° = 110°$$

$$180° - 110° = 70°$$

The triangle has two angles with the same measure. It is an isosceles triangle.

55. Begin by finding the missing angle. To find the missing angle, add the two given measurements and then subtract that from $180°$.

$$130° + 25° = 155°$$

$$180° - 155° = 25°$$

The triangle has two angles with the same measure. It is an isosceles triangle.

57. To find the missing angle, add the two given measurements and then subtract that from $180°$.

$$120° + 30° = 150°$$

$$180° - 150° = 30°$$

The missing angle is $30°$.
Since the triangle has two angles with the same measure, it is an isosceles triangle.

59. To find the missing angle, add the two given measurements and then subtract that from $180°$.

$$90° + 45° = 135°$$

$$180° - 135° = 45°$$

The missing angle is $45°$.
Since the triangle has two angles with the same measure, it is an isosceles triangle.

61. To find the missing angle, add the two given measurements and then subtract that from $180°$.

$$67° + 46° = 113°$$

$$180° - 113° = 67°$$

The missing angle is $67°$.
Since the triangle has two angles with the same measure, it is an isosceles triangle.

63. To find the indicated angle, add the two given measurements and then subtract that from $180°$.

$$82° + 61° = 143°$$

$$180° - 143° = 37°$$

$$m\angle C = 37°$$

65. To find the indicated angle, add the two given measurements then subtract that from $180°$.

$$39° + 18° = 57°$$

$$180° - 57° = 123°$$

$$m\angle A = 123°$$

67. To find the indicated angle, add the two given measurements and then subtract that from $180°$.

$$90° + 63° = 153°$$

$$180° - 153° = 27°$$

$$m\angle B = 27°$$

69. Since the triangle is isosceles, the two missing angles have equal measures. To find the measure of each of the missing angles, subtract the measure of the given angle from 180° and then divide that by two.
$$180° - 110° = 70°$$
$$70° \div 2 = 35°$$
$$m\angle A = 35°$$
$$m\angle C = 35°$$

71. False

73. True

75. False

77. **(a)** The needle travels 10° for each 1 volt increase $\left(\dfrac{100°}{10}\right)$. Therefore, the angular distance travelled by the needle in moving from 0 V to 5 V $= 5 \times 10° = 50°$.
(b) (Measured voltage) × 10 = 85°
Measured Voltage = 8.5 V

79. $180° - 25° = 155°$, which is the total of the two remaining equal angles. The angle at the base of the isosceles triangle $155° \div 2 = 77.5°$.

81. Begin by finding the missing angle of the triangle.
$$83° + 44° = 127°$$
$$180° - 127° = 53°$$
Subtract this angle from 180°, since there are 180° in a straight angle.
$$180° - 53° = 127°$$
The exterior angle measures 127°.

83. Above and Beyond

85. Above and Beyond

87. Above and Beyond

Exercises 8.4

< Objective 1 >

1. Yes; quadrilateral; not regular

3. No

5. Yes; rectangle; not regular

7. Yes; hexagon; regular

9. No

11. Yes; trapezoid; not regular

13. No

15. No

< Objective 2 >

17. $P = 5 \text{ ft} + 7 \text{ ft} + 6 \text{ ft} + 4 \text{ ft} = 22 \text{ ft}$

19. $P = 6 \text{ yd} + 8 \text{ yd} + 7 \text{ yd} = 21 \text{ yd}$

21. $P = 3 \text{ m} + 3 \text{ m} + 10 \text{ m} + 10 \text{ m} = 26 \text{ m}$

23. $P = 1 \text{ mm} + 0.7 \text{ mm} + 0.4 \text{ mm} = 2.1 \text{ mm}$

25. $P = \dfrac{3}{4} \text{in.} + \dfrac{3}{4} \text{in.} + \dfrac{3}{4} \text{in.} + \dfrac{3}{4} \text{in.} + \dfrac{3}{4} \text{in.} + \dfrac{3}{4} \text{in.}$
$= \dfrac{18}{4} \text{in.} = \dfrac{9}{2} \text{in.} = 4\dfrac{1}{2} \text{in.}$

27. $P = 16 \text{ km} + 13 \text{ km} + 26 \text{ km} + 13 \text{ km}$
$= 68 \text{ km}$

< Objective 3 >

29. $A = L \times W = 8 \text{ in.} \times 11 \text{ in.} = 88 \text{ in.}^2$

31. $A = L \times W = 12 \text{ ft} \times 7 \text{ ft} = 84 \text{ ft}^2$

33. $A = (s)^2 = (6 \text{ mm})^2 = 36 \text{ mm}^2$

35. $A = (s)^2 = (14 \text{ yd})^2 = 196 \text{ yd}^2$

37. $A = b \cdot h = 100 \text{ ft} \times 58 \text{ ft} = 5,800 \text{ ft}^2$

39. $A = b \cdot h = 15 \text{ m} \times 24 \text{ m} = 360 \text{ m}^2$

41. $A = \dfrac{1}{2} \cdot b \cdot h = \dfrac{1}{2} \times 14 \text{ yd} \times 14 \text{ yd} = 98 \text{ yd}^2$

43. $A = \dfrac{1}{2} \cdot b \cdot h = \dfrac{1}{2} \times 21 \text{ cm} \times 7 \text{ cm} = \dfrac{147}{2} \text{ cm}^2$
$= 73.5 \text{ cm}^2$

45. $A = \dfrac{1}{2} h (b_1 + b_2)$
$= \dfrac{1}{2} \times 70 \text{ ft} \times (75 \text{ ft} + 140 \text{ ft})$
$= \dfrac{1}{2} \times 70 \text{ ft} \times (215 \text{ ft}) = 7,525 \text{ ft}^2$

47. $A = \dfrac{1}{2} h (b_1 + b_2)$
$= \dfrac{1}{2} \times 8 \text{ mm} \times (6.5 \text{ mm} + 9.5 \text{ mm})$
$= \dfrac{1}{2} \times 8 \text{ mm} \times (16 \text{ mm}) = 64 \text{ mm}^2$

< Objective 4 >

49. $15 \text{ ft}^2 = 15 \cancel{\text{ft}}^2 \left(\dfrac{144 \text{ in.}^2}{1 \cancel{\text{ft}}^2} \right) = 2,160 \text{ in.}^2$

51. $360 \text{ ft}^2 = 360 \cancel{\text{ft}}^2 \left(\dfrac{1 \text{ yd}^2}{9 \cancel{\text{ft}}^2} \right) = 40 \text{ yd}^2$

53. $8 \text{ km}^2 = 8 \cancel{\text{km}}^2 \left(\dfrac{1,000,000 \text{ m}^2}{1 \cancel{\text{km}}^2} \right)$
$= 8,000,000 \text{ m}^2$

55. $30 \text{ m}^2 = 30 \cancel{\text{m}}^2 \left(\dfrac{10,000 \text{ cm}^2}{1 \cancel{\text{m}}^2} \right)$
$= 300,000 \text{ cm}^2$

57. $16 \text{ in.}^2 = 16 \cancel{\text{in.}}^2 \left(\dfrac{6.45 \text{ cm}^2}{1 \cancel{\text{in.}}^2} \right) = 103.2 \text{ cm}^2$

59. $18,000 \text{ acres} = 18,000 \cancel{\text{acres}} \left(\dfrac{1 \text{ mi}^2}{640 \cancel{\text{acres}}} \right)$
$= 28.1 \text{ mi}^2$

61. $12,000,000 \text{ m}^2$
$= 12,000,000 \cancel{\text{m}}^2 \left(\dfrac{1 \text{ ha}}{10,000 \cancel{\text{m}}^2} \right) = 1,200 \text{ ha}$

63. $2,500 \text{ ha}$
$= 2,500 \cancel{\text{ha}} \left(\dfrac{10,000 \cancel{\text{m}}^2}{1 \cancel{\text{ha}}} \right) \left(\dfrac{1 \text{ km}^2}{1,000,000 \cancel{\text{m}}^2} \right)$
$= 25 \text{ km}^2$
$25 \text{ km}^2 = 25 \cancel{\text{km}}^2 \left(\dfrac{0.38 \cancel{\text{mi}}^2}{1 \cancel{\text{km}}^2} \right) \left(\dfrac{640 \text{ acres}}{1 \cancel{\text{mi}}^2} \right)$
$= 6,080 \text{ acres}$

65. $A = \dfrac{1}{2} \cdot b \cdot h = \dfrac{1}{2} \times 12 \text{ in.} \times 12 \text{ in.} = 72 \text{ in.}^2$
$12 \times 72 \text{ in.}^2 = 864 \text{ in.}^2$
864 in.^2 of material is needed.

67. $A = \dfrac{1}{2} \cdot b \cdot h = 0.5 \times b \times h$
$A = 0.5 \times 1.6 \text{ in.} \times 0.85 \text{ in.} = 0.68 \text{ in.}^2$
The area of the hole to be punched is 0.68 in.^2

69. **(a)** $A = L \times W = 180 \text{ ft} \times 265 \text{ ft}$
$= 47,700 \text{ ft}^2$
$47,700 \text{ ft}^2 = 47,700 \cancel{\text{ft}}^2 \left(\dfrac{1 \text{ acre}}{43,560 \cancel{\text{ft}}^2} \right)$
$= 1.1 \text{ acres}$
1.1 acres make up the garden.
(b) $P = 2L + 2W = 2(180 \text{ ft}) + 2(265 \text{ ft})$
$= 360 \text{ ft} + 530 \text{ ft} = 890 \text{ ft}$
It takes 890 ft of fencing to enclose the garden.

71. (a) $A = L \times W = 160 \text{ yd} \times 160 \text{ yd}$
$= 25{,}600 \text{ yd}^2$

$25{,}600 \text{ yd}^2 = 25{,}600 \text{ yd}^2 \left(\dfrac{1 \text{ acre}}{4{,}840 \text{ yd}^2} \right)$

$= 5.3 \text{ acres}$

(rounded to the nearest tenth acre)
The area of the square lot is 5.3 acres.

(b)
5.3 acres

$= 5.3 \text{ acres} \left(\dfrac{1 \text{ mi}^2}{640 \text{ acres}} \right) \left(\dfrac{2.59 \text{ km}^2}{1 \text{ mi}^2} \right)$

$= 0.021 \text{ km}^2$

$= 0.021 \text{ km}^2 \left(\dfrac{1{,}000{,}000 \text{ m}^2}{1 \text{ km}^2} \right) \left(\dfrac{1 \text{ ha}}{10{,}000 \text{ m}^2} \right)$

$= 2.1 \text{ ha}$
The area of the square lot is 2.1 ha.

73. False

75. True

77. always

79. (a) $A = L \times W = 2.3 \text{ in.} \times 4.8 \text{ in.} = 11 \text{ in.}^2$
(rounded to the nearest square inch)
The area to be covered is 11 in.^{2}

(b) $500 \text{ ft}^2 \left(\dfrac{144 \text{ in.}^2}{1 \text{ ft}^2} \right) = 72{,}000 \text{ in.}^2$

No. of parts $= 72{,}000 \text{ in.}^2 \left(\dfrac{1 \text{ part}}{11 \text{ in.}^2} \right)$

$= 6{,}545.45$

Therefore, one can of sealer covers 6,545 whole parts.

81. $A = b \times h = 4.75 \text{ in.} \times 3 \text{ in.} = 14.25 \text{ in.}^2$
The area of each piece of strapping material is 14.25 in.^{2}

83. $A = L \times W = 4 \text{ cm} \times 7 \text{ cm} = 28 \text{ cm}^2$
The area of the rectangular tire patch is 28 cm^2.

85. $A = L \times W = 4 \text{ in.} \times 14 \text{ in.} = 56 \text{ in.}^2$
$56 \times 0.048 = 2.688 \text{ lb}$
Therefore, the weight of the plate is 2.688 lb.

87. Above and beyond

89. Above and beyond

Exercises 8.5

< Objective 1 >

1. $C = 2\pi r = 2 \times 3.14 \times 9 \text{ ft} \approx 56.5 \text{ ft}$
(rounded to one decimal place)

3. $C = \pi d \approx 3.14 \times 8.5 \text{ in.} \approx 26.7 \text{ in.}$
(rounded to one decimal place)

5. $C = \pi d = 3.14 \times 17.5 \text{ cm} \approx 55 \text{ cm}$

7. $C = 2\pi r = 2 \times 3.14 \times \dfrac{3}{4} \text{ cm} \approx 4.7 \text{ cm}$

(rounded to one decimal place)

< Objective 2 >

9. $A = \pi r^2 = 3.14 \times (7 \text{ in.})^2 \approx 153.9 \text{ in.}^2$
(rounded to one decimal place)

11. $A = \pi r^2 = 3.14 \times \left(\dfrac{7}{2} \text{ yd} \right)^2 \approx 38.5 \text{ yd}^2$
(rounded to one decimal place)

13. $A = \pi r^2 = 3.14 \times \left(\dfrac{3.5}{2} \text{ m} \right)^2 = 3.14 \times (1.75)^2$

$= 3.14 \times 3.0625 \approx 9.6 \text{ m}^2$
(rounded to one decimal place)

15. $A = \pi r^2 = 3.14 \times \left(\dfrac{5.5}{2} \text{ ft} \right)^2 = 3.14 \times (2.75)^2$

$= 3.14 \times 7.5625 \approx 23.7 \text{ ft}^2$
(rounded to one decimal place)

17. Circumference of the circular lake
$= 3.14 \times 1,000$ yd
Robert jogs around the lake 3 times, so
$3 \times 3,140$ yd $= 9,420$ yd .
Therefore, he has run about 9,420 yd.

19. The area of the circular lawn is
$A = 3.14 \times (28 \text{ ft})^2 \approx 2,462 \text{ ft}^2$.
Yes, you do have enough fertilizer to cover the lawn.

21. The area of the circular terrace is
$A = 3.14 \times (6 \text{ ft})^2 = 113.04 \text{ ft}^2$.
$\$4.50 \times 113.04 = \$508.68$
It will cost about $508.68 to cover the terrace.

< Objective 3 >

23. $P = 5 + 2 + 2 + 6 + 3 + 6 + 2 = 26$ m

25. $P = 1.7 + 1.0 + 1.0 + 0.7 + 0.4 + 0.4 + 0.4$
$\qquad + 0.6$
$= 6.2$ m

27. $P = \dfrac{3}{2} + \dfrac{3}{4} + 1 + 1 + \dfrac{5}{2} + \dfrac{7}{4} = \dfrac{17}{2} = 8\dfrac{1}{2}$ yd

29. Circumference (upper part)
$\approx \dfrac{1}{2} \times 3.14 \times 9$ ft ≈ 14.1 ft
(rounded to one decimal place)
Perimeter (lower part) $= 7 + 9 + 7 = 23$ ft
The perimeter of the figure is
14.1 ft $+ 23$ ft $= 37.1$ ft .

31. Circumference (upper part)
$\approx \dfrac{1}{2} \times 2 \times 3.14 \times 4$ ft ≈ 12.6 ft
(rounded to one decimal place)
Perimeter (lower part) $= 7 + 8 + 7 = 22$ ft
The perimeter of the figure is
12.6 ft $+ 22$ ft $= 34.6$ ft .

< Objective 4 >

33.

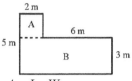

$A = L \times W$
$A_A = 2 \text{ m} \times 2 \text{ m} = 4 \text{ m}^2$
$A_B = 3 \text{ m} \times 8 \text{ m} = 24 \text{ m}^2$
$A_A + A_B = 4 \text{ m}^2 + 24 \text{ m}^2 = 28 \text{ m}^2$
Therefore, the area of the figure is 28 m^2.

35.

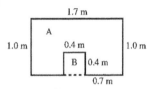

$A = L \times W$
$A_A = 1.7 \text{ m} \times 1.0 \text{ m} = 1.7 \text{ m}^2$
$A_B = 0.4 \text{ m} \times 0.4 \text{ m} = 0.16 \text{ m}^2$
$A_A - A_B = 1.7 \text{ m}^2 - 0.16 \text{ m}^2 = 1.54 \text{ m}^2$
Therefore, the area of the figure is 1.54 m^2.

37.

$A = L \times W$

Area of the larger rectangle is

$$A_A = \left(1\frac{3}{4} \times 2\frac{1}{2}\right) yd^2$$

Area of the smaller rectangle is

$$A_B = \left(\frac{3}{4} \times 1\right) yd^2$$

$$A_A - A_B = \left(1\frac{3}{4} \times 2\frac{1}{2}\right) - \left(\frac{3}{4} \times 1\right)$$

$$= \left(\frac{7}{4} \times \frac{5}{2}\right) - \frac{3}{4} = \frac{35}{8} - \frac{3}{4}$$

$$= \frac{29}{8} = 3\frac{5}{8}$$

Therefore, the area of the figure is $3\frac{5}{8}$ yd^2.

39. Area of the outer circle $= 3.14 \times (5 \text{ ft})^2$.

$$= 78.5 \text{ ft}^2$$

Area of the inner circle $= 3.14 \times (3 \text{ ft})^2$.

$$= 28.3 \text{ ft}^2$$

(rounded to one decimal place)
The area of the shaded part is
$78.5 \text{ ft}^2 - 28.3 \text{ ft}^2 = 50.2 \text{ ft}^2$.

41. Area of the square $= 20 \text{ ft} \times 20 \text{ ft} = 400 \text{ ft}^2$

Area of the circle $= 3.14 \times (10 \text{ ft})^2 = 314 \text{ ft}^2$

The area of the shaded part is
$400 \text{ ft}^2 - 314 \text{ ft}^2 = 86 \text{ ft}^2$.

43. Circumference of the circle, $C = 2\pi r$
Entering into the calculator:

$2 \boxed{\times} \boxed{2\text{nd}} \boxed{[\pi]} \boxed{\times} 5 \boxed{=}$

We get 31.416.
The circumference of the circle is
31.416 in.
Area of the circle $A = \pi r^2$

$\boxed{2\text{nd}} \boxed{[\pi]} \boxed{\times} 5 \boxed{x^2} \boxed{=}$

We get 78.54.
The area of the circle is 78.54 in.^{2}

45. Circumference of the circle, $C = 2\pi r$
Entering into the calculator:

$2 \boxed{\times} \boxed{2\text{nd}} \boxed{[\pi]} \boxed{\times} 6.5 \boxed{=}$

We get 40.841.
The circumference of the circle is
40.841 in.
Area of the circle, $A = \pi r^2$

$\boxed{2\text{nd}} \boxed{[\pi]} \boxed{\times} 6.5 \boxed{x^2} \boxed{=}$

We get 132.732.
(rounded to the nearest thousandth)
The area of the circle is 132.732 in.^{2}

47. Circumference of the circle, $C = \pi d$
Entering into the calculator:

$\boxed{2\text{nd}} \boxed{[\pi]} \boxed{\times} 11 \boxed{=}$

We get 34.558.
The circumference of the circle is 34.558 ft.

Radius $(r) = \dfrac{11}{2} = 5.5$

Area of the circle, $A = \pi r^2$

$\boxed{2\text{nd}} \boxed{[\pi]} \boxed{\times} 5.5 \boxed{x^2} \boxed{=}$

We get 95.033.
(rounded to the nearest thousandth)
The area of the circle is 95.033 ft.^{2}

49. $C = \pi d = 3.14 \times 2.5 \approx 7.9$ cm
The head circumference of the fetus is
7.9 cm.

51. First, find the radius.

Radius $= \dfrac{d}{2} = \dfrac{1.7}{2} = 0.85$ cm

$A = \pi r^2 = 3.14 \times (0.85)^2 = 3.14 \times (0.7225)$

$\approx 2.3 \text{cm}^2$

The area of the circular fluid pocket is 2.3 cm^2.

53. For the perimeter, you first need to find the circumference.

$C = 2\pi r = 2 \times 3.14 \times 6 \approx 37.68$ mm

Next, divide the circumference by 4, since the shape is a quarter of a circle.

$37.68 \div 4 = 9.42$ mm

Finally, add the two radii and the quarter circle side.

$6 \text{ mm} + 6 \text{ mm} + 9.42 \text{ mm} = 21.42$ mm

Next, find the area.

$A = \pi r^2 = 3.14 \times (6)^2 = 3.14 \times 36$

$= 113.04 \text{ mm}^2$

Finally, divide the area by 4, since the shape is a quarter of a circle.

$113.04 \div 4 = 28.26 \text{ mm}^2$

55. Above and Beyond

57. Above and Beyond

Exercises 8.6

< Objectives 1–3 >

1. $L = 8$ in., $H = 6$ in., $W = 5$ in.

(a) LSA $= 2LH + 2WH$

$= 2(8 \text{ in.})(6 \text{ in.}) + 2(5 \text{ in.})(6 \text{ in.})$

$= 96 \text{ in.}^2 + 60 \text{ in.}^2 = 156 \text{ in.}^2$

(b) TSA $= 2(LW + LH + WH)$

$= 2\left(\begin{array}{l} 8 \text{ in.} \times 6 \text{ in.} + 8 \text{ in.} \\ \times 5 \text{ in.} + 6 \text{ in.} \times 5 \text{ in.} \end{array} \right)$

$= 2(48 \text{ in.}^2 + 40 \text{ in.}^2 + 30 \text{ in.}^2)$

$= 236 \text{ in.}^2$

(c) $V = LWH = 8 \text{ in.} \times 6 \text{ in.} \times 5 \text{ in.}$

$= 240 \text{ in.}^3$

3. $L = 3$ cm, $W = 12$ cm, $H = 6$ cm

(a) LSA $= 2LH + 2WH$

$= 2(3 \text{ cm})(6 \text{ cm}) + 2(12 \text{ cm})(6 \text{ cm})$

$= 36 \text{ cm}^2 + 144 \text{ cm}^2$

$= 180 \text{ cm}^2$

(b) TSA $= 2(LW + LH + WH)$

$= 2\left(\begin{array}{l} 3 \text{ cm} \times 12 \text{ cm} + 3 \text{ cm} \\ \times 6 \text{ cm} + 12 \text{ cm} \times 6 \text{ cm} \end{array} \right)$

$= 2(36 \text{ cm}^2 + 18 \text{ cm}^2 + 72 \text{ cm}^2)$

$= 252 \text{ cm}^2$

(c) $V = LWH = 3 \text{ cm} \times 12 \text{ cm} \times 6 \text{ cm}$

$= 216 \text{ cm}^3$

5. $L = 30 \text{ yd}$, $W = 20 \text{ yd}$, $H = 10 \text{ yd}$ yd

   (a) $\text{LSA} = 2LH + 2WH$

$$= 2(30 \text{ yd})(10 \text{ yd}) + 2(20 \text{ yd})(10 \text{ yd})$$

$$= 600 \text{ yd}^2 + 400 \text{ yd}^2 = 1,000 \text{ yd}^2$$

   (b) $\text{TSA} = 2(LW + LH + WH)$

$$= 2\begin{pmatrix} 30 \text{ yd} \times 20 \text{ yd} + 30 \text{ yd} \\ \times 10 \text{ yd} + 20 \text{ yd} \times 10 \text{ yd} \end{pmatrix}$$

$$= 2(600 \text{ yd}^2 + 300 \text{ yd}^2 + 200 \text{ yd}^2)$$

$$= 2,200 \text{ yd}^2$$

   (c) $V = LWH = 30 \text{ yd} \times 20 \text{ yd} \times 10 \text{ yd}$

$$= 6,000 \text{ yd}^3$$

7. $L = 12.5 \text{ m}$, $H = 7.8 \text{ m}$, $W = 3.4 \text{ m}$

   (a) $\text{LSA} = 2LH + 2WH$

$$= 2(12.5 \text{ m})(7.8 \text{ m})$$
$$+ 2(7.8 \text{ m})(3.4 \text{ m})$$

$$= 195 \text{ m}^2 + 53.04 \text{ m}^2$$

$$= 248.04 \text{ m}^2$$
(rounded to two decimal places)

   (b) $\text{TSA} = 2(LW + LH + WH)$

$$= 2\begin{pmatrix} 12.5 \text{ m} \times 7.8 \text{ m} + 12.5 \text{ m} \\ \times 3.4 \text{ m} + 7.8 \text{ m} \times 3.4 \text{ m} \end{pmatrix}$$

$$= 2(97.5 \text{ m}^2 + 42.5 \text{ m}^2 + 26.52 \text{ m}^2)$$

$$= 333.04 \text{ m}^2$$
(rounded to two decimal places)

   (c) $V = LWH = 12.5 \text{ m} \times 7.8 \text{ m} \times 3.4 \text{ m}$

$$= 331.5 \text{ m}^3$$
(rounded to two decimal places)

9. $L = \dfrac{3}{4} \text{ in}$, $W = \dfrac{3}{4} \text{ in.}$, $H = \dfrac{1}{2} \text{ in.}$

   (a)

$\text{LSA} = 2LH + 2WH$

$$= 2\left(\frac{3}{4} \text{ in.}\right)\left(\frac{1}{2} \text{ in.}\right) + 2\left(\frac{3}{4} \text{ in.}\right)\left(\frac{1}{2} \text{ in.}\right)$$

$$= \frac{3}{4} \text{ in.}^2 + \frac{3}{4} \text{ in.}^2$$

$$= \frac{3}{2} \text{ in.}^2 = 1\frac{1}{2} \text{ in.}^2$$

   (b) $\text{TSA} = 2(LW + LH + WH)$

$$= 2\begin{pmatrix} \dfrac{3}{4}\text{in.} \times \dfrac{3}{4}\text{in.} + \dfrac{3}{4}\text{in.} \\ \times \dfrac{1}{2}\text{in.} + \dfrac{3}{4}\text{in.} \times \dfrac{1}{2}\text{in.} \end{pmatrix}$$

$$= 2\left(\frac{9}{16}\text{in.}^2 + \frac{3}{8}\text{in.}^2 + \frac{3}{8}\text{in.}^2\right) = 2\frac{5}{8}\text{in.}^2$$

   (c) $V = LWH = \dfrac{3}{4}\text{in.} \times \dfrac{3}{4}\text{in.} \times \dfrac{1}{2}\text{in.}$

$$= \frac{9}{32}\text{in.}^3$$

11. Side (s) of the cube = 8 mm

   (a) $\text{LSA} = 4s^2 = 4(8 \text{ mm})^2 = 256 \text{ mm}^2$

   (b) $\text{TSA} = 6s^2 = 6(8 \text{ mm})^2 = 384 \text{ mm}^2$

   (c) $V = s^3 = (8 \text{ mm})^3 = 512 \text{ mm}^3$

13. Side (s) of the cube $= \dfrac{1}{2}$ in.

   (a) $\text{LSA} = 4s^2 = 4\left(\dfrac{1}{2} \text{ in.}\right)^2 = 1 \text{ in.}^2$

   (b) $\text{TSA} = 6s^2 = 6\left(\dfrac{1}{2} \text{ in.}\right)^2 = \dfrac{3}{2} \text{ in.}^2$

$$= 1\frac{1}{2} \text{ in.}^2$$

   (c) $V = s^3 = \left(\dfrac{1}{2} \text{ in.}\right)^3 = \dfrac{1}{8} \text{ in.}^3$

15. $r = 3$ in., $h = 10$ in. (cylinder)

   (a) LSA $= 2\pi rh = 2 \times 3.14 \times 3$ in. $\times 10$ in.

$$= 188.4 \text{ in.}^2$$

   (b) TSA $= 2\pi rh \;+\; 2\pi r^2$

$$= 2 \times 3.14 \times 3 \text{ in.} \times 10 \text{ in.}$$
$$+ 2 \times 3.14 \times (3 \text{ in.})^2$$
$$= 188.4 \text{ in.}^2 + 56.52 \text{ in.}^2$$
$$= 244.92 \text{ in.}^2$$

   (c) $V = \pi r^2 h = 3.14 \times (3 \text{ in.})^2 \times 10 \text{ in.}$

$$= 282.6 \text{ in.}^3$$

17. $r = 16$ m , $r = 35$ m (cylinder)

   (a) LSA $= 2\pi rh = 2 \times 3.14 \times 16$ m $\times 35$ m

$$= 3{,}516.8 \text{ m}^2$$

   (b) TSA $= 2\pi rh + 2\pi r^2$

$$= 2 \times 3.14 \times 16 \text{ m} \times 35 \text{ m} + 2 \times 3.14$$
$$\times (16 \text{ m})^2$$
$$= 3{,}516.8 \text{ m}^2 + 1{,}607.68 \text{ m}^2$$
$$= 5{,}124.48 \text{ m}^2$$

   (c) $V = \pi r^2 h = 3.14 \times (16 \text{ m})^2 \times 35 \; m$

$$= 28{,}134.4 \text{ m}^3$$

19. $r = \dfrac{d}{2} = \dfrac{8 \text{ cm}}{2} = 4$ cm , $h = 12$ cm (cylinder)

   (a) LSA $= 2\pi rh = 2 \times 3.14 \times 4$ cm $\times 12$ cm

$$= 301.44 \text{ cm}^2$$

   (b) TSA $= 2\pi rh + 2\pi r^2$

$$= 2 \times 3.14 \times 4 \text{ cm} \times 12 \text{ cm} + 2$$
$$\times 3.14 \times (4 \text{ cm})^2$$
$$= 301.44 \text{ cm}^2 + 100.48 \text{ cm}^2$$
$$= 401.92 \text{ in.}^2$$

   (c) $V = \pi r^2 h = 3.14 \times (4 \text{ cm})^2 \times 12$ cm

$$= 602.88 \text{ cm}^3$$

< Objective 4 >

21. $r = 15$ in. (sphere)

   (a) TSA $= 4\pi r^2 = 4 \times 3.14 \times (15 \text{ in.})^2$

$$= 2{,}826 \text{ in.}^2$$

   (b) $V = \dfrac{4}{3}\pi r^3 = \dfrac{4}{3} \times 3.14 \times (15 \text{ in.})^3$

$$= 14{,}130 \text{ in.}^3$$

23. $r = 2$ cm (sphere)

   (a) TSA $= 4\pi r^2 = 4 \times 3.14 \times (2 \text{ cm})^2$

$$= 50.24 \text{ cm}^2$$

   (b) $V = \dfrac{4}{3}\pi r^3 = \dfrac{4}{3} \times 3.14 \times (2 \text{ cm})^3$

$$= 33.49 \text{ cm}^3$$

25. $r = 20$ yd (sphere)

   (a) TSA $= 4\pi r^2 = 4 \times 3.14 \times (20 \text{ yd})^2$

$$= 5{,}024 \text{ yd}^2$$

   (b) $V = \dfrac{4}{3}\pi r^3 = \dfrac{4}{3} \times 3.14 \times (20 \text{ yd})^3$

$$= 33{,}493.33 \text{ yd}^3$$

27. The figure is composed of 2 half spheres and 1 cylinder.

$$r = \frac{d}{2} = \frac{37 \text{ in.}}{2} = 18.5 \text{ in.}, \quad h = 83 \text{ in.}$$

(a) Surface area for the figure would be

$$TSA = LSA_{cyl} + TSA_{sph}$$

$$LSA_{cyl} = 2\pi rh = 2 \times 3.14 \times 18.5 \text{ in.} \times 83 \text{ in.}$$

$$= 9,642.94 \text{ in.}^2$$

$$TSA_{sph} = 4\pi r^2 = 4 \times 3.14 \times (18.5 \text{ in.})^2$$

$$= 4,298.66 \text{ in.}^2$$

$$TSA = 9,642.94 \text{ in.}^2 + 4298.66 \text{ in.}^2$$

$$= 13,941.6 \text{ in.}^2$$

(b) Volume of the figure would be

$$V = V_{cyl} + V_{sph}$$

$$V_{cyl} = \pi r^2 h = 3.14 \times (18.5 \text{ in.})^2 \times 83 \text{ in.}$$

$$= 89,197.20 \text{ in.}^3$$

$$V_{sph} = \frac{4}{3}\pi r^3 = \frac{4}{3} \times 3.14 \times (18.5 \text{ in.})^3$$

$$= 26,508.40 \text{ in.}^3$$

$$V = 89,197.20 \text{ in.}^3 + 26,508.40 \text{ in.}^3$$

$$= 115,705.6 \text{ in.}^3$$

< Objective 5 >

29. $25 \text{ ft}^3 = 25 \text{ ft}^3 \left(\dfrac{1,728 \text{ in.}^3}{1 \text{ ft}^3} \right) = 43,200 \text{ in.}^3$

31. $16 \text{ yd}^3 = 16 \text{ yd}^3 \left(\dfrac{27 \text{ ft}^3}{1 \text{ yd}^3} \right) = 432 \text{ ft}^3$

33. $6 \text{ yd}^3 = 6 \text{ yd}^3 \left(\dfrac{27 \text{ ft}^3}{1 \text{ yd}^3} \right) \left(\dfrac{1,728 \text{ in.}^3}{1 \text{ ft}^3} \right)$

$$= 279,936 \text{ in.}^3$$

35. $25 \text{ cm}^3 = 25 \text{ cm}^3 \left(\dfrac{1,000 \text{ mm}^3}{1 \text{ cm}^3} \right)$

$$= 25,000 \text{ mm}^3$$

37. $12 \text{ m}^3 = 12 \text{ m}^3 \left(\dfrac{1,000,000 \text{ cm}^3}{1 \text{ m}^3} \right)$

$$= 12,000,000 \text{ cm}^3$$

39. $18 \text{ in.}^3 = 18 \text{ in.}^3 \left(\dfrac{16.387 \text{ cm}^3}{1 \text{ in.}^3} \right)$

$$= 294.97 \text{ cm}^3$$

(rounded to two decimal places)

41. $42 \text{ m}^3 = 42 \text{ m}^3 \left(\dfrac{34.966 \text{ ft}^3}{1 \text{ m}^3} \right) = 1,468.57 \text{ ft}^3$

(rounded to two decimal places)

43. $10 \text{ m}^3 = 10 \text{ m}^3 \left(\dfrac{34.966 \text{ ft}^3}{1 \text{ m}^3} \right) \left(\dfrac{1728 \text{ in.}^3}{1 \text{ ft}^3} \right)$

$$= 604,212.48 \text{ in.}^3$$

(rounded to two decimal places)

45. Dimensions of room, $L = 12 \text{ ft}$, $W = 16 \text{ ft}$, $H = 8 \text{ ft}$

(a) Volume of the room

$$V = LWH = 12 \text{ ft} \times 16 \text{ ft} \times 8 \text{ ft} = 1,536 \text{ ft}^3$$

(b) For 1 cubic foot of air to be heated, 3 BTUs are needed. BTUs needed to heat the room is $1,536 \times 3 = 4,608$ BTUs

47. $r = \dfrac{d}{2} = \dfrac{20 \text{ ft}}{2} = 10 \text{ ft}$

(Since a silo is composed of cylindrical part and a hemispherical part)

Height of cylinder $(h) = 60 \text{ ft} - 10 \text{ ft} = 50 \text{ ft}$

(a) Volume of silo

$$V = V_{\text{cyl}} + V_{\text{hemisphere}}$$

$$V_{\text{cyl}} = \pi r^2 h = 3.14 \times (10 \text{ ft})^2 \times 50 \text{ ft}$$

$$= 15,700 \text{ ft}^3$$

$$V_{\text{hemisphere}} = \dfrac{\dfrac{4}{3}\pi r^3}{2} = \dfrac{\dfrac{4}{3} \times 3.14 \times (10 \text{ ft})^3}{2}$$

$$= 2,093.33 \text{ ft}^3$$

$$V = 15,700 \text{ ft}^3 + 2,093.33 \text{ ft}^3 \approx 17,933 \text{ ft}^3$$

Silo will hold $17,933 \text{ ft}^3$ of silage.

(b) Surface area of the silo

$$\text{TSA} = \text{LSA}_{\text{cyl}} + \text{TSA}_{\text{hemi}}$$

$$\text{LSA}_{\text{cyl}} = 2\pi rh = 2 \times 3.14 \times 10 \text{ ft} \times 50 \text{ ft}$$

$$= 3,140 \text{ ft}^2$$

$$\text{TSA}_{\text{hemi}} = \dfrac{4\pi r^2}{2} = \dfrac{4 \times 3.14 \times (10 \text{ ft})^2}{2}$$

$$= 628 \text{ ft}^2$$

$$\text{TSA} = 3,140 \text{ ft}^2 + 628 \text{ ft}^2 = 3,768 \text{ ft}^2$$

49. False

51. sometimes

53. $r = 7$ in. (sphere)

(a) $\text{TSA} = 4\pi r^2$

Entering into the calculator:

$\boxed{4}\,\boxed{\times}\,\boxed{\text{2nd}}\,\boxed{[\pi]}\,\boxed{\times}\,\boxed{7}\,\boxed{x^2}\,\boxed{=}$

We get 615.752.

Total surface area $= 615.752$ in.^{2} (rounded to three decimal places)

(b) $V = \dfrac{4}{3}\pi r^3$

Entering into the calculator:

$\boxed{4}\,\boxed{\times}\,\boxed{\text{2nd}}\,\boxed{[\pi]}\,\boxed{\times}\,\boxed{7}\,\boxed{x^3}\,\boxed{\div}\,\boxed{3}\,\boxed{=}$

We get 1436.755.

Volume $= 1,436.755$ in.^{3} (rounded to three decimal places)

55. $r = 18$ cm (sphere)

(a) $\text{TSA} = 4\pi r^2$

Entering into the calculator:

$\boxed{4}\,\boxed{\times}\,\boxed{\text{2nd}}\,\boxed{[\pi]}\,\boxed{\times}\,\boxed{18}\,\boxed{x^2}\,\boxed{=}$

We get 4071.504.

Total surface area $= 4,071.504$ cm^2 (rounded to three decimal places)

(b) $V = \dfrac{4}{3}\pi^3$

Entering into the calculator:

$\boxed{4}\,\boxed{\times}\,\boxed{\text{2nd}}\,\boxed{[\pi]}\,\boxed{\times}\,\boxed{18}\,\boxed{x^3}\,\boxed{\div}\,\boxed{3}\,\boxed{=}$

We get 24429.024.

Volume $= 24,429.024$ cm^3 (rounded to three decimal places)

57. $r = 3$ cm, $h = 8$ cm (cylinder)

(a) $\text{TSA} = 2\pi rh + 2\pi r^2$

Entering into the calculator:

$\boxed{2}\,\boxed{\times}\,\boxed{\text{2nd}}\,\boxed{[\pi]}\,\boxed{\times}\,\boxed{3}\,\boxed{\times}\,\boxed{8}\,\boxed{=}$

We get 150.796.

$\boxed{2}\,\boxed{\times}\,\boxed{\text{2nd}}\,\boxed{[\pi]}\,\boxed{\times}\,\boxed{3}\,\boxed{x^2}\,\boxed{=}$

We get 56.549.

Total surface area

$150.796 + 56.549 = 207.345$ cm^2 (rounded to three decimal places)

(b) $V = \pi r^2 h$

Entering into the calculator:

$\boxed{\text{2nd}}\,\boxed{[\pi]}\,\boxed{\times}\,\boxed{3}\,\boxed{x^2}\,\boxed{\times}\,\boxed{8}\,\boxed{=}$

We get 226.195.

Volume $= 226.195$ cm^3 (rounded to three decimal places)

59. For the steel cylinder

$$r = \dfrac{d}{2} = \dfrac{\dfrac{1}{8}}{2} \text{ in.} = 0.0625 \text{ in.}$$

Height, $h = 1\dfrac{1}{2}$ in. $= 2.5$ in.

Volume of the steel stock

$$V = \pi r^2 h = 3.14 \times (0.0625 \text{ in.})^2 \times 2.5 \text{ in.}$$

$$= 0.03 \text{ in.}^3$$

(rounded to two decimal places)

The volume of the steel stock is 0.03 in.^{3}

61. $L = 5$ in., $W = 12$ in., $H = 3$ in.
Volume
$V = LWH = 5$ in.$\times 12$ in.$\times 3$ in.$= 180$ in.^{3}
Weight of the metal is
180 in.$^3 \times 0.092 \, \dfrac{\text{lb}}{\text{in.}^3} = 16.56$ lb

63. $L = 12$ in., $W = 30$ in., $H = 18$ in.
Volume
$V = LWH = 12$ in.$\times 30$ in.$\times 18$ in.
$\quad = 6,480$ in.^{3}
Now
$\qquad 1 \text{ gal} = 231 \text{ in.}^3$

$\qquad 1 \text{ in.}^3 = \dfrac{1}{231} \text{ gal}$

$6,480 \text{ in.}^3 = 6,480 \text{ in.}^3 \left(\dfrac{1 \text{ gal}}{231 \text{ in.}^3} \right) = 28 \text{ gal}$

The tank holds 28 gal.

Summary Exercises

1. $11 \text{ ft} = 11 \text{ ft} \left(\dfrac{12 \text{ in.}}{1 \text{ ft}} \right) = 132 \text{ in.}$

3. $6 \text{ gal} = 6 \text{ gal} \left(\dfrac{4 \text{ qt}}{1 \text{ gal}} \right) = 24 \text{ qt}$

5. $4 \text{ lb} = 4 \text{ lb} \left(\dfrac{16 \text{ oz}}{1 \text{ lb}} \right) = 64 \text{ oz}$

7. $8,000 \text{ lb} = 8,000 \text{ lb} \left(\dfrac{1 \text{ ton}}{2,000 \text{ lb}} \right) = 4 \text{ tons}$

9. $3 \text{ ft } 23 \text{ in.} = 3 \text{ ft} + 1 \text{ ft} + 11 \text{ in.} = 4 \text{ ft } 11 \text{ in.}$

11. $\quad 3 \text{ lb} \quad 9 \text{ oz}$
$\underline{+ 5 \text{ lb } 10 \text{ oz}}$
$\quad 8 \text{ lb } 19 \text{ oz} = 9 \text{ lb } 3 \text{ oz}$

13. $\quad 7 \text{ ft } 11 \text{ in.}$
$\underline{- 2 \text{ ft} \quad 4 \text{ in.}}$
$\quad 5 \text{ ft} \quad 7 \text{ in.}$

15. $\quad 1 \text{ hr } 25 \text{ min}$
$\underline{\times \qquad\qquad 3}$
$\quad 3 \text{ hr } 75 \text{ min} = 4 \text{ hr } 15 \text{ min}$

17. $\dfrac{10 \text{ lb } 12 \text{ oz}}{2} = 5 \text{ lb } 6 \text{ oz}$

19. $\quad 6 \text{ hr } 15 \text{ min}$
$\quad 8 \text{ hr}$
$\quad 5 \text{ hr } 50 \text{ min}$
$\quad 7 \text{ hr } 30 \text{ min}$
$\underline{+ 6 \text{ hr} \qquad\qquad}$
$\quad 32 \text{ hr } 95 \text{ min} = 33 \text{ hr } 35 \text{ min}$
Total hours worked were 33 hr 35 min.

21. A marathon race: **(a)** 40 km.

23. The diameter of a penny: **(c)** 19 mm.

25. A small Post-it is 39 mm wide.

27. A 1-lb coffee can has a diameter of 10 cm.

29. $2 \text{ km} = 2 \text{ km} \left(\dfrac{1,000 \text{ m}}{1 \text{ km}} \right) = 2,000 \text{ m}$

31. $3,000 \text{ mm} = 3,000 \text{ mm} \left(\dfrac{1 \text{ m}}{1,000 \text{ mm}} \right) = 3 \text{ m}$

33. $6 \text{ cm} = 6 \text{ cm} \left(\dfrac{1 \text{ m}}{100 \text{ cm}} \right) = 0.06 \text{ m}$

35. $\quad 8.3 \text{ m} = 8.3 \text{ m} \left(\dfrac{100 \text{ cm}}{1 \text{ m}} \right) = 830 \text{ cm}$

$830 \text{ cm} = 830 \text{ cm} \left(\dfrac{1 \text{ in.}}{2.54 \text{ cm}} \right) = 326.77 \text{ in.}$

37. $65 \text{ cm} = 65 \text{ cm} \left(\dfrac{1 \text{ in.}}{2.54 \text{ cm}} \right) = 25.35 \text{ in.}$

39. $19 \text{ yd} = 19 \text{ yd} \left(\dfrac{0.91 \text{ m}}{1 \text{ yd}} \right) \left(\dfrac{100 \text{ cm}}{1 \text{ m}} \right)$
$\qquad = 1,729 \text{ cm}$

41. Evaluate $F = \dfrac{9C}{5} + 32$ with $C = 17$.

$$F = \frac{9C}{5} + 32 = \frac{9 \times 17}{5} + 32 = 30.6 + 32$$
$$= 62.6$$
$$17°C = 62.6°F$$

43. Evaluate $C = \dfrac{5(F-32)}{9}$ with $F = 98.6$.

$$C = \frac{5(98.6-32)}{9} = \frac{333}{9} = 37$$
$$98.6°F = 37°C$$

45. Evaluate $C = \dfrac{5(F-32)}{9}$ with $F = 59$.

$$C = \frac{5(59-32)}{9} = \frac{135}{9} = 15$$
$$59°F = 15°C$$

47. Evaluate $F = \dfrac{9C}{5} + 32$ with $C = 5$.

$$F = \frac{9C}{5} + 32 = \frac{9 \times 5}{5} + 32 = 9 + 32 = 41$$
$$5°C = 41°F$$

49. The vertex of the angle is O, and the angle begins at A and ends at B, so we would name the angle $\angle AOB$ or $\angle BOA$. $\angle BOA$ is an acute angle, since its measure is between $0°$ and $90°$. Using the protractor, we find $m\angle BOA = 70°$.

51. The vertex of the angle is O, and angle begins at A and ends at C, so we could name the angle $\angle AOC$ or $\angle COA$. $\angle AOC$ is an obtuse angle, since its measure is between $90°$ and $180°$. Using the protractor, we find $m\angle AOC = 100°$.

53. The vertex of the angle is Y, and angle begins at X and ends at Z, so we could name the angle $\angle XYZ$ or $\angle ZYX$. $\angle XYZ$ is a straight angle. The measure of $m\angle XYZ = 180°$

55. $\dfrac{3}{8} \times 360 = 135$

The measure of $\angle A$ is $135°$.

57. $m\angle x$ + its complement = $90°$, so the complement $= 90° - m\angle x = 90° - 43°$
$$= 47°.$$

59. $m\angle x + 17° = 90°$ (complementary angles), so $m\angle x = 90° - 17° = 73°$.

61. Since the measure of the given angle is $79°$, $m\angle x = 79°$ (vertical angles).

63. Since the measure of the given angle is $67°$, $m\angle x = 67°$ (alternate interior angles).

65. To find the missing angle, add the two given measurements and then subtract that from $180°$.
$$60° + 60° = 120°$$
$$180° - 120° = 60°$$
The missing angle is $60°$.
Since the triangle has three angles with the same measure, it is an equilateral triangle.

67. To find the missing angle, add the two given measurements then subtract that from $180°$.
$$90° + 45° = 135°$$
$$180° - 135° = 45°$$
The missing angle is $45°$.
Since the triangle has two angles with the same measure, it is an isosceles triangle.

69. $\text{Area}(A) = b \cdot h = 30 \text{ ft} \times 25 \text{ ft} = 750 \text{ ft}^2$

71. $P = 2L + 2W = 2(18 \text{ in.}) + 2(12 \text{ in.})$
$$= 36 + 24 = 60 \text{ in.}$$
$$A = l \times w = 18 \times 12 = 216 \text{ in.}^2$$

73. $P = 16 + 25 + 16 + 36 = 93 \text{ ft}$
$$A = \frac{1}{2}h(b_1 + b_2) = \frac{1}{2}(15)(25+36)$$
$$= \frac{1}{2}(15)(61) = 457.5 \text{ ft}^2$$

75. $A = 10 \text{ ft} \times 18 \text{ ft} = 180 \text{ ft}^2$
180 ft² of floor covering will be needed.

$$\overset{20}{\cancel{180 \text{ ft}^2}} \times \frac{1 \text{ yd}^2}{\underset{1}{\cancel{9 \text{ ft}^2}}} = 20 \text{ yd}^2$$

20 yd² of floor covering will be needed.

77. $C = \pi d \approx 3.14 \times 12 \text{ in.} = 37.7 \text{ in.}$
(rounded to the nearest tenth)
$A = \pi r^2 \approx 3.14(6)^2 \approx 3.14(36) \approx 113.0 \text{ in.}^2$
(rounded to the nearest tenth)

79. $P = 10 + 14 + 14 + 4 + 3 + 4 + 4 + 3 = 56 \text{ ft}$

$A = l \times w$
$A_A = 10 \times 14 = 140 \text{ ft}^2$
$A_B = 3 \times 4 = 12 \text{ ft}^2$
$A = A_A - A_B = 140 - 12 = 128 \text{ ft}^2$

81. $L = 2 \text{ in.}, \ W = 4 \text{ in.}, \ H = 3 \text{ in.}$
   (a) $\text{LSA} = 2LH + 2WH$
   $$= 2(2 \text{ in.})(3 \text{ in.}) + 2(4 \text{ in.})(3 \text{ in.})$$
   $$= 12 \text{ in.}^2 + 24 \text{ in.}^2 = 36 \text{ in.}^2$$
   (b) $\text{TSA} = 2(LW + LH + WH)$
   $$= 2\begin{pmatrix} 2 \text{ in.} \times 4 \text{ in.} + 2 \text{ in.} \\ \times 3 \text{ in.} + 4 \text{ in.} \times 3 \text{ in.} \end{pmatrix}$$
   $$= 2\left(8 \text{ in.}^2 + 6 \text{ in.}^2 + 12 \text{ in.}^2\right)$$
   $$= 52 \text{ in.}^2$$
   (c) $V = LWH = 2 \text{ in.} \times 4 \text{ in.} \times 3 \text{ in.}$
   $$= 24 \text{ in.}^3$$

83. $s = 8\dfrac{1}{2} \text{ in.} = \dfrac{17}{2} \text{ in.}$ (cube)

   (a) $\text{LSA} = 4s^2 = 4\left(\dfrac{17}{2} \text{ in.}\right)^2 = 289 \text{ in.}^2$

   (b) $\text{TSA} = 6s^2 = 6\left(\dfrac{17}{2} \text{ in.}\right)^2 = \dfrac{867}{2} \text{ in.}^2$
   $$= 433\dfrac{1}{2} \text{ in.}^2$$

   (c) $V = s^3 = \left(\dfrac{17}{2} \text{ in.}\right)^3 = \dfrac{4973}{8} \text{ in.}^3$
   $$= 614\dfrac{1}{8} \text{ in.}^3$$

85. $r = 11.5 \text{ cm}$ (sphere)
   (a) $\text{TSA} = 4\pi r^2 = 4 \times 3.14 \times (11.5 \text{ cm})^2$
   $$= 1,661.06 \text{ cm}^2$$
   (b) $V = \dfrac{4}{3}\pi r^3 = \dfrac{4}{3} \times 3.14 \times (11.5 \text{ cm})^3$
   $$= 6,367.4 \text{ cm}^3$$

87. $12 \text{ ft}^3 = 12 \cancel{\text{ft}^3}\left(\dfrac{1,728 \text{ in.}^3}{1 \cancel{\text{ft}^3}}\right) = 20,736 \text{ in.}^3$

89. $30 \text{ in.}^3 = 30 \cancel{\text{in.}^3}\left(\dfrac{16.387 \text{ cm}^3}{1 \cancel{\text{in.}^3}}\right) = 491.61 \text{ cm}^3$

Chapter Test 8

1. $8 \text{ ft} = 8 \cancel{\text{ft}}\left(\dfrac{12 \text{ in.}}{1 \cancel{\text{ft}}}\right) = 96 \text{ in.}$

3. $40 \text{ in} = 40 \cancel{\text{in.}}\left(\dfrac{2.54 \text{ cm}}{1 \cancel{\text{in.}}}\right) = 101.6 \text{ cm}$

5. $12 \text{ in} = 12 \cancel{\text{in.}}\left(\dfrac{2.54 \text{ cm}}{1 \cancel{\text{in.}}}\right) = 30.5 \text{ cm}$

7. Evaluate $F = \dfrac{9C}{5} + 32$ with $C = 24$.

$$F = \dfrac{9C}{5} + 32 = \dfrac{9 \times 24}{5} + 32 = 43.2 + 32$$
$$= 75.2$$
$$24°C = 75.2°F$$

9. 2 days 47 hr 72 min = 2 days 48 hr 12 min
$$= 4 \text{ days } 12 \text{ min}$$

11.
$$
\begin{array}{r}
3 \text{ hr} \quad 50 \text{ min} \\
\times \qquad\qquad 4 \\
\hline
12 \text{ hr } 200 \text{ min} = 15 \text{ hr } 20 \text{ min}
\end{array}
$$

13. The speed limit on a freeway:
(b) 100 km/hr.

15. **(a)** The lines are parallel, since the lines never intersect.
(b) The lines are neither parallel nor perpendicular
(c) The lines are perpendicular, since the lines intersect to form four equal angles.
(d) The lines are neither parallel nor perpendicular.

17. The triangle has an obtuse angle. It is an obtuse triangle.

19. Using the protractor, we find $m\angle BOA = 50°$.

21. $m\angle x + 43° = 180°$ (supplementary angles), so $m\angle x = 180° - 43° = 137°$.

23. To find the indicated angle, add the two given measurements and then subtract that from 180°.
$$75° + 38° = 113°$$
$$180° - 113° = 67°$$
$$m\angle A = 67°$$

25. $8 \text{ yd}^3 = 8 \text{ yd}^3 \left(\dfrac{27 \text{ ft}^3}{1 \text{ yd}^3} \right) = 216 \text{ ft}^3$

27. $P = 2l + 2w = 2(4) + 2(2) = 8 + 4 = 12$ in.

29. $P = 30 + 12 + 20 = 62$ m

31. To find the perimeter of this figure, we will add all sides as well as half the circumference of the circle.
$$C = \pi d \approx 3.14 \times 10 \approx 31.4 \text{ ft}$$
$$31.4 \div 2 = 15.7 \text{ ft}$$
$$P = 15.7 + 21 + 16 + 7 + 10 + 6 + 10 + 6$$
$$= 91.7 \text{ ft}$$

33. $A = b \times h = 4 \times 1.75 = 7$ in.^{2}

35. $A = \dfrac{1}{2} \times b \times h = \dfrac{1}{2} \times 12 \times 9 = 54$ m^2

37.

To find the area of this figure, you will need to find the area of the large rectangle, then subtract the area of the small rectangle and half the area of the circle
$$A_A = l \times w = 16 \times 27 = 432$$
$$A_B = l \times w = 6 \times 10 = 60$$
$$A_C = \pi r^2 \approx 3.14(5)^2 \approx 3.14(25) \approx 78.5$$
$$78.5 \div 2 = 39.25$$
Therefore, the area of the figure is
$$A = A_A - A_B - A_C = 432 - 60 - 39.25$$
$$= 332.75 \text{ ft}^2 = 332.8 \text{ ft}^2$$
(rounded to the nearest tenth)

39. $r = 1$ in., $h = 4$ in. (cylinder)

(a) LSA $= 2\pi rh = 2 \times 3.14 \times 1$ in.$\times 4$ in.

$= 25.12$ in.^{2}

(b) TSA $= 2\pi rh + 2\pi r^2$

$= 2 \times 3.14 \times 1$ in.$\times 4$ in.

$+ 2 \times 3.14 \times (1 \text{ in.})^2$

$= 25.12$ in.$^2 + 6.28$ in.^{2}

$= 31.4$ in.^{2}

(c) $V = \pi r^2 h = 3.14 \times (1 \text{ in.})^2 \times 4$ in.

$= 12.56$ in.^{3}

41. $c^2 = a^2 + b^2 = (39)^2 + (52)^2 = 1{,}521 + 2{,}704$

$= 4{,}225$

$c = \sqrt{4{,}225} = 65$

The hypotenuse is 65 m.

43. Volume of the room is

$9 \text{ ft} \times 11 \text{ ft} \times 8 \text{ ft} = 792 \text{ ft}^3$.

BTUs needed for 1 cubic feet is 3, hence for this room the required quantity is

$792 \times 3 = 2{,}376$ BTUs.

Cumulative Review: Chapters 1–8

1. The place value of 6 is ten thousands.

3. First, find the area of the room.

$7 \text{ yd} \times 8 \text{ yd} = 56 \text{ yd}^2$

Now, multiply by the cost per square yard.

$56 \times \$16 = \$896$

The cost of the carpeting is \$896.

5. $(-12)(-13) = 156$

7. $x - 16 = 41$ Check: $57 - 16 \overset{?}{=} 41$

$x - 16 + 16 = 41 + 16$ $41 = 41$

$x = 57$

9. $630 = 2 \times 3 \times 3 \times 5 \times 7$

11. $20 = 2 \times 2 \times 5$

$24 = 2 \times 2 \times 2 \times 3$

LCM $= 2 \times 2 \times 2 \times 3 \times 5 = 120$

13. $2\dfrac{5}{8} \div \dfrac{7}{12} = \dfrac{21}{8} \div \dfrac{7}{12} = \dfrac{\overset{3}{\cancel{21}}}{\underset{2}{\cancel{8}}} \times \dfrac{\overset{3}{\cancel{12}}}{\underset{1}{\cancel{7}}} = \dfrac{9}{2} = 4\dfrac{1}{2}$

15. $270 \text{ miles} \div 4\dfrac{1}{2} \text{ hours} = \dfrac{270}{1} \div \dfrac{9}{2} = \dfrac{270}{1} \times \dfrac{2}{9}$

$= \dfrac{540}{9}$

$= 60$ miles per hour

Your average speed is 60 mi/hr.

17. $\left(-\dfrac{1}{8}\right) - \left(-\dfrac{7}{8}\right) = \dfrac{-1 + 7}{8} = \dfrac{6}{8} = \dfrac{3}{4}$

19. $-\dfrac{2}{3} + \left(-\dfrac{1}{6}\right) = -\dfrac{4}{6} + \left(-\dfrac{1}{6}\right) = -\dfrac{5}{6}$

21. $7\dfrac{3}{8} - 3\dfrac{5}{6} = \dfrac{59}{8} - \dfrac{23}{6} = \dfrac{177}{24} - \dfrac{92}{24} = \dfrac{85}{24} = 3\dfrac{13}{24}$

23. The place value of 3 is hundredths.

25. $32.67 + (-21.5) = 11.17$

27. $P = 4.8 + 7.3 + 4.8 + 7.3 = 24.2$ ft

Area of the figure is $4.8 \times 7.3 = 35.04$ ft^2.

29. $\dfrac{5}{16} = 5 \div 16 = 0.3125$

31. $\dfrac{35}{14} = \dfrac{10}{w}$

$35w = 140$

$\dfrac{35w}{35} = \dfrac{140}{35}$

$w = 4$

33. $8.5\% = 8.5\left(\dfrac{1}{100}\right) = 0.085$

35. $\dfrac{27}{40} = 27 \div 40 = 0.675 = 67.5\%$

37. $\dfrac{35}{100} = \dfrac{525}{x}$

$35x = 52,500$

$\dfrac{35x}{35} = \dfrac{52,500}{35}$

$x = 1,500$

35% of 1,500 is 525

39. $5 \text{ days} = 5 \cancel{\text{ days}} \left(\dfrac{24 \text{ hr}}{1 \cancel{\text{ day}}} \right) = 120 \text{ hr}$

41.   8 lb 14 oz

$\underline{+ \ 12 \text{ lb } 13 \text{ oz}}$

20 lb 27 oz = 21 lb 11 oz

43. $43 \text{ cm} = 43 \cancel{\text{ cm}} \left(\dfrac{1 \text{ m}}{100 \cancel{\text{ cm}}} \right) = 0.43 \text{ m}$

45. $8.3 \text{ mi} = 8.3 \cancel{\text{ mi}} \left(\dfrac{1.6 \text{ km}}{1 \cancel{\text{ mi}}} \right) = 13.3 \text{ km}$

47. Using the protractor, we find $m\angle AOB = 45°$.

49. $A = L \times W = 22.5 \text{ cm} \times 15.2 \text{ cm} = 342 \text{ cm}^2$

51. $A = \dfrac{1}{2} \times b \times h = \dfrac{1}{2} \times 14.1 \text{ ft} \times 8.6 \text{ ft}$

$= 60.63 \text{ ft}^2$

53. $s = 5 \text{ cm (cube)}$

  (a) $\text{LSA} = 4s^2 = 4(5 \text{ cm})^2 = 100 \text{ cm}^2$

  (b) $\text{TSA} = 6s^2 = 6(5 \text{ cm})^2 = 150 \text{ cm}^2$

  (c) $V = s^3 = (5 \text{ cm})^3 = 125 \text{ cm}^3$

55. Volume of the fish tank is

$V = LWH = 24 \text{ in.} \times 48 \text{ in.} \times 30 \text{ in.}$

$= 34,560 \text{ in.}^3$

given 1 gal is 231 in.^{3}

$1 \text{ in.}^3 = \dfrac{1}{231} \text{ gal}$

$34,560 \text{ in.}^3 = \dfrac{34,560}{231} \text{ gal} \approx 150 \text{ gal}$

Chapter 9
Introductions to Statistics and Graphs

Prerequisite Check

1. $\dfrac{20}{25} = \dfrac{4}{5}$

3. $\dfrac{3+5+7+9}{4} = \dfrac{24}{4} = 6$

5. $\dfrac{1}{4} \times 360 = 90$

7. $\dfrac{5}{8} = 0.625 = 62.5\%$

9. $0, 1, 2, 5, 7, 10, 11, 13$

11. $1\dfrac{1}{8}$ inches

13.
$$2 - 5x = 12$$
$$\underline{-2 \qquad\quad -2}$$
$$\dfrac{-5x}{-5} = \dfrac{10}{-5}$$
$$x = -2$$

15.
$$2x - 2 = 6$$
$$\underline{+2 \quad +2}$$
$$\dfrac{2x}{2} = \dfrac{8}{2}$$
$$x = 4$$

17.
$$6 - 3x = 8$$
$$\underline{-6 \qquad\quad -6}$$
$$\dfrac{-3x}{-3} = \dfrac{2}{-3}$$
$$x = -\dfrac{2}{3}$$

19. $C = \dfrac{5}{9}(F - 32) = \dfrac{5}{9}(113 - 32) = \dfrac{5}{9} \times 81$
$$= 45°C$$

Exercises 9.1

< Objective 1 >

1. Step 1: Add the numbers in the set.
$6 + 9 + 10 + 8 + 12 = 45$
Step 2: Divide that sum by the number of items in the set.
$45 \div 5 = 9$
The mean of this set of numbers is 9.

3. Step 1: Add the numbers in the set.
Step 2: Divide that sum by the number of items in the set.
$$113 \div 6 = 18\dfrac{5}{6}$$
The mean of this set of numbers is $18\dfrac{5}{6}$.

5. Step 1: Add the numbers in the set.
$12 + 14 + 15 + 16 + 16 + 16 + 17 + 22$
$\qquad\qquad + 25 + 27$
$= 180$
Step 2: Divide that sum by the number of items in the set.
$180 \div 10 = 18$
The mean of this set of numbers is 18.

7. Step 1: Add the numbers in the set.
$8 + (-4) + (-1) + 12 + (-11) = 4$
Step 2: Divide that sum by the number of items in the set.
$4 \div 5 = 8$
The mean of this set of numbers is 8.

9. Step 1: Add the numbers in the set.
$9 + 8 + 11 + 14 + 9 = 51$
Step 2: Divide that sum by the number of items in the set.
$51 \div 5 = 10.2$
The mean of this set of numbers is 10.2.

< Objective 2 >

11. Step 1: Rewrite the numbers in order from smallest to largest (in this case the data set is sorted already).
2, 3, 5, 6, 10
Step 2: There is an odd number of data points (5), so select the middle value; this is the median. The middle value is 5. Therefore, the median is 5.

13. Step 1: Rewrite the numbers in order from smallest to largest.
23, 24, 27, 31, 36, 38
Step 2: There is an even number of data points (6), so select the two middle values (27 and 31).
Step 3: Find the mean of the pair of middle values.
$\dfrac{27 + 31}{2} = 29$
Therefore, the median is 29.

15. Step 1: Rewrite the numbers in order from smallest to largest.
−22, −8, 1, 1, 5, 11
Step 2: There is an even number of data points (6), so select the two middle values (1 and 1).
Step 3: Find the mean of the pair of middle values.
$\dfrac{1 + 1}{2} = 1$
Therefore, the median is 1.

< Objective 3 >

17. (a) Mean:
Step 1: Add the numbers in the set.
$45 + 60 + 70 + 38 + 54 + 64 + 70 = 401$
Step 2: Divide that sum by the number of items in the set.
$\dfrac{401}{7} = 57.29$
The mean of this set of numbers is 57.29.
(b) Median:
Rewrite the numbers in order from smallest to largest
38, 45, 54, 60, 64, 70, 70.
Step 2: There is an odd number of data points (5), so select the middle value; this is the median. The middle value is 60. Therefore, the median is 60.
(c) Mean

19. (a) Mean:
Step 1: Add the numbers in the set.
$209,000 + 224,900 + 249,900 + 215,000$
$\qquad + 289,900 + 265,000 + 274,900$
$\qquad + 749,900$
$= 2,478,500$
Step 2: Divide that sum by the number of items in the set.
$\dfrac{2,478,500}{8} = 309,812.50$
The mean home price of the agent's listings is $309,812.50.
(b) Median:
Step 1: Rewrite the numbers in order from smallest to largest.
209,000; 215,000' 224,900; 249,900;
$\qquad$ 265,000; 274,900; 289,900; 749,900
Step 2: There is an even number of data points (8), so select the two middle values (249,900 and 265,000).
Step 3: Find the mean of the pair of middle values.
$\dfrac{249,900 + 265,000}{2} = \dfrac{514,900}{2} = 257,450$
Therefore, the median home price of the agent's listing is $257,450.
(c) Median

21. (a) Mean:
Step 1: Add the numbers in the set.
$66 + 60 + 62 + 64 + 64 + 58 + 63 + 68 + 65$

$\qquad +64 + 63 + 65$

$= 762$
Step 2: Divide that sum by the number of items in the set.
$\dfrac{762}{12} = 63.5$

The mean height of her female patients is 63.5.
(b) Median:
Step 1: Rewrite the numbers in order from smallest to largest.
58, 60, 62, 63, 63, 64, 64, 64, 65, 65, 66, 68
Step 2: There is an even number of data points (12), so select the two middle values (64 and 64).
Step 3: Find the mean of the pair of middle values.
$\dfrac{64 + 64}{2} = \dfrac{128}{2} = 64$

Therefore, the median height of her female patients is 64.
(c) Mean

< Objective 4 >

23. The mode, 17, is the number that appears most frequently in the data set.

25. The mode, 44, is the number that appears most frequently in the data set.

27. Since each numbers occurs only once, there is no mode in the data set.

29. The modes, 18 and 36, are the numbers that appear most frequently in the data set. The data set is bimodal.

31. Green color is the mode, because it is the color that appears most frequently in the data set.

33. Step 1: Add the numbers in the set.
$86 + 91 + 92 + 103 + 98 = 470$
Step 2: Divide that sum by the number of items in the set.
$470 \div 5 = 94$
The mean of this set of numbers is 94. Therefore, the mean high temperature was 94°F.

35. Step 1: Add the numbers in the set.
$43 + 29 + 51 + 36 + 33 + 42 + 32 = 266$
Step 2: Divide that sum by the number of items in the set.
$266 \div 7 = 38$
The mean of this set of numbers is 38. Therefore, the mean rating is 38 mi/gal.

37. Since you must have a mean of 90 on five tests to get an A in history, the total number of points you need to score is $90 \times 5 = 450$ points. So far you have scored $83 + 93 + 88 + 91 = 355$ points. To receive an A you need $450 - 355 = 95$ points. Therefore, you must have 95 points on the final test to receive an A.

39. To find Louis's mean, add his scores on all five tests.
$87 + 82 + 93 + 89 + 84 = 435$
Then, divide by 5, the total number of tests.
$435 \div 5 = 87$
To find Tamika's mean, add her scores on all five tests.
$92 + 83 + 89 + 94 + 87 = 445$
Then, divide the sum by 5.
$445 \div 5 = 89$
Therefore, Tamika had the higher mean by $89 - 87 = 2$ points.

41. To find the mean number of kilowatt-hours used each month by the four families for heating their homes, add the number of kilowatt-hours used by each of the four families for electric heat. Then, divide the sum by 4.
$1,200 + 1,086 + 1,103 + 975 = 4,364$

$4,364 \div 4 = 1,091$

Therefore, the mean number of kilowatt-hours used each month by the four families for heating their homes is 1,091 kWh.

43. To find the mean number of kilowatt-hours used per appliance by the McCarthy family, add the number of kilowatt-hours used by each of the McCarthy's seven appliances. Then, divide the sum by 7.

$115 + 1,086 + 386 + 154 + 99 + 117 + 45$

$= 2,002$

$2,002 \div 7 = 286$

Therefore, the mean number of kilowatt-hours used per appliance by the McCarthy family is 286 kWh.

45. False

47. sometimes

49. Entering into the calculator:

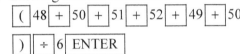

Display: 50
Therefore, the mean of this set of numbers is 50.

51. Entering into the calculator:

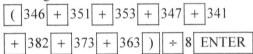

Display: 357
Therefore, the mean of this set of numbers is 357.

53. Entering into the calculator:

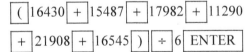

Display: 16607
Therefore, the mean of this set of numbers is 16,607.

55. Entering into the calculator:

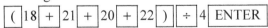

Display: 20.25
Therefore, the mean of this set of numbers is 20.25.

57. Entering into the calculator:

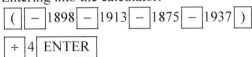

Display: −1905.75
Therefore, the mean of this set of numbers is −1,905.75.

59. Entering into the calculator:

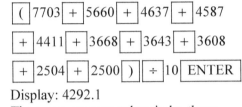

Display: 4292.1
The mean revenue taken in by these companies is $4,292.1 million, or $4,292,100,000.

61. To find the mean number of work stoppages per year from 2003 to 2010 add the stoppages. Then, divide the sum by 8.

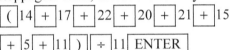

Display: 15.625
Therefore, the mean number of work stoppages per year from 2003 to 2010 was $15\frac{5}{8}$ stoppages.

63. (a) To find the mean number of response times from your computer to a local router using ping, add the ten response times. Then, divide the sum by 10.

$2.2 + 2.3 + 1.9 + 2.0 + 2.5 + 2.5 + 2.4 + 2.2$
$+2.4 + 2.5 = 22.9$

$22.9 \div 10 = 2.29$

The mean response time is 2.29 ms.

(b) Rewrite the numbers in order from smallest to largest.

1.9, 2.0, 2.2, 2.2, 2.3, 2.4, 2.4, 2.5, 2.5, 2.5

There is an even number of data points (10), so select the two middle values (2.3 and 2.4). Next, find the mean of the pair of middle values.

$\dfrac{2.3 + 2.4}{2} = 2.35$

The median is 2.35 ms.

(c) 2.5 ms is the mode, because it is the response time that appeared most frequently in the data set.

65. (a) To find the mean price of regular gasoline, add the twenty prices. Then, divide the sum by 20.

$3.14 + 2.99 + 3.48 + 3.33 + 3.18 + 2.93$

$+3.36 + 3.31 + 3.55 + 3.19 + 3.16 + 3.25$

$+3.04 + 3.22 + 3.38 + 3.19 + 3.02 + 3.15$

$+3.24 + 3.33$

$= 64.4$

$64.4 \div 20 = 3.22$

The mean gas price was $3.22.

(b) Rewrite the numbers in order from smallest to largest.

2.93, 2.99, 3.02, 3.04, 3.14, 3.15, 3.16, 3.18, 3.19, 3.19, 3.22, 3.24, 3.25, 3.31, 3.33, 3.33, 3.36, 3.38, 3.48, 3.55

There is an even number of data points (20), so select the two middle values (3.19 and 3.22). Next, find the mean of the pair of middle values.

$\dfrac{3.19 + 3.22}{2} = 3.21.$

The median gas price was $ 3.21.

(c) Mean

67. (a) To find the mean of Fred's monthly utility bills, add the twelve utility bills. Then, divide the sum by 12.

$153 + 151 + 143 + 137 + 132 + 129 + 134$
$+141 + 158 + 155 + 149 + 158$

$= 1,740$

$1,740 \div 12 = 145$

The mean of Fred's monthly utility bills was $145.

(b) Rewrite the numbers in order from smallest to largest.

129, 132, 134, 137, 141, 143, 149, 151, 153, 155, 158, 158

There is an even number of data points (20), so select the two middle values (143 and 149). Next, find the mean of the pair of middle values.

$\dfrac{143 + 149}{2} = 146$

The median of Fred's monthly utility bills was $146.

(c) Above and Beyond

69. Above and Beyond

71. Answers will vary.

73. Answers will vary.

Exercises 9.2

< Objective 1 >

1. (a) Looking at the cell that is in the row labeled North America and the column labeled 1950, we find a population listed as 221. Because the population figures are given in millions, the population was 221,000,000.
(b) Looking at the cell that is in the row labeled North America and the column labeled percent of Earth's total land area, we find a percentage listed as 16.2. Thus, the land area of North America as a percent of Earth's total land area is 16.2%.

3. **(a)** Note that the population in Asia in 1950 was 1,591,000,000, and the population in Asia in 1900 was 932,000,000. To find the percent increase, begin by subtracting to find the actual increase.

$1,591,000,000 - 932,000,000 = 659,000,000$

Use the population in 1900, 932,000,000 as the base of our comparison. Now, to find the rate, we have $\dfrac{659,000,000}{932,000,000} \approx 0.7071 \approx 70.7\%$. The population increased at a rate of 70.7%.

(b) Note that the population in Asia in 2000 was 4,028,000,000, and the population in Asia in 1950 was 1,591,000,000. To find the percent increase, begin by subtracting to find the actual increase.

$4,028,000,000 - 1,591,000,000$

$= 2,427,000,000$

Use the population in 1950, 1,591,000,000, as the base of our comparison. Now, to find the rate, we have

$\dfrac{2,427,000,000}{1,591,000,000} \approx 1.532 \approx 153.2\%$. The

population increased at a rate of 153.2%.

(c) The population of Asia in 1950 was 1,591,000,000 people. The land area of Asia is 17,400 thousand mi^2, or 17,400,000 square miles. The population per mi^2 of Asia in 1950

was $\dfrac{1,591,000,000}{17,400,000} \approx 91.4$. Therefore, the

population per square mile in Asia in 1950 was 91.4 people.

(d) The population of Asia in 2000 was 4,028,000,000 people. The land area is 17,400 thousand mi^2, or 17,400,000 mi^2. The population per mi^2 of Asia in 2000 was

$\dfrac{4,028,000,000}{17,400,000} \approx 231.5$. Therefore, the

population per square mile in Asia in 200 was 231.5 people.

5. Begin by finding the total population of five habitable continents except Asia in 1950.

$2,556,000,000 - 1,591,000,000 = 965,000,000$

Divide by 5, the total number of continents except Asia.

$\dfrac{965,000,000}{5} = 193,000,000$.

Therefore, the average for the five habitable continents besides Asia in 1950 was 193,000,000 people per continent.

7. **(a)** Note that 90.7% of the Earth's land is inhabitable. Therefore, the percent of the Earth's inhabitable land in North America is

$\dfrac{16.2}{90.7} \approx 0.1786 \approx 17.9\%$. (since the land area

in North America makes up 16.2% of the Earth)

(b) The population in North America was 305,000,000 people in 2000. The world's population was 6,279,000,000 people in 2000. Therefore, the percentage of world's population in the year 2000 in North America

was $\dfrac{305,000,000}{6,279,000,000} \approx 0.049 \approx 4.9\%$.

9. **(a)** The world's total population in 1950 was 2,556,000,000 and the world's total land area is 57,900 thousand mi^2, or 57,900,000 mi^2.

$$\frac{2,556,000,000}{57,900,000} \approx 44.1$$

Therefore, the number of people per square mile for the entire world in 1950 was 44.1 people.

(b) Similarly, the world's total population in 2000 was 6,279,000,000. The number of people per mi^2 in 2000 was

$$\frac{6,279,000,000}{57,900,000} \approx 108.4 .$$

(c) To find the percent increase in the number of people per square mile for the entire world from 1950 to 2000, begin by subtracting to find the actual increase.

$108.4 - 44.1 = 64.3$

Use the number of people per mi^2 in 1950, 44.1, as the base of our comparison. Now to find the rate, we have $\frac{64.3}{44.1} \approx 1.458 \approx 145.8\%$.

The percent of increase in the number of people per square mile for the entire world from 1950 to 2000 was 145.8%.

11. **(a)** Looking at the cell that is in the row labeled 2000 and the column labeled regular, we find the mean cost to be $1.484.

(b) Looking at the cell that is in the row labeled 2005 and the column labeled premium, we find the mean cost to be $2.468.

13. **(a)** The price of regular gas in 1996 was $1.199. The price of regular gas in 2010 was $2.782. Therefore, the increase in the price of regular gas from 1996 to 2010 was $2.782 - \$1.199 = \$1.583$.

(b) The price of premium gas in 1996 was $1.381. The price of regular gas in 2010 was $3.022. Therefore, the increase in the price of premium gas from 1996 to 2010 was $3.022 - \$1.381 = \$1.641$.

< Objective 2 >

15. The production in 2008 was 8,700,000 vehicles.

17. The number of vehicles produced in each year was the following:
2001: 11.4 million
2002: 12.4 million
2003: 12.1 million
2004: 12 million
2005: 12 million
2006: 11.4 million
2007: 11 million
2008: 8.7 million
2009: 5.8 million
2010: 7.9 million
Begin by arranging the production values in order from smallest to largest.
5.8, 7.9, 8.7, 11, 11.4, 11.4, 12, 12, 12.1, 12.4
Count from both ends to find the middle values. 11.4 and 11.4 are the middle values.

The median is $\frac{11.4 + 11.4}{2} = 11.4$. The median number of cars produced in the 10 years was 11.4 million cars.

19. The attendance on August 4 was 2,800 people

21. August 3 had the lowest attendance.

23. We have to estimate our answer when reading the bar graph. In this case 729,000 would be a good estimate. Therefore, the sale of wagons in 2008 was about 729,000.

25. The sales in 2001 were 450,000 wagons. The sales in 2010 were 710,000. Begin by finding the increase in sales.
Increase: $710,000 - 450,000 = 260,000$
Use the sales in 2001, 450,000 as the base of our comparison. Now to find the rate, we have $\frac{260,000}{450,000} = 0.5777 \approx 0.578 = 57.8\%$. The percent increase in sales from 2001 to 2010 was 57.8%.

27. Assuming that you consume 1 cup of soup, you have consumed $2 \times 110 = 220$ calories, since, as indicated by the label, there are 110 calories per $\frac{1}{2}$ cup of soup.

29. Assuming that you consume 1 cup of soup, you have consumed $2 \times 4 = 8\%$ of the daily value of fiber, since, as indicated by the label, 4% of the daily value of fiber is in $\frac{1}{2}$ cup of soup.

31. Tomato soup has the least fat, since it contains 0 g of fat.

33. Tomato soup has the least sodium, since it contains 760 mg of sodium.

35. Begin by adding the amount of calories for each type of soup.
$110 + 130 + 180 + 100 = 520$
Divide the sum by 4, the total number of soups.
$520 \div 4 = 130$
The mean number of calories in the soups is 130 calories.

< Objective 3 >

37.

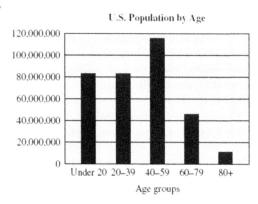

39.

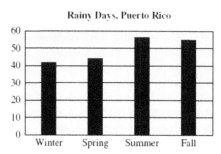

41. **(a)** The diameter for the 12-gauge wire is 0.0808 inches.
(b) The diameter for the 14-gauge is 0.0640 in. and the diameter for the 10-gauge is 0.1019 in. Therefore, the difference in diameter between 14-gauge and 10-guage wire is $0.1019 \text{ in.} - 0.0640 \text{ in.} = 0.0379 \text{ in.}$

43. **(a)** The density of titanium is 4.507 g/cm³.
(b) The melting point for copper is 1,084.9°C. The melting point for iron is 1,538°C. The difference in melting point between copper and iron is $1,538°C - 1,084.9°C = 453.1°C$.
(c) Titanium has the highest melting point.

45.

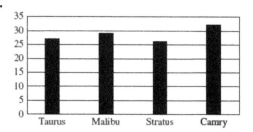

47. General Motors sold 100,000 cars in the U.S. in March 2012.

49. The difference in cars sold between Toyota and GM is $125,000 - 100,000 = 25,000$. Toyota sold 25,000 more cars than General Motors.

51. The income brought in at Toronto, Canada conference was $270,000.

53. Looking at the bar graph, at Salt Lake City, UT the expenses exceeded the income.

55. Above and Beyond

Exercises 9.3

< Objective 1 >

1. Look across the bottom until finding 2010. Look straight up until you see the line of the graph. Following across to the left, we see that annual electricity cost is $1,220.

3. The cost of electricity in 2009 was $1,240. The cost of electricity in 2010 was $1,220. The decrease in the cost of electricity from 2009 to 2010 was $1,240 - $1,220 = 20.

5. Look across the bottom until finding 2010. Look straight up until you see the line of the graph. Following across to the left, we see that natural gas cost is $680.

7. The cost of natural gas in 2009 was $750. The cost of natural gas in 2010 was $680. The decrease in the cost of natural gas from 2009 to 2010 was $750 - $680 = 70.

9. Looking at the graph, the greatest number of robberies occurred in December.

11. The number of robberies in August was 500. The number of robberies in September was 300. The decrease in robberies from August to September was $500 - 300 = 200$.

< Objective 2 >

13. The electricity cost for this family grew from $1,250 to $1,300 between 2011 and 2012. It is not unreasonable to expect at least this much growth between 2012 and 2013.
$1,300 - $1,250 = 50
Therefore, at least another $50 rise in electricity cost can be expected by 2013.
$1,300 + $50 = $1,350$
This gives $1,350, as the electricity cost in 2013.

15. The number of beneficiaries grew from 9.7 million to 10.2 million between 2009 and 2010. It is not unreasonable to expect at least this much growth between 2010 and 2011.
$10.2 - 9.7 = 0.5$
Therefore, at least another 0.5 million rise in the number of beneficiaries can be expected by 2011.
$10.2 + 0.5 = 10.7$
This gives 10.7 million beneficiaries in 2011.

< Objective 3 >

17.

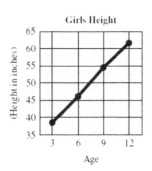

19.

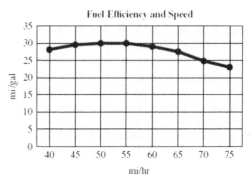

21.

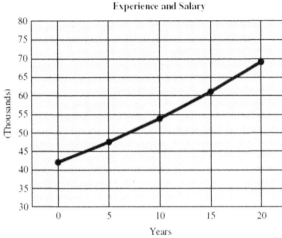

The rise in the salary for the number of years of service from 15 to 20 years is
$68.7 - 60.7 = 8$. Therefore, at least another 8 thousand rise in the salary can be expected in next five years.
$68.7 + 8 = 76.7$.
The expected salary of a professor with 25 years of experience is $76,700.

< Objective 4>

23. 31% of a typical household's electricity usage goes toward heating.

25. From the pie chart, 31% of household's electricity usage goes toward heating. Household usage of electricity was 11,600 kWh. Hence, kWh used for heating is 31% of 11,600kWh.

$\dfrac{31}{100} \times 11,600 = 3,596$ k Wh

27. 19% of the year's expenses went toward supplies.

29. From the pie chart, 19% of the year's expenses went toward supplies. The year's expenses totaled $35,000. Hence, expenses for supplies = 19% of $35,000.

$= \dfrac{19}{100} \times 35,000 = \$6,650$

< Objective 5 >

31.

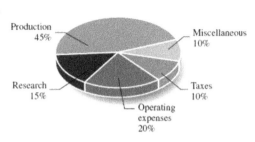

33.

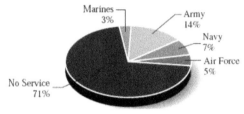

35.

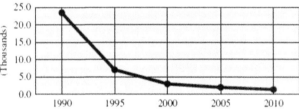

The decrease in number of recorded polio cases from 2005 to 2010 is $1.4 - 2 = 0.6$ (thousands). Therefore, decrease in at least another 0.6 polio cases can be expected between 2010 and 2020.
$1.4 - 0.6 = 0.8$ (thousands)
The expected number of recorded polio cases by 2020 is 800.

37.

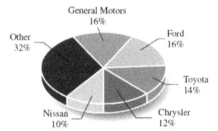

39. Looking at the income line, Chicago brought in the most income.

41. Looking at the income and the expenses line, largest distance between two lines is at Chicago. Hence, the conference at Chicago produced the largest difference between the income and the expenses.

43. Looking at the line graph, the income and the expense line meet at Phoenix, AZ. Hence, at Phoenix, AZ conference, the income and the expenses were approximately equal.

45. The average high temperature in May can be predicted by taking mean temperature for April and June.
$$\frac{64.9 + 84.8}{2} = 74.85°F$$
Therefore, the average temperature in May is predicted to be 74.85°F.

47. Continuing the extrapolation as used in problem 46
High temperature in September:
$97.4°F + 6.3°F = 103.7°F$
High temperature in October:
$103.7°F + 6.3°F = 110°F$
High temperature in November:
$110°F + 6.3°F = 116.3°F$
High temperature in December:
$116.3°F + 6.3°F = 122.6°F$
Hence, the high temperature in December is predicted to be 122.6°F.

49. The kitten's weight can be predicted by extrapolation. The difference in weight in 10 and 8 months is $11\text{ lb} - 9\text{ lb} = 2\text{ lb}$. The kitten's weight at 12 months is $11\text{ lb} + 2\text{ lb} = 13\text{ lb}$. Yes, this seems reasonable.

Exercises 9.4

< Objectives 1 and 2 >

1. $x + y = 6$

(4, 2): $(4) + (2) \overset{?}{=} 6$
$6 = 6$
(−2, 4): $(-2) + (4) \overset{?}{=} 6$
$2 \neq 6$
(0, 6): $(0) + (6) \overset{?}{=} 6$
$6 = 6$
(−3, 9): $(-3) + (9) \overset{?}{=} 6$
$6 = 6$
(4, 2), (0, 6), and (−3, 9) are solutions.

3. $2x - y = 8$

(5, 2): $2(5) - (2) \overset{?}{=} 8$
$8 = 8$
(4, 0): $2(4) - (0) \overset{?}{=} 8$
$8 = 8$
(0, 8): $2(0) - (8) \overset{?}{=} 8$
$-8 \neq 8$
(6, 4): $2(6) - (4) \overset{?}{=} 8$
$8 = 8$
(5, 2), (4, 0), and (6, 4) are solutions.

5. $4x + y = 8$

(2, 0): $4(2) + (0) \overset{?}{=} 8$
$8 = 8$
(2, 3): $4(2) + (3) \overset{?}{=} 8$
$11 \neq 8$
(0, 2): $4(0) + (2) \overset{?}{=} 8$
$2 \neq 8$
(1, 4): $4(1) + (4) \overset{?}{=} 8$
$8 = 8$
(2, 0) and (1, 4) are solutions.

7. $2x - 3y = 6$

$(0, 2)$: $2(0) - 3(2) \overset{?}{=} 6$

$-6 \neq 6$

$(3, 0)$: $2(3) - 3(0) \overset{?}{=} 6$

$6 = 6$

$(6, 2)$: $2(6) - 3(2) \overset{?}{=} 6$

$6 = 6$

$(0, -2)$: $2(0) - 3(-2) \overset{?}{=} 6$

$6 = 6$

$(3, 0)$, $(6, 2)$, and $(0, -2)$ are solutions.

9. $3x - 2y = 12$

$(4, 0)$: $3(4) - 2(0) \overset{?}{=} 12$

$12 = 12$

$\left(\dfrac{2}{3}, -5\right)$: $3\left(\dfrac{2}{3}\right) - 2(-5) \overset{?}{=} 12$

$12 = 12$

$(0, 6)$: $3(0) - 2(6) \overset{?}{=} 12$

$-12 \neq 12$

$\left(5, \dfrac{3}{2}\right)$: $3(5) - 2\left(\dfrac{3}{2}\right) \overset{?}{=} 12$

$12 = 12$

$(4, 0)$, $\left(\dfrac{2}{3}, -5\right)$, and $\left(5, \dfrac{3}{2}\right)$ are solutions.

11. $3x + 5y = 15$

$(0, 3)$: $3(0) + 5(3) \overset{?}{=} 15$

$15 = 15$

$\left(1, \dfrac{12}{5}\right)$: $3(1) + 5\left(\dfrac{12}{5}\right) \overset{?}{=} 15$

$15 = 15$

$(5, 3)$: $3(5) + 5(3) \overset{?}{=} 15$

$30 \neq 15$

$(0, 3)$ and $\left(1, \dfrac{12}{5}\right)$ are solutions.

13. $x = 3$

$(3, 5)$, $(3, 0)$, and $(3, 7)$ are solutions.

15.

$x + y = 12$

$(4) + y = 12$

$y = 8; (4, 8)$

$x + (5) = 12$

$x = 7; (7, 5)$

$(0) + y = 12$

$y = 12; (0, 12)$

$x + (0) = 12$

$x = 12; (12, 0)$

17.

$3x + y = 9$

$3(3) + y = 9$

$y = 0; (3, 0)$

$3x + (9) = 9$

$3x = 0$

$x = 0; (0, 9)$

$3x + (-3) = 9$

$3x = 12$

$x = 4; (4, -3)$

$3(0) + y = 9$

$y = 9; (0, 9)$

19.

$5x - y = 15$

$5x - (0) = 15$

$x = 3; (3, 0)$

$5(2) - y = 15$

$-y = 5$

$y = -5; (2, -5)$

$5(4) - y = 15$

$-y = -5$

$y = 5; (4, 5)$

$5x - (-5) = 15$

$5x = 10$

$x = 2; (2, -5)$

21.

$$4x - 2y = 16$$
$$4x - 2(0) = 16$$
$$x = 4; (4, 0)$$
$$4x - 2(-6) = 16$$
$$4x = 4$$
$$x = 1; (1, -6)$$
$$4(2) - 2y = 16$$
$$-2y = 8$$
$$y = -4; (2, -4)$$
$$4x - 2(6) = 16$$
$$4x = 28$$
$$x = 7; (7, 6)$$

23.

$$y = 3x + 9$$
$$(0) = 3x + 9$$
$$-9 = 3x$$
$$-3 = x; (-3, 0)$$
$$y = 3\left(\frac{2}{3}\right) + 9 = 11; \left(\frac{2}{3}, 11\right)$$
$$y = 3(0) + 9$$
$$y = 9; (0, 9)$$
$$y = 3\left(-\frac{2}{3}\right) + 9$$
$$y = 7; \left(-\frac{2}{3}, 7\right)$$

25.

$$y = 3x - 4 = 3(0) - 4 = -4; (0, -4)$$
$$(5) = 3x - 4$$
$$9 = 3x$$
$$x = 3; (3, 5)$$
$$(0) = 3x - 4$$
$$3x = 4$$
$$x = \frac{4}{3}; \left(\frac{4}{3}, 0\right)$$
$$y = 3\left(\frac{5}{3}\right) - 4 = 1; \left(\frac{5}{3}, 1\right)$$

27. $x - y = 10$
$(0, -10)$, $(10, 0)$, $(5, -5)$, and $(12, 2)$ are solutions.

29. $2x - y = 6$
$(0, -6)$, $(3, 0)$, $(6, 6)$, and $(9, 12)$ are solutions.

31. $x + 4y = 8$
$(8, 0)$, $(-4, 3)$, and $(4, 1)$ are solutions.

33. $5x - 2y = 10$
$(0, -5)$, $(4, 5)$, $(-6, -20)$, and $(2, 0)$ are solutions.

35. $\frac{1}{3}x - \frac{1}{4}y = 1$
$(0, 3)$, $(1, 5)$, $(2, 7)$, and $(3, 9)$ are solutions.

37. $x = -5$
$(-5, 0)$, $(-5, 1)$, $(-5, 2)$, and $(-5, 3)$ are solutions.

39. $\frac{1}{2}x + \frac{1}{3}y = 1$

$\frac{1}{2}x + \frac{1}{3}(0) = 1$	$\frac{1}{2}(0) + \frac{1}{3}y = 1$
$\frac{1}{2}x = 1$	$\frac{1}{3}y = 1$
$x = 2$	$y = 3$
$(2, 0)$	$(0, 3)$

41. $0.3x + 0.5y = 2$

$0.3x + 0.5(0) = 2$ $0.3x = 2$ $x = \dfrac{20}{3}$	$0.3(0) + 0.5y = 2$ $0.5y = 2$ $y = 4$
$\left(\dfrac{20}{3}, 0\right)$	$(0, 4)$

43. $\dfrac{3}{4}x - \dfrac{2}{5}y = 6$

$\dfrac{3}{4}x - \dfrac{2}{5}(0) = 6$ $\dfrac{3}{4}x = 6$ $x = 8$	$\dfrac{3}{4}(0) - \dfrac{2}{5}y = 6$ $-\dfrac{2}{5}y = 6$ $y = -15$
$(8, 0)$	$(0, -15)$

45. $0.4x - 0.7y = 3$

$0.4x - 0.7(0) = 3$ $0.4x = 3$ $x = \dfrac{15}{2}$	$0.4(0) - 0.7y = 3$ $-0.7y = 3$ $y = -\dfrac{30}{7}$
$\left(\dfrac{15}{2}, 0\right)$	$\left(0, -\dfrac{30}{7}\right)$

< Objective 3 >

47. $y = 0.75x + 8$

$y = 0.75(2) + 8 = 9.5$

$y = 0.75(5) + 8 = 11.75$

$y = 0.75(10) + 8 = 15.5$

$y = 0.75(15) + 8 = 19.25$

$y = 0.75(20) + 8 = 23$

The hourly wages for producing 2, 5, 10, 15, and 20 units per hour are $9.50, $11.75, $15.50, $19.25, and $23.00, respectively

49. $A = 4s$

$A = 4(5) = 20$

$A = 4(10) = 40$

$A = 4(12) = 48$

$A = 4(15) = 60$

For squares whose sides are 5 cm, 10 cm, 12 cm, and 15 cm, their perimeters are 20 cm, 40 cm, 48 cm, and 60 cm, respectively.

51.

$p = \dfrac{-x}{2} + 75$

$p = \dfrac{-2}{2} + 75 = 74$

$p = \dfrac{-7}{2} + 75 = 71.5$

$p = \dfrac{-9}{2} + 75 = 70.5$

$p = \dfrac{-11}{2} + 75 = 69.5$

When 2, 7, 9, and 11 units are sold, the price for each unit is $74, $71.50, $70.50, and $69.50, respectively.

53. $y = 162x + 4,365$

$y = 162(1) + 4,365 = 4,527$

$y = 162(2) + 4,365 = 4,689$

$y = 162(3) + 4,365 = 4,851$

$y = 162(4) + 4,365 = 5,013$

$y = 162(6) + 4,365 = 5,337$

x	1	2	3	4	6
y	4,527	4,689	4,851	5,013	5,337

55. False

57. False

59. True

61. sometimes

63.

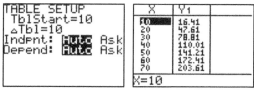

The ordered pairs are (10, 16. 41),
(20, 47.61), (30, 78.81), (40, 110.01),
and (50, 141.21).

65. $d = 7.5w$

(a) $d = 7.5(30 \text{ kg}) = 225 \text{ mg}$

(b) $150 \text{ mg} = 7.5w$
$20 \text{ kg} = w$

67. $b = \dfrac{8.25}{144} L$

(a) $b = \dfrac{8.25}{144}(12 \text{ ft}) = 0.6875 \approx 0.69 \text{ bd ft}$

(b) $b = \dfrac{8.25}{144}(16 \text{ ft}) = 0.91\overline{6} \approx 0.92 \text{ bd ft}$

(c) $b = \dfrac{8.25}{144}(20 \text{ ft}) = 1.14583\overline{3} \approx 1.15 \text{ bd ft}$

69. $F = kx$

(a) $F = 72x = 72(3 \text{ ft}) = 216 \text{ lb}$

(b) $F = 72x = 72(5 \text{ ft}) = 360 \text{ lb}$

71. $x + y + z = 0 \qquad (2, -3, \ )$
$2 + (-3) + z = 0$
$2 - 3 + z = 0$
$-1 + z = 0$
$z = 1$
$(2, -3, 1)$ is the ordered-triple solution.

73. $x + y + z = 0 \qquad (1, \ , 5)$
$1 + y + 5 = 0$
$y + 6 = 0$
$y = -6$
$(1, -6, 5)$ is the ordered-triple solution.

75. $2x + y + z = 2 \qquad (-2, \ , 1)$
$2(-2) + y + 1 = 2$
$-4 + y + 1 = 2$
$y - 3 = 2$
$y = 5$
$(-2, 5, 1)$ is the ordered-triple solution.

77. Above and Beyond

Exercises 9.5

< Objective 1 >

1. $A(5, 6)$; I

3. $C(2, 0)$; x-axis

5. $E(-4, -5)$; III

7. $S(-5, -3)$; III

9. $U(-3, 5)$; II

< Objective 2 >

11.

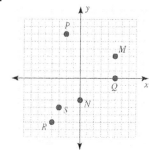

13.

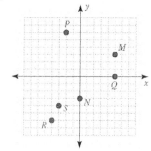

15.

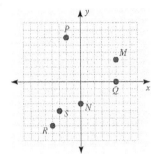

17.

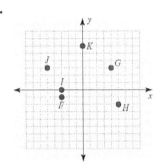

19.

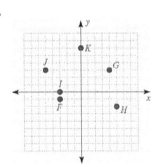

21.

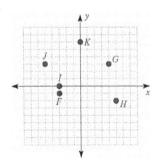

23. (4, 5) is in quadrant I.

25. (−6, −8) is in quadrant III.

27. (5, 0) is on the *x*-axis.

29. (−4, 7) is in quadrant II.

31. (0, −4) is on the *y*-axis.

33. $\left(5\frac{3}{4}, -3\right)$ is in quadrant IV.

< Objective 3 >

35. (a)-(c) The ordered pairs are *A*(1500, 350), *B*(2300, 430), and *C*(1200, 320).

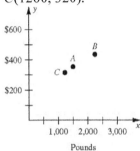

37. The ordered pairs are (1, 4), (2, 14), (3, 26), (4, 33), (5, 42), and (6, 51).

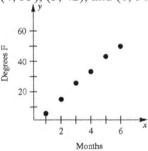

39. The ordered pairs are (1, 11), (2, 9), (3, 2), (4, 4), and (5, 3).

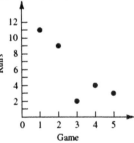

41. The ordered pairs are (1, 30), (2, 45), (3, 60), (4, 60), (5, 75), (6, 90), and (7, 95).

43. The ordered pairs are (7, 100), (15, 70), (20, 80), (30, 70), (40, 50), (50, 40), (60, 30), (70, 40), and (80, 25).

45. True

47. sometimes

49. always

51.

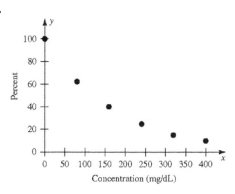

53.

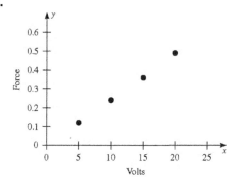

55.

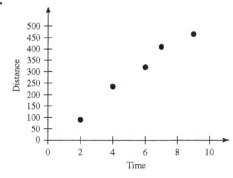

57.

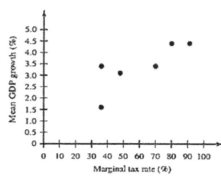

59. The points lie on a line; another point on the line is the point (1, 2).

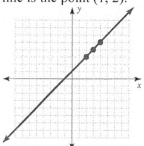

61. **(a)** $(-2, -4), (1, 2), (3, 6)$
    (b) The y-value is twice the x-value.
    (c) $y = 2x$

63. **(a)** $(-2, 6), (-1, 3), (1, -3)$
    (b) The y-value is -3 times the x-value.
    (c) $y = -3x$

65. .Above and Beyond

67. Above and Beyond

Exercises 9.6

< Objectives 1 and 2 >

1. $x + y = 6$

Two solutions are (0, 6) and (6, 0). The graph is the line through both points.

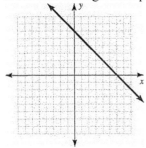

3. $x - y = -3$

Two solutions are (−3, 0) and (0, 3). The graph is the line through both points.

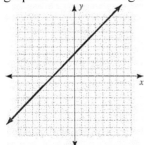

5. $2x + y = 2$

Two solutions are (0, 2) and (1, 0). The graph is the line through both points.

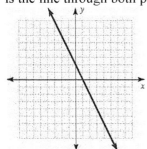

7. $3x + y = 0$

Two solutions are (0, 0) and (1, −3). The graph is the line through both points.

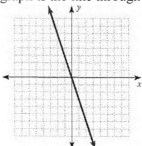

9. $x + 4y = 8$

Two solutions are (0, 2) and (4, 1). The graph is the line through both points.

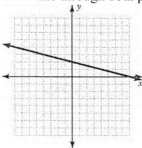

11. $y = 5x$

Two solutions are (0, 0) and (1, 5). The graph is the line through both points.

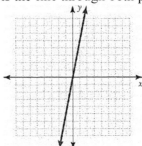

13. $y = 2x - 1$

Two solutions are (0, −1) and (3, 5). The graph is the line through both points.

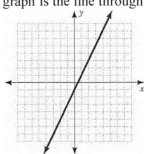

15. $y = -3x + 1$

Two solutions are (0, 1) and (2, −5). The graph is the line through both points.

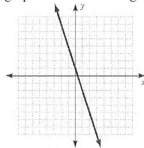

17. $y = \frac{1}{3}x$

Two solutions are (0, 0) and (3, 1). The graph is the line through both points.

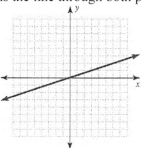

19. $y = \frac{2}{3}x - 3$

Two solutions are (0, −3) and (6, 1). The graph is the line through both points.

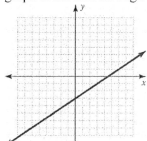

21. $x = 5$

Two solutions are (5, 0) and (5, 6). The graph is the line through both points.

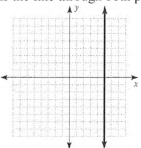

23. $y = 1$

Two solutions are (0, 1) and (4, 1). The graph is the line through both points.

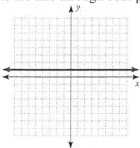

< Objective 3 >

25. $x - 2y = 4$

The x-intercept is (4, 0) and the y-intercept is (0, −2).

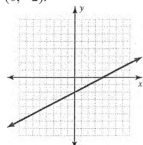

27. $5x + 2y = 10$

The x-intercept is $(2, 0)$ and the y-intercept is $(0, 5)$.

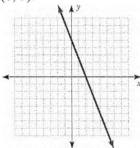

29. $3x + 5y = 15$

The x-intercept is $(5, 0)$ and the y-intercept is $(0, 3)$.

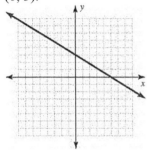

31. $3x - 2y = -6$

The x-intercept is $(-2, 0)$ and the y-intercept is $(0, 3)$.

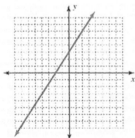

33. $5x + 2y = -10$

The x-intercept is $(-2, 0)$ and the y-intercept is $(0, -5)$.

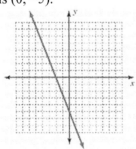

< Objective 4 >

35. $x + 3y = 6$

$$3y = 6 - x$$

$$y = -\frac{x}{3} + 2$$

Two solutions are $(0, 2)$ and $(6, 0)$. The graph is the line through both points.

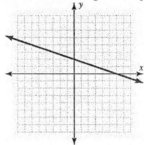

37. $3x + 4y = 12$

$$4y = 12 - 3x$$

$$y = -\frac{3}{4}x + 3$$

Two solutions are $(0, 3)$ and $(4, 0)$. The graph is the line through both points.

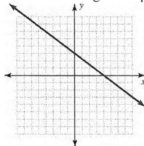

39. $5x - 4y = 20$

$$-4y = 20 - 5x$$

$$y = \frac{5}{4}x - 5$$

Two solutions are $(4, 0)$ and $(0, -5)$. The graph is the line through both points.

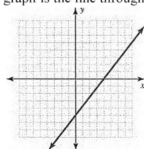

41. $y = 0.10x + 200$

Two solutions are (0, 200) and (2000, 400). The graph is the line through both points.

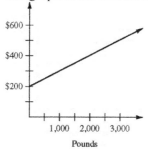

Pounds

43. (a) Two solutions of the equation $y = 11x - 100$ are (0, −100) and (10, 10). The graph is the line through both points.

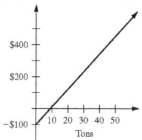

Tons

(b) Solve $0 = 11x - 100$

$x = \dfrac{100}{11} \approx 9$ tons

(c) $y = 11(16) - 100 = 76$

The class will earn $76.

(d) $y = 17x - 125$

45. (a) $C = 15x + 200$

(b)

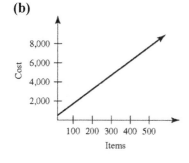

Items

47. $C = 0.04n + 8$

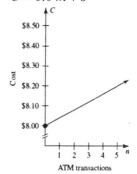

ATM transactions

49. $y = 2x$

Two solutions are (0, 0) and (3, 6). The graph is the line through both points.

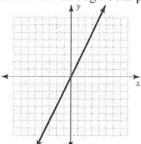

51. $y = 3x - 3$

Two solutions are (0, −3) and (1, 0). The graph is the line through both points.

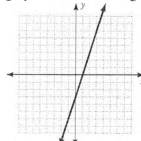

53. $x - 4y = 12$

Two solutions are (0, −3) and (4, −2). The graph is the line through both points.

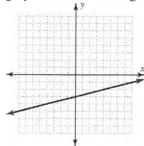

55. True

57. always

59. never

61. $w = -1.75d + 25$
The graph contains the points (0, 25) and (4, 18).

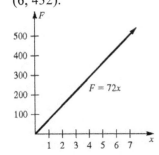

63. $F = 72x$
The graph contains the points (0, 0) and (6, 432).

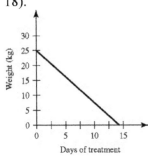

65. $s = \dfrac{3}{4}L + 1$

The graph contains the points (0, 1) and (20, 16).

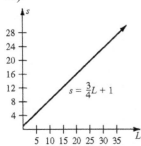

67. (3, 1) is the point of intersection.

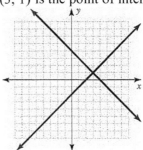

69. The lines do not intersect. The y-intercepts are (0, 0), (0, 4), and (0, −5).

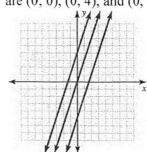

71. **(a)**

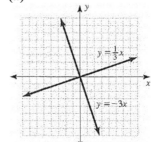

(b) Above and Beyond

73. (a)

Point	x	y
A	5	13
B	6	15
C	7	17
D	8	19
E	9	21

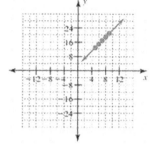

(b) The y-coordinate increases by 2 each time the x-coordinate increases by 1.
(c) Yes
(d) grows by 2 units

75. $y = 3x - 2$

(b) The y-coordinate increases by 3 each time the x-coordinate increases by 1.
(c) Yes
(d) grows by 3 units

77. $y = -4x + 50$

(b) The y-coordinate decreases by 4 each time the x-coordinate increases by 1.
(c) Yes
(d) decreases by 4 units

Summary Exercises

1. Step 1: Add the numbers in the set.
$8 + 6 + 7 + 4 + 5 = 30$
Step 2: Divide that sum by the number of items in the set.
$30 \div 5 = 6$
The mean of this set of numbers is 6.

3. Step 1: Add the numbers in the set.
$117 + 121 + 122 + 118 + 115 + 125 + 123 + 119$
$= 960$
Step 2: Divide that sum by the number of items in the set.
$960 \div 8 = 120$
The mean of this set of numbers is 120.

5. Step 1: Add the numbers in the set.
$(-12) + (-3) + (-9) + (-15) + (-18) + (-9)$
$\quad\quad + (-6) + (-12)$
$= -84$
Step 2: Divide that sum by the number of items in the set.
$-84 \div 8 = 10.5$
The mean of this set of numbers is 10.5.

7. Step 1: Add the numbers in the set.
$89 + 71 + 93 + 87 = 340$
Step 2: Divide that sum by the number of items in the set.
$340 \div 4 = 85$
The mean of this set of numbers is 85.
Therefore, Elmer's mean test score was 85.

9. First, find the median.
Step 1: Rewrite the numbers in order from smallest to largest.
16, 18, 19, 20, 20
Step 2: There is an odd number of data points (5), so select the middle value; this is the median. The middle value is 19. Therefore, the median is 19. Next, find the mode. The mode, 20, is the number that appears most frequently.

11. First, find the median.
Step 1: Rewrite the numbers in order from smallest to largest.
26, 27, 28, 28, 28, 30, 30, 31, 31, 35
Step 2: There is an even number of data points (10), so select the two middle values (28 and 30).
Step 3: Find the mean of the pair of middle values.
$(28 + 30) \div 2 = 58 \div 2 = 29$

29 is the median. Next find the mode. The mode, 28, is the number that appears most frequently.

13. First, find the median.
Step 1: Rewrite the numbers in order from smallest to largest.
$-18, -15, -12, -12, -9, -9, -6, -3$
Step 2: There is an even number of data points (8), so select the two middle values (-12 and -9).
Step 3: Find the mean of the pair of middle values.
$$\left[-12 + (-9)\right] \div 2 = -10.5$$
-10.5 is the median. Next, find the mode. The modes, -12 and -9, are the numbers that appear most frequently.

15. In order for Anita to have a mean of 90 on five tests, the total number of points she needs is $90 \times 5 = 450$ points. So far she has scored $88 + 91 + 86 + 93 = 358$ points. To get a mean of 90, Anita needs $450 - 358 = 92$ points on her fifth test.

17. The motor vehicle production in Japan in 1960 was 482,000. In 2010, the motor vehicle production in Japan was 9,629,000.

19. The decrease in motor vehicle production in United States from 1960 to 2010 is
$7,905,000 - 7,763,000 = 142,000$.
Use the motor vehicle in 1960 as the base of our comparison.
$$\frac{142,000}{7,905,000} \approx .018 = 1.8\%$$
Hence, the percent decrease in production was 1.8%.

21. The motor vehicle production in Japan in 2010 was 9,629,000. The percentage of total world production that occurred in Japan in 2010 is $\dfrac{9,269,000}{77,629,000} \approx 0.124 = 12.4\%$.

23. The motor vehicle production occurred outside United States and Japan can be calculated by subtracting the production in United States and Japan from the total world production.
$77,629,000 - (9,629,000 + 7,763,000)$
$= 77,629,000 - 17,392,000 = 60,237,000$
The percentage of total world production that occurred outside United States and Japan is $\dfrac{60,237,000}{77,629,000} \approx 0.776 = 77.6\%$.

25. Students enrolled in 2000 were 5,000. Students enrolled in 2010 were 10,000.
Increase in enrollment:
$10,000 - 5,000 = 5,000$
5,000 more students were enrolled in 2010 than in 2000.

27.

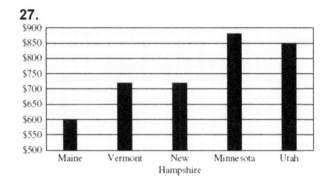

29. Look across the bottom until finding 2009. Look straight up until you see the line of the graph. Following across to the left, you see that PCs sold were 50,000. Similarly, in 2012, PCs sold were 300,000.
$300,000 - 50,000 = 250,000$
250,000 more PCs were sold in 2012 than in 2009.

31. The number of PCs that they will sell in 2013 can be found by extrapolation.
The increase in sales of number of PCs between 2012 and 2011 is
$300,000 - 200,000 = 100,000$
Therefore, an increase in sales of number of PCs by 100,000 can be expected in 2013.
$300,000 + 100,000 = 400,000$
The expected number of PCs that they will sell in 2013 is 400,000.

33.

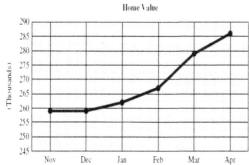

35. The home's value in November is $259 thousand.
The home's value in April is $286 thousand.
The percent increase in the home's value in 6 months
$$\frac{286-259}{259}=\frac{27}{259}\approx.104=10.4\%$$

37. Looking at the pie chart, Côte d'Ivore was the largest Cocoa bean producer in 2011. Côte d'Ivore produced 30% of world's cocoa beans.

39.

Grade	Count	Percent	Degrees
A	7	17.5%	63°
B	12	30%	108°
C	13	32.5%	117°
D	5	12.5%	45°
E	3	7.5%	27°

Grade Distribution

17.5% A
30% B
7.5% F
12.5% D
32.5% C

41. $7(x)+2=16$, $x=2$

$7(2)+2\overset{?}{=}16$

$16=16$

Yes, 2 is a solution of $7x+2=16$.

43. $7x-2=2x+8$, $x=2$

$7(2)-2\overset{?}{=}2(2)+8$

$14-2\overset{?}{=}4+8$

$12=12$

Yes, 2 is a solution of $7x-2=2x+8$.

45. $x+5+3x=2+x+23$, $x=6$

$6+5+3(6)\overset{?}{=}2+6+23$

$29\neq31$

No, 6 is not a solution of
$x+5+3x=2+x+23$.

47. $x-y=6$

$(6, 0)$: $6-0\overset{?}{=}6$

$6=6$

$(3, 3)$: $3-3\overset{?}{=}6$

$0\neq6$

$(3, -3)$: $3-(-3)\overset{?}{=}6$

$0\neq6$

$(0, -6)$: $0-(-6)\overset{?}{=}6$

$6=6$

The solutions are (6, 0), (3, -3), and (0, −6).

49. $2x+3y=6$

$(3, 0)$: $2(3)+3(0)\overset{?}{=}6$

$6=6$

$(6, 2)$: $2(6)+3(2)\overset{?}{=}6$

$18\neq6$

$(-3, 4)$: $2(-3)+3(4)\overset{?}{=}6$

$6=6$

$(0, 2)$: $2(0)+3(2)\overset{?}{=}6$

$6=6$

The solutions are (3, 0), (−3, 4), and (0, 2).

51.

$$x + y = 8$$
$$4 + y = 8$$
$$y = 4; (4, 4)$$
$$x + 8 = 8$$
$$x = 0; (0, 8)$$
$$8 + y = 8$$
$$y = 0; (8, 0)$$
$$6 + y = 8$$
$$y = 2; (6, 2)$$

53.

$$2x + 3y = 6$$
$$2(3) + 3y = 6$$
$$6 + 3y = 6$$
$$y = 0; (3, 0)$$
$$2(6) + 3y = 6$$
$$12 + 3y = 6$$
$$y = -2; (6, -2)$$
$$2x + 3(-4) = 6$$
$$2x - 12 = 6$$
$$x = 9; (9, -4)$$
$$2(-3) + 3y = 6$$
$$-6 + 3y = 6$$
$$x = 4; (-3, 4)$$

55. $x + y = 10$

Answers will vary. Some possible solutions are (0, 10), (2, 8), (4, 6), and (6, 4).

57. $2x - 3y = 6$

Answers will vary. Some possible solutions are (0, −2), (3, 0), (6, 2), and (9, 4).

59. $A(4, 6)$

61. $E(-1, -5)$

63.

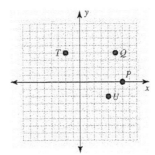

65.

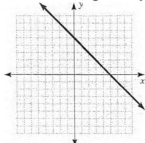

67. $x + y = 5$

Two solutions are (0, 5) and (5, 0). The graph is the line through both points.

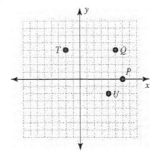

69. $y = 2x$

Two solutions are (0, 0) and (2, 4). The graph is the line through both points.

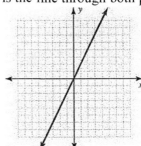

71. $y = \dfrac{3}{2}x$

Two solutions are (0, 0) and (2, 3). The graph is the line through both points.

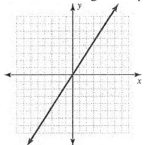

73. $y = 2x - 3$

Two solutions are (0, −3) and (2, 1). The graph is the line through both points.

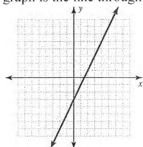

75. $y = \dfrac{2}{3}x + 2$

Two solutions are (−3, 0) and (0, 2). The graph is the line through both points.

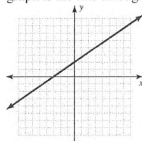

77. $3x - y = 3$

Two solutions are (0, −3) and (1, 0). The graph is the line through both points.

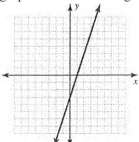

79. $3x + 2y = 12$

Two solutions are (0, 6) and (4, 0). The graph is the line through both points.

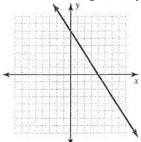

81. $x = 3$

Two solutions are (3, 0) and (3, 3). The graph is the line through both points.

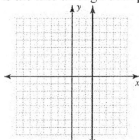

83. $5x - 3y = 15$

Two solutions are (0, −5) and (3, 0). The graph is the line through both points.

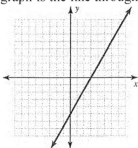

85. $2x + y = 6$

$y = -2x + 6$

Two solutions are $(0, 6)$ and $(1, 4)$. The graph is the line through both points.

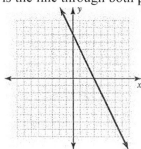

Chapter Test 9

1. Step 1: Add the numbers in the set.
$12 + 19 + 15 + 20 + 11 + 13 = 90$
Step 2: Divide that sum by the number of items in the set.
$90 \div 6 = 15$
The mean of this set of numbers is 15.

3. The mode, 6, is the number that appears most frequently.

5. Since you must have a mean of 90 on four tests to get an A, the total number of points you need to score is $90 \times 4 = 360$ points. So far you have scored $87 + 89 + 91 = 267$ points. To get an A, you need $360 - 267 = 93$ points on the final test.

7. Looking at the table, the world's total cocoa production can be calculated by adding the individual production of each nation.
$2,706 + 1,608 + 1,078 + 462 + 462 + 363$
$\qquad + 286 + 70 + 755$
$= 7,790$
Hence, the world's total cocoa production is 7,790,000,000lb.

9. For the current year Côte d'Ivoire production was 2,706 million pounds. For the subsequent year, Côte d'Ivoire's production fell to 2,688 million pounds. Therefore, the percent decrease can be calculated as
$$\frac{2,706 - 2,688}{2,706} = \frac{18}{2,706} \approx .0067 = 0.67\%.$$

11.

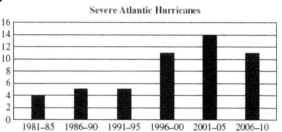

13. The number of severe hurricanes in the period 1996-2000 = 11
The number of severe hurricanes in the period 2001-2005 = 14
The percent increase in hurricanes is calculated as $\frac{14 - 11}{11} = \frac{3}{11} \approx 0.27 = 27\%$.

15.

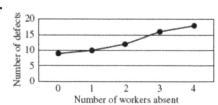

17. 45% was shipped by truck.

19. Truck shipped 45% of items. As the company shipped 1,200 items in a month, the number items shipped by truck was $\frac{45}{100} \times 1,200 = 540$.

21.

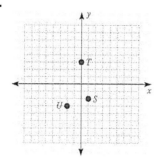

23.

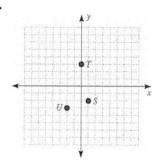

25. $B(-4, 6)$

27. Answers will vary.

29. $x + y = 9$

$(3, 6)$ $3 + 6 \overset{?}{=} 9$

$9 = 9$

$(9, 0)$ $9 + 0 \overset{?}{=} 9$

$9 = 9$

$(3, 2)$ $3 + 2 \overset{?}{=} 9$

$5 \neq 9$

The solutions are $(3, 6)$ and $(9, 0)$.

31.

$$4x + 3y = 12$$
$$4(3) + 3y = 12$$
$$3y = 0$$
$$y = 0; (3, 0)$$
$$4x + 3(4) = 12$$
$$4x = 0$$
$$x = 0; (0, 4)$$
$$4x + 3(3) = 12$$
$$4x = 12 - 9 = 3$$
$$x = \frac{3}{4}; \left(\frac{3}{4}, 3\right)$$

33. $2x + 5y = 10$

Two solutions are $(0, 2)$ and $(5, 0)$. The graph is the line through both points.

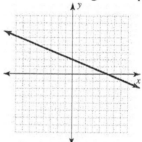

35. $x + y = 4$

Two solutions are $(0, 4)$ and $(4, 0)$. The graph is the line through both points.

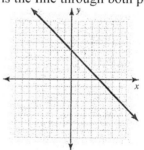

37. $y = \dfrac{3}{4}x - 4$

Two solutions are (0, –4) and (4, –1). The graph is the line through both points.

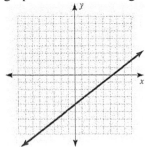

Cumulative Review: Chapters 1–9

1. The place value of 6 in the numeral 126,489 is thousands.

3.
$$74,9\overset{7}{\cancel{8}}{}^{1}3$$
$$\underline{-\ 35,6\ 9\ 5}$$
$$8$$

$$74,\overset{8}{\cancel{9}}\ \overset{17}{\cancel{8}}{}^{1}3$$
$$\underline{-\ 35,\ 6\ 9\ 5}$$
$$2\ 8\ 8$$

$$\overset{6}{\cancel{7}}{}^{1}4,\overset{8}{\cancel{9}}{}^{1}\ \overset{7}{\cancel{8}}{}^{1}3$$
$$\underline{-\ 3\ 5,\ 6\ 9\ 5}$$
$$3\ 9,\ 2\ 8\ 8$$

5.
$$\begin{array}{r} 308 \\ 27\overline{)8,322} \\ \underline{81} \\ 22 \\ \underline{\ 0} \\ 222 \\ \underline{216} \\ 6 \end{array}$$
We have $8322 \div 27 = 308$ r6.

7. $2.45(-30.7) = -(2.45 \times 30.7)$

$$\begin{array}{r} 2.45 \\ \underline{\times\ 30.7} \\ 1715 \\ 000 \\ \underline{735} \\ 75.215 \end{array}$$

$2.45(-30.7) = -(2.45 \times 30.7) = -75.215$

9. $\left(-\dfrac{11}{15}\right) \div \left(-\dfrac{121}{90}\right) = \dfrac{11}{15} \times \dfrac{90}{121} = \dfrac{\overset{1}{\cancel{11}} \times \overset{6}{\cancel{90}}}{\underset{1}{\cancel{15}} \times \underset{11}{\cancel{121}}}$

$$= \dfrac{6}{11}$$

11. $\dfrac{4}{7} = \dfrac{8}{x}$

$$7x\left(\dfrac{4}{7}\right) = 7x\left(\dfrac{8}{x}\right)$$

$$4x = 56$$

$$\dfrac{4x}{4} = \dfrac{56}{4}$$

$$x = 14$$

13. $18\% = 18\left(\dfrac{1}{100}\right) = 0.18$

$18\% = 18\left(\dfrac{1}{100}\right) = \dfrac{18}{100} = \dfrac{9}{50}$

15.
$$\begin{array}{r} 7\ \text{lb}\ \ 9\ \text{oz} \\ \underline{+\ 3\ \text{lb}\ 12\ \text{oz}} \\ 10\ \text{lb}\ 21\ \text{oz} = 11\ \text{lb}\ 5\ \text{oz} \end{array}$$
Since 21 oz = 1 lb 5 oz, the final result is 11 lb 5 oz.

17. $8\ \text{km} = 8\ \cancel{\text{km}}\left(\dfrac{1{,}000\ \text{m}}{1\ \cancel{\text{km}}}\right) = 8{,}000\ \text{m}$

19. $8\ \text{in.} = 8\ \cancel{\text{in.}}\left(\dfrac{1\ \text{cm}}{0.3937\ \cancel{\text{in.}}}\right) = 20.32\ \text{cm}$

21. Looking at the line graph, the increase in benefits between 2008 and 2009 was $80 - 40 = 40$. Similarly, the increase in benefits between 2010 and 2011 was $160 - 80 = 80$. Hence, the greatest increase in benefits occurred between 2010 and 2011.

23. Mean:
Step 1: Add the numbers in the set.
$$11 + 9 + 3 + 6 + 7 + 9 + 8 + 11 + 12 + 13 + 11$$
$$+ 11 + 4 + 8 + 12$$
$$= 135$$
Step 2: Divide that sum by the number of items in the set.
$$135 \div 5 = 9$$
The mean of this set of numbers is 9.
Median:
Step 1: Rewrite the numbers in order from smallest to greatest.
3, 4, 6, 7, 8, 8, 9, 9, 11, 11, 11, 11, 12, 12, 13
Step 2: There is an odd number of data points (15), so select the middle value; this is the median. The middle value is 9.
Therefore, the median is 9.
Mode:
The mode, 11, is the number that appears most frequently.

25. $\dfrac{1760 \text{ ft}}{20 \text{ s}} = \dfrac{1760}{20} \dfrac{\text{ft}}{\text{s}} = 88 \dfrac{\text{ft}}{\text{s}}$

27. $12 \text{ ft} = 12 \text{ ft} \left(\dfrac{1 \text{ yd}}{3 \text{ ft}} \right) = 4 \text{ yd}$

$18 \text{ ft} = 18 \text{ ft} \left(\dfrac{1 \text{ yd}}{3 \text{ ft}} \right) = 6 \text{ yd}$

Area of the floor of room $= 4 \text{ yd} \times 6 \text{ yd}$
The price of the carpet is $17 per square yard.
Therefore,
$$\left(4 \text{ yd} \times 6 \text{ yd} \right) \times \dfrac{\$17}{\text{yd}^2} = 24 \text{ yd}^2 \times \dfrac{\$17}{\text{yd}^2}$$
$$= \$408.$$
The carpet will cost $408.

29. Perimeter of rectangle is
$$P = 2L + 2W$$
$$2L = 2 \times 8\frac{3}{5} = \frac{2}{1} \times \frac{43}{5} = \frac{86}{5}$$
$$2W = 2 \times 5\frac{7}{10} = \frac{2}{1} \times \frac{57}{10} = \frac{57}{5}$$

Therefore,
$$P = 2L + 2W = \frac{86}{5} + \frac{57}{5} = \frac{86 + 57}{5} = \frac{143}{5}$$
$$= 28\frac{3}{5} \text{ cm.}$$

The perimeter of the rectangle is $28\dfrac{3}{5}$ cm.

31. $\dfrac{A}{1,400} = \dfrac{9\frac{1}{2}}{100}$

$$100A = 1,400 \times \frac{19}{2} = 13,300$$

$$A = \frac{13,300}{100} = 133$$

$9\dfrac{1}{2}\%$ of 1,400 is 133.

33. $\dfrac{111}{B} = \dfrac{60}{100}$

$$60B = 111 \times 100$$
$$B = \frac{11,100}{60} = 185$$
111 is 60% of 185.

35. $\dfrac{2,968}{B} = \dfrac{106}{100}$
$$106B = 2,968 \cdot 100$$
$$B = \frac{296,800}{106}$$
$$B = 2,800$$
2,800 students attended the school last year.

37.

$$4x - 3y = 24$$
$$4(0) - 3y = 24$$
$$-3y = 24$$
$$y = -8; (0, -8)$$
$$4x - 3(8) = 24$$
$$4x = 48$$
$$x = 12; (12, 8)$$
$$4(3) - 3y = 24$$
$$-3y = 24 - 12 = 12$$
$$y = -4; (3, -4)$$
$$4x - 3(-8) = 24$$
$$4x = 0$$
$$x = 0; (0, -8)$$

39. $y = -7$

Two solutions are (0, –7) and (4, –7). The graph is the line through both points.

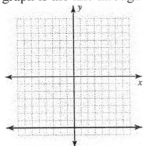

Chapter 10
An Introduction to Polynomials

Prerequisite Check

1. $5^4 = 5 \cdot 5 \cdot 5 \cdot 5 = 625$

3. $-3^4 = -(3)(3)(3)(3) = -81$

5. $2.3 \times 10^5 = 230,000$

7. $8^0 = 1$

9. $360 = 2 \cdot 2 \cdot 2 \cdot 3 \cdot 3 \cdot 5 = 2^3 \cdot 3^2 \cdot 5$

11. $5x - 2(3x - 4) = 5x - 6x + 8 = -x + 8$

13. $7x^2 + 4x - 3 = 7(-1)^2 + 4(-1) - 3$
$\qquad = 7(1) + 4(-1) - 3 = 7 - 4 - 3$
$\qquad = 7 - 7 = 0$

15. Let x be the first odd integer, and $x + 2$ be the second odd integer.
$$3(x) = 2(x + 2) + 5$$
$$3x = 2x + 4 + 5 = 2x + 9$$
$$3x - 2x = 2x - 2x + 9$$
$$x = 9; x + 2 = 11$$
The two numbers are 9 and 11.

Exercises 10.1

< Objectives 1 and 2 >

1. $x^4 \cdot x^5 = x^{4+5} = x^9$

3. $x^5 \cdot x^3 \cdot x^2 = x^{5+3+2} = x^{10}$

5. $3^5 \cdot 3^2 = 3^{5+2} = 3^7$

7. $(-2)^3 (-2)^5 = (-2)^{3+5} = (-2)^8 = 2^8$

9. $4 \cdot x^2 \cdot x^4 \cdot x^7 = 4x^{2+4+7} = 4x^{13}$

11. $\left(\dfrac{1}{2}\right)^2 \left(\dfrac{1}{2}\right)^3 \left(\dfrac{1}{2}\right) = \left(\dfrac{1}{2}\right)^{2+3+1} = \left(\dfrac{1}{2}\right)^6$

13. $(-2)^2 (-2)^3 (x^4)(x^5) = (-2)^{2+3} (x^{4+5})$
$\qquad = (-2)^5 x^9$

15. $(2x)^2 (2x)^3 (2x)^4 = (2x)^{2+3+4} = (2x)^9$

17. $(x^2 y^3)(x^4 y^2) = (x^2 x^4)(y^3 y^2) = x^6 y^5$

19. $(x^3 y^2)(x^4 y^2)(x^2 y^3) = (x^3 x^4 x^2)(y^2 y^2 y^3)$
$\qquad = x^9 y^7$

21. $(2x^4)(3x^3)(-4x^3) = [2 \cdot 3 \cdot (-4)](x^4 x^3 x^3)$
$\qquad = -24x^{10}$

23. $(5x^2)(3x^3)(x)(-2x^3)$
$\qquad = [5 \cdot 3 \cdot 1 \cdot (-2)](x^2 \cdot x^3 \cdot x \cdot x^3) = -30x^9$

25. $(5xy^3)(2x^2 y)(3xy)$
$\qquad = (5 \cdot 2 \cdot 3)(x \cdot x^2 \cdot x)(y^3 \cdot y \cdot y) = 30x^4 y^5$

27. $(x^2 yz)(x^3 y^5 z)(x^4 yz)$
$\qquad = (x^2 \cdot x^3 \cdot x^4)(y \cdot y^5 \cdot y)(z \cdot z \cdot z) = x^9 y^7 z^3$

29. $\dfrac{x^{10}}{x^7} = x^{10-7} = x^3$

31. $\dfrac{x^7 y^{11}}{x^4 y^3} = x^{7-4} y^{11-3} = x^3 y^8$

33. $\dfrac{x^5 y^4 z^2}{xy^2 z} = x^{5-1} y^{4-2} z^{2-1} = x^4 y^2 z$

35. $\dfrac{21 x^4 y^5}{7xy^2} = \dfrac{21}{7} x^{4-1} y^{5-2} = 3x^3 y^3$

37. $(-3x)(5x^5) = (-3 \cdot 5)(x \cdot x^5) = -15 x^6$

39. $(2x)^3 = 2^3 x^3 = 8x^3$

41. $\left(x^3\right)^7 = x^{3 \cdot 7} = x^{21}$

43. $(3x)(-2x)^3 = (3x)(-2)^3 x^3 = 3(-8)(x \cdot x^3)$
$\qquad\qquad = -24 x^4$

45. $\left(2x^3\right)^5 = 2^5 x^{3 \cdot 5} = 32 x^{15}$

47. $\left(-2x^2\right)^3 \left(3x^2\right)^3 = (-2)^3 x^{2 \cdot 3} (3)^3 x^{2 \cdot 3}$
$\qquad\qquad\qquad = (-8)(27)\left(x^6 \cdot x^6\right)$
$\qquad\qquad\qquad = -216 x^{12}$

49. $\left(3x^3\right)^2 \left(x^2\right)^4 = 3^2 x^{3 \cdot 2} x^{2 \cdot 4} = 9x^6 x^8 = 9 x^{14}$

51. $\left(\dfrac{3}{4}\right)^2 = \dfrac{3^2}{4^2} = \dfrac{9}{16}$

53. $\left(\dfrac{x}{5}\right)^3 = \dfrac{x^3}{5^3} = \dfrac{x^3}{125}$

55. $\left(\dfrac{m^3}{n^2}\right)^3 = \dfrac{\left(m^3\right)^3}{\left(n^2\right)^3} = \dfrac{m^9}{n^6}$

57. $\left(\dfrac{a^3 b^2}{c^4}\right)^2 = \dfrac{\left(a^3 b^2\right)^2}{\left(c^4\right)^2} = \dfrac{\left(a^3\right)^2 \left(b^2\right)^2}{c^8} = \dfrac{a^6 b^4}{c^8}$

59. $\left(\dfrac{2x^5}{y^3}\right)^2 = \dfrac{\left(2x^5\right)^2}{\left(y^3\right)^2} = \dfrac{2^2 \left(x^5\right)^2}{y^6} = \dfrac{4 x^{10}}{y^6}$

61. $(-8x^2 y)(-3x^4 y^5)^4$
$\quad = (-8x^2 y)\left[(-3)^4 \left(x^4\right)^4 \left(y^5\right)^4\right]$
$\quad = (-8x^2 y)\left(81 x^{16} y^{20}\right)$
$\quad = (-8 \cdot 81)\left(x^2 \cdot x^{16}\right)\left(y \cdot y^{20}\right) = -648 x^{18} y^{21}$

63. $\left(\dfrac{3x^4 y^9}{2x^2 y^7}\right)\left(\dfrac{x^6 y^3}{x^3 y^2}\right)^2$
$\quad = \left(\dfrac{3}{2} x^{4-2} y^{9-7}\right)\left(x^{6-3} y^{3-2}\right)^2$
$\quad = \left(\dfrac{3}{2} x^2 y^2\right)\left(x^3 y\right)^2 = \left(\dfrac{3}{2} x^2 y^2\right)\left(x^6 y^2\right)$
$\quad = \dfrac{3x^8 y^4}{2}$

65. False

67. False

69. With a scientific calculator:
4 $\boxed{x^y}$ 3 $\boxed{=}$
With a graphing calculator:
4 $\boxed{\wedge}$ 3 $\boxed{\text{ENTER}}$
Display: 64
The result is 64.

71. With a scientific calculator:
$\boxed{(}$ 3 $\boxed{+/-}$ $\boxed{)}$ $\boxed{x^y}$ 4 $\boxed{=}$
With a graphing calculator:
$\boxed{(}$ $\boxed{(-)}$ 3 $\boxed{)}$ $\boxed{\wedge}$ 4 $\boxed{\text{ENTER}}$
Display: 81
The result is 81.

73. With a scientific calculator:

2 $\boxed{x^y}$ 3 $\boxed{\times}$ 3 $\boxed{x^y}$ 5 $\boxed{=}$

With a graphing calculator:

$\boxed{(}$ 2 $\boxed{\wedge}$ 3 $\boxed{)}$ $\boxed{\times}$ $\boxed{(}$ 2 $\boxed{\wedge}$ 5 $\boxed{)}$ $\boxed{\text{ENTER}}$

Display: 256
The result is 256.

75. $\left(3x^2\right)\left(2x^4\right) = \left[3(2)^2\right]\left[2(2)^4\right] = 384$

77. $\left(2x^4\right)\left(4x^2\right) = \left[2(-2)^4\right]\left[4(-2)^2\right] = 512$

79. $\left(-2x^3\right)\left(-3x^5\right) = \left[-2(2)^3\right]\left[-3(2)^5\right] = 1,536$

81. If $P = \$2,000$, $r = 0.05$, and $t = 8$ then
$$A = P(1+r)^t = (2,000)(1+0.05)^8$$
$$= 2,000 \cdot 1.05^8 = 2,954.91$$
There will be \$2,954.91 in the account.

83. (a) If $m = 12$ and $v = 4.9t^2$, then
$$KE = \frac{1}{2}mv^2 = \frac{1}{2}(12)\left(4.9t^2\right)^2 = \frac{1}{2}(12)\left(24.01t^4\right)$$
$$= 144.06t^4$$

The kinetic energy of a 12-kg falling object is given by $KE = 144.06t^4$.

(b) $KE = 144.06t^4 = 144.06(4)^4$
$$= 144.06(256) = 36,879.36 \text{ joules}$$
The kinetic energy of a 12-kg object 4s after it is dropped is 36,879.36 joules.

85. Let D be the depth of the beam; L be the length of the beam; and M be the moment of inertia of the beam.

Then $D = 4L^3$ and $M = \frac{1}{9}D^3$.

So, $M = \frac{1}{9}D^3 = \frac{1}{9}\left(4L^3\right)^3 = \frac{1}{9}\left(64L^9\right) = \frac{64}{9}L^9$

The moment of inertia of a 4-in-wide beam is
$M = \frac{64}{9}L^9$.

87. Above and Beyond

89. $x^{12} = x^{2 \cdot 6} = \left(x^2\right)^6$

91. $a^{16} = a^{2 \cdot 8} = \left(a^2\right)^8$

93. $2^{12} = 2^{3 \cdot 4} = \left(2^3\right)^4 = 8^4$
$$2^{18} = 2^{3 \cdot 6} = \left(2^3\right)^6 = 8^6$$
$$\left(2^5\right)^3 = 2^{15} = 2^{3 \cdot 5} = \left(2^3\right)^5 = 8^5$$
$$\left(2^7\right)^6 = 2^{42} = 2^{3 \cdot 14} = \left(2^3\right)^{14} = 8^{14}$$

95. $-8x^6y^9z^{15} = (-2)^3 x^{2 \cdot 3} y^{3 \cdot 3} z^{5 \cdot 3} = \left(-2x^2y^3z^5\right)^3$
$-2x^2y^3z^5$ raised to the third power is $-8x^6y^9z^{15}$.

97. (a) If a country has a 2% growth rate for 35 years, then it takes that long for its population to double. Therefore, over a period of 105 years, $\frac{105}{35}$ doublings will occur. This means there will be 3 doublings.
$$G = (1+R)^y = (1+0.02)^{105} = (1.02)^{105}$$
$$= 7.998 \approx 8$$
The country's population will be 8 times as large.

(b) The total population of a less developed country in 2010 is 5.9 billion. If the average growth rate remains unchanged in the next 105 years, the population will be 8 times as large. Therefore the country's population after 105 years is: $5.9 \times 8 = 47.2$ billion.

99. Above and Beyond

Exercises 10.2

< Objectives 1 and 2 >

1. $x^{-5} = \frac{1}{x^5}$

3. $5^{-2} = \dfrac{1}{5^2} = \dfrac{1}{25}$

5. $(-5)^{-2} = \dfrac{1}{(-5)^2} = \dfrac{1}{25}$

7. $(-2)^{-3} = \dfrac{1}{(-2)^3} = \dfrac{1}{-8} = -\dfrac{1}{8}$

9. $\left(\dfrac{2}{3}\right)^{-3} = \left(\dfrac{3}{2}\right)^3 = \dfrac{3^3}{2^3} = \dfrac{27}{8}$

11. $3x^{-2} = 3 \cdot \dfrac{1}{x^2} = \dfrac{3}{x^2}$

13. $-5x^{-4} = -5 \cdot \dfrac{1}{x^4} = -\dfrac{5}{x^4}$

15. $(-3x)^{-2} = \dfrac{1}{(-3x)^2} = \dfrac{1}{9x^2}$

17. $\dfrac{1}{x^{-3}} = x^3$

19. $\dfrac{2}{5x^{-3}} = \dfrac{2}{5} \cdot \dfrac{1}{x^{-3}} = \dfrac{2}{5}x^3$

21. $\dfrac{x^{-3}}{y^{-4}} = \dfrac{y^4}{x^3}$

23. $x^5 \cdot x^{-3} = x^{5+(-3)} = x^2$

25. $a^{-9} \cdot a^6 = a^{-9+6} = a^{-3} = \dfrac{1}{a^3}$

27. $z^{-2} \cdot z^{-8} = z^{-2+(-8)} = z^{-10} = \dfrac{1}{z^{10}}$

29. $a^{-5} \cdot a^5 = a^{-5+5} = a^0 = 1$

31. $\dfrac{x^{-5}}{x^{-2}} = x^{-5-(-2)} = x^{-3} = \dfrac{1}{x^3}$

33. $(x^5)^3 = x^{5 \cdot 3} = x^{15}$

35. $(2x^{-3})(x^2)^4 = 2 \cdot x^{-3} \cdot x^{2 \cdot 4} = 2 \cdot x^{-3+8} = 2x^5$

37. $(3a^{-4})(a^3)(a^2) = 3 \cdot a^{-4+3+2} = 3 \cdot a^1 = 3a$

39. $(x^4 y)(x^2)^3 (y^3)^0 = (x^4 y)x^{2 \cdot 3} \cdot y^{3 \cdot 0}$
$$= (x^4 y) \cdot x^6 \cdot y^0$$
$$= (x^4 \cdot x^6)(y \cdot y^0) = x^{10} y$$

41. $(ab^2 c)(a^4)^4 (b^2)^3 (c^3)^4$
$$= a^{1+4 \cdot 4} \cdot b^{2+2 \cdot 3} \cdot c^{1+3 \cdot 4} = a^{17} b^8 c^{13}$$

43. $(x^5)^{-3} = x^{5(-3)} = x^{-15} = \dfrac{1}{x^{15}}$

45. $(b^{-4})^{-2} = b^{(-4)(-2)} = b^8$

47. $(x^5 y^{-3})^2 = (x^5)^2 (y^{-3})^2 = x^{5 \cdot 2} \cdot y^{-3 \cdot 2}$
$$= x^{10} \cdot y^{-6} = \dfrac{x^{10}}{y^6}$$

49. $(x^{-4} y^{-2})^{-3} = x^{(-4)(-3)} \cdot y^{(-2)(-3)} = x^{12} y^6$

51. $(2x^{-3} y^0)^{-5} = 2^{-5} \cdot x^{(-3)(-5)} \cdot y^{0(-5)} = \dfrac{x^{15} y^0}{2^5}$
$$= \dfrac{x^{15}}{32}$$

53. $\dfrac{x^{-2}}{y^{-4}} = \dfrac{y^4}{x^2}$

55. $\dfrac{x^{-4}}{y^{-2}} = \dfrac{y^2}{x^4}$

57. $(4x^{-2})^2 (3x^{-4}) = 4^2 \cdot 3 \cdot x^{-2 \cdot 2 + (-4)} = 16 \cdot 3x^{-8}$
$$= \dfrac{48}{x^8}$$

59. $(2x^5)^4 (x^3)^2 = 2^4 (x^5)^4 x^{3 \cdot 2} = 16 \cdot x^{5 \cdot 4 + 6}$
$$= 16x^{26}$$

61. $\left(2x^{-3}\right)^3\left(3x^3\right)^2 = 2^3\left(x^{-3}\right)^3 \cdot 3^2\left(x^3\right)^2$

$\qquad = 8 \cdot 9 \cdot x^{-3\cdot3} \cdot x^{3\cdot2} = 72 \cdot x^{-9+6}$

$\qquad = 72 \cdot x^{-3} = \dfrac{72}{x^3}$

63. $\left(xy^5z\right)^4\left(xyz^2\right)^8\left(x^6yz\right)^5$

$\qquad = \left(x^4y^{5\cdot4}z^4\right)\left(x^8y^8z^{2\cdot8}\right)\left(x^{6\cdot5}y^5z^5\right)$

$\qquad = x^{4+8+30} \cdot y^{20+8+5} \cdot z^{4+16+5} = x^{42}y^{33}z^{25}$

65. $\left(3x^{-2}\right)\left(5x^2\right)^2 = \left(3x^{-2}\right)\left(25x^4\right) = 75 \cdot x^{-2+4}$

$\qquad = 75x^2$

67. $\left(2w^3\right)^4\left(3w^{-5}\right)^2 = 2^4 \cdot w^{3\cdot4} \cdot 3^2 \cdot w^{-5\cdot2}$

$\qquad = 16 \cdot 9 \cdot w^{12-10} = 144w^2$

69. $\dfrac{3x^6}{2y^9} \cdot \dfrac{y^5}{x^3} = \dfrac{3}{2} \cdot x^{6-3} \cdot y^{5-9} = \dfrac{3}{2} \cdot x^3 \cdot y^{-4}$

$\qquad = \dfrac{3x^3}{2y^4}$

71. $\left(-7x^2y\right)\left(-3x^5y^6\right)^4$

$\qquad = \left(-7x^2y\right)\left[(-3)^4 x^{5\cdot4}y^{6\cdot4}\right]$

$\qquad = -7 \cdot 81 \cdot x^{2+20} \cdot y^{1+24} = -567x^{22}y^{25}$

73. $\left(2x^2y^{-3}\right)\left(3x^{-4}y^{-2}\right) = 6 \cdot x^{2+(-4)} \cdot y^{-3+(-2)}$

$\qquad = 6 \cdot x^{-2} \cdot y^{-5} = \dfrac{6}{x^2y^5}$

75. $\dfrac{\left(x^{-3}\right)\left(y^2\right)}{y^{-3}} = x^{-3} \cdot y^{2-(-3)} = x^{-3} \cdot y^5 = \dfrac{y^5}{x^3}$

77. $\dfrac{15x^{-3}y^2z^{-4}}{20x^{-4}y^{-3}z^2} = \dfrac{15}{20} \cdot x^{-3-(-4)} \cdot y^{2-(-3)} \cdot z^{-4-2}$

$\qquad = \dfrac{3}{4} \cdot x^1 \cdot y^5 \cdot z^{-6} = \dfrac{3xy^5}{4z^6}$

79. $\dfrac{x^{-5}y^{-7}}{x^0y^{-4}} = x^{-5-0} \cdot y^{-7-(-4)} = x^{-5} \cdot y^{-3} = \dfrac{1}{x^5y^3}$

81. $\dfrac{x^{-2}y^2}{x^3y^{-2}} \cdot \dfrac{x^{-4}y^2}{x^{-2}y^{-2}} = \dfrac{x^{-2+(-4)} \cdot y^{2+2}}{x^{3+(-2)} \cdot y^{-2+(-2)}} = \dfrac{x^{-6} \cdot y^4}{x \cdot y^{-4}}$

$\qquad = x^{-6-1} \cdot y^{4-(-4)} = x^{-7} \cdot y^8$

$\qquad = \dfrac{y^8}{x^7}$

< Objective 3 >

83. $9 \times 10^4 = 90{,}000$

85. $1.21 \times 10^1 = 12.1$

87. $8 \times 10^{-3} = 0.008$

89. $2.8 \times 10^{-5} = 0.000028$

91. $3{,}420{,}000 = 3.42 \times 10^6$

93. $8 = 8 \times 10^0$

95. $0.0005 = 5 \times 10^{-4}$

97. $0.00037 = 3.7 \times 10^{-4}$

99. $93{,}000{,}000 = 9.3 \times 10^7$

101. $130{,}000{,}000{,}000 = 1.3 \times 10^{11}$

103. $1.98 \times 10^{30} = 198 \times 10^{28}$
28 zeros follow the digit 8.

105. $\left(2 \times 10^5\right)\left(4 \times 10^4\right) = (2 \times 4)\left(10^{5+4}\right) = 8 \times 10^9$

107. $\dfrac{6 \times 10^9}{3 \times 10^7} = 2 \times 10^{9-7} = 2 \times 10^2$

109. $\dfrac{\left(3.3 \times 10^{15}\right)\left(6 \times 10^{15}\right)}{\left(1.1 \times 10^8\right)\left(3 \times 10^6\right)} = \left(\dfrac{3 \times 3 \times 6}{1.1 \times 3}\right)10^{15+15-8-6}$

$\qquad = 6 \times 10^{16}$

111. $\left(4 \times 10^{-3}\right)\left(2 \times 10^{-5}\right) = (4 \times 2)\left(10^{-3} \times 10^{-5}\right)$

$\qquad = 8 \times 10^{-3+(-5)} = 8 \times 10^{-8}$

113. $\dfrac{9 \times 10^3}{3 \times 10^{-2}} = 3 \times 10^{3-(-2)} = 3 \times 10^5$

< Objective 4 >

115. $\dfrac{6.6 \times 10^{17}}{10^{16}} = 6.6 \times 10^{17-16} = 6.6 \times 10^1 = 66$

It takes light approximately 66 years to travel from Megrez to Earth.

117. $15,500 \times 10^{19} = 1.55 \times 10^4 \times 10^{19} = 1.55 \times 10^{23}$
The amount of water on Earth is approximately 1.55×10^{23} liters.

119. $\left(7.1 \times 10^9\right)\left(6.57 \times 10^5\right) \approx 46.6 \times 10^{14}$
$$= 4.66 \times 10^{15}$$
There is 4.66×10^{15} L of freshwater on the Earth.

121. False

123. True

125. **(a)** $12 \times 10^4 \ \Omega = 120,000 \ \Omega$
(b) $1.2 \times 10^5 \ \Omega$

127. $c = \dfrac{A}{\varepsilon d} = \dfrac{0.254}{\left(3.6 \times 10^3\right)(1.4)} = \dfrac{2.54 \times 10^{-1}}{5.04 \times 10^3}$
$$\approx 0.504 \times 10^{-1-3} = 0.504 \times 10^{-4}$$
$$= 5.04 \times 10^{-5}$$
The concentration is approximately 5.04×10^{-5} moles per liter.

129. $x^{2n} \bullet x^{3n} = x^{2n+3n} = x^{5n}$

131. $\dfrac{x^{n+3}}{x^{n+1}} = x^{(n+3)-(n+1)} = x^2$

133. $\left(y^n\right)^{3n} = y^{n \bullet 3n} = y^{3n^2}$

135. $\dfrac{x^{2n} \bullet x^{n+2}}{x^{3n}} = x^{2n+(n+2)-3n} = x^2$

137. $P = 7 \times 2^{(2025-2010)/35} \approx 9.42$
(rounded to two decimal places)
Earth's population in 2025 will be approximately 9.42 billion.

139. $P = 310 \times 2^{(2025-2010)/66} \approx 363$
The U.S. population in 2025 will be approximately 363 million, rounded to the nearest million.

141. **(a)** 30 cuts gives 2^{30} pieces of paper.
$$\left(2^{30}\right)(0.002) \approx 2,147,484 \text{ in.} = 178,957 \text{ ft}$$
$$\approx 33.9 \text{ mi}$$
The stack would be approximately 33.9 miles high.
(b) $\left(2^{50}\right)(0.002) \approx 2.2517998 \times 10^{12}$ in
$$\approx 1.8764998 \times 10^{11} \text{ ft}$$
$$\approx 35,539,770 \text{ mi}$$
The stack would be approximately 35.5 million mi high.

143. Above and Beyond

145. Plan 2:

Day	Amount paid on that day(p)	Total for all days up to and including that day (t)
1	0.01	0.01
2	0.02	0.03 or $0.01(2^2-1)$
3	0.04	0.07 or $0.01(2^3-1)$
4	0.08	0.15 or $0.01(2^4-1)$
5	0.16	0.31 or $0.01(2^5-1)$
$\vdots$	$\vdots$	$\vdots$
28	$0.01(2)^{27}$	$0.01(2^{28}-1)=2{,}684{,}354.55$
$\vdots$	$\vdots$	$\vdots$
n	$0.01(2)^{n-1}$	$0.01(2^n-1)$

After 28 days, Plan 1 is worth $4,000,000 while Plan 2 is worth $2,684,354.55. Plan 1 is the better choice. The formula for the amount that you make on *n*th day is

$$p = 0.01(2)^{n-1} = \frac{2^{n-1}}{100}.$$

The formula for the total after *n* days is

$$t = 0.01(2^n-1) = \frac{2^n-1}{100} = \frac{2^n}{100} - \frac{1}{100}.$$

Exercises 10.3

< Objective 1 >

1. $7x^3$ is a polynomial.

3. $2x^5y^3 - 4x^2y^4$ is a polynomial.

5. -7 is a polynomial.

7. $\dfrac{3+x}{x^2}$ is not a polynomial.

9. Terms: $2x^2, -3x$
Coefficients: 2, -3

11. Terms: $4x^3, -3x, 2$
Coefficients: 4, -3, 2

13. $4x^3 - 2x^2$ is a binomial.

15. $7y^2 + 4y + 5$ is a trinomial.

17. $2x^4 - 3x^2 + 5x - 2$ is not classified.

19. $7x^{10}$ is a monomial.

21. $x^5 - \dfrac{3}{x^2}$ is not a polynomial.

< Objectives 2 and 3 >

23. $4x^5 - 3x^2$ has degree 5. The leading coefficient is 4.

25. $-5x^9 + 7x^7 + 4x^3$ has degree 9. The leading coefficient is -5.

27. $-9x$ has degree 1. The leading coefficient is -9.

29. $x^6 - 3x^5 + 5x^2 - 7$ has degree 6. The leading coefficient is 1.

31. $x = 1$
$9x - 2 = 9(1) - 2 = 7$
     $x = -1$
$9x - 2 = 9(-1) - 2 = -11$

33. $x = 2$
$x^3 - 2x = (2)^3 - 2(2) = 8 - 4 = 4$
     $x = -2$
$x^3 - 2x = (-2)^3 - 2(-2) = -8 + 4 = -4$

35. $x = 4$
$3x^2 + 4x - 2 = 3(4)^2 + 4(4) - 2$
$= 3\cdot16 + 16 - 2 = 62$
     $x = -4$
$3x^2 + 4x - 2 = 3(-4)^2 + 4(-4) - 2$
$= 3\cdot16 - 16 - 2 = 30$

37.
$$x = 3$$
$$-x^2 - x + 12 = -(3)^2 - (3) + 12 = -9 - 3 + 12$$
$$= 0$$
$$x = -4$$
$$-x^2 - x + 12 = -(-4)^2 - (-4) + 12$$
$$= -16 + 4 + 12 = 0$$

39. The polynomial to describe the cost of typing x number of pages is $5x + 35$. For 50-page paper: $5(50) + 35 = 285$. The cost of typing a 50-page paper is $285.

41. $4(12)^2 - 95 = 481$

The revenue, when 12 pairs of slippers are sold, is $481.

43. always

45. sometimes

47. sometimes

49. sometimes

51. $Wh = 58t + 144 = 58(0) + 144 = 144$

In the 24-hour period of non-operation, the TV consumed 144 Wh.

53. $g = 0.472t^3 - 5.298t^2 + 11.802t + 93.143$
$$= 0.472(5)^3 - 5.298(5)^2 + 11.802(5)$$
$$+ 93.143$$
$$= 78.703$$

55. $P(1) = (1)^3 - 2(1)^2 + 5 = 4$

57. $Q(2) = 2(2)^2 + 3 = 11$

59. $P(3) = (3)^3 - 2(3)^2 + 5 = 14$

61. $P(0) = (0)^3 - 2(0)^2 + 5 = 5$

63. $P(2) + Q(-1)$
$$= \left[(2)^3 - 2(2)^2 + 5\right] + \left[2(-1)^2 + 3\right] = 5 + 5$$
$$= 10$$

65. $P(3) - Q(-3) \div Q(0)$
$$= \left[(3)^3 - 2(3)^2 + 5\right] - \left[2(-3)^2 + 3\right]$$
$$\div \left[2(0)^2 + 3\right]$$
$$= 14 - (21 \div 3) = 7$$

67. $|Q(4)| - |P(4)|$
$$= \left|2(4)^2 + 3\right| - \left|(4)^3 - 2(4)^2 + 5\right| = |35| - |37|$$
$$= -2$$

Exercises 10.4

< Objective 1 >

1. $(5a - 7) + (4a + 11) = 5a - 7 + 4a + 11$
$$= (5a + 4a) + (-7 + 11)$$
$$= 9a + 4$$

3. $(8b^2 - 11b) + (5b^2 - 7b)$
$$= 8b^2 - 11b + 5b^2 - 7b = 13b^2 - 18b$$

5. $(3x^2 - 2x) + (-5x^2 + 2x) = -2x^2$

7. $(2x^2 + 5x - 3) + (3x^2 - 7x + 4)$
$$= 2x^2 + 5x - 3 + 3x^2 - 7x + 4 = 5x^2 - 2x + 1$$

9. $(3b^2 - 7) + (2b - 7) = 3b^2 - 7 + 2b - 7$
$$= 3b^2 + 2b - 14$$

11. $(8y^3 - 5y^2) + (5y^2 - 2y) = 8y^3 - 2y$

13. $(2a^2 - 4a^3) + (3a^3 + 2a^2) = -a^3 + 4a^2$

15. $\left(7x^2 - 5 + 4x\right) + \left(8 - 5x - 9x^2\right)$

$= -2x^2 - x + 3$

17. $-\left(4a + 5b\right) = -4a - 5b$

19. $5a - \left(2b - 3c\right) = 5a - 2b + 3c$

21. $9r - \left(3r + 5s\right) = 9r - 3r - 5s = 6r - 5s$

23. $5p - \left(-3p + 2q\right) = 5p + 3p - 2q = 8p - 2q$

< Objective 2 >

25. $\left(2x - 3\right) - \left(x + 4\right) = 2x - 3 - x - 4 = x - 7$

27. $\left(4m^2 - 5m\right) - \left(3m^2 - 2m\right)$

$= 4m^2 - 5m - 3m^2 + 2m = m^2 - 3m$

29. $\left(4y^2 + 5y\right) - \left(6y^2 + 5y\right)$

$= 4y^2 + 5y - 6y^2 - 5y = -2y^2$

31. $\left(3x^2 - 5x - 2\right) - \left(x^2 - 4x - 3\right)$

$= 3x^2 - 5x - 2 - x^2 + 4x + 3 = 2x^2 - x + 1$

33. $\left(8a^2 - 9a\right) - \left(3a + 7\right) = 8a^2 - 9a - 3a - 7$

$= 8a^2 - 12a - 7$

35. $\left(5b - 2b^2\right) - \left(4b^2 - 3b\right)$

$= 5b - 2b^2 - 4b^2 + 3b = -6b^2 + 8b$

37. $\left(7x^2 - 11x + 31\right) - \left(5x^2 + 19 - 11x\right)$

$= 7x^2 - 11x + 31 - 5x^2 - 19 + 11x = 2x^2 + 12$

39. $\left(2x^3 - 3x + 1\right) - \left(4x^2 + 8x + 6\right)$

$= 2x^3 - 3x + 1 - 4x^2 - 8x - 6$

$= 2x^3 - 4x^2 - 11x - 5$

41. $\left[\left(5b - 4\right) + \left(3b + 8\right)\right] - \left(2b + 5\right)$

$= \left(8b + 4\right) - \left(2b + 5\right) = 8b + 4 - 2b - 5$

$= 6b - 1$

43. $\left[\left(x^2 + 5x - 2\right) + \left(2x^2 + 7x - 8\right)\right]$

$\qquad - \left(3x^2 + 2x - 1\right)$

$= \left(3x^2 + 12x - 10\right) - \left(3x^2 + 2x - 1\right)$

$= 3x^2 + 12x - 10 - 3x^2 - 2x + 1 = 10x - 9$

45. $\left[\left(4x^2 - 5\right) + \left(2x - 7\right)\right] - \left(2x^2 - 3x\right)$

$= \left(4x^2 + 2x - 12\right) - \left(2x^2 - 3x\right)$

$= 4x^2 + 2x - 12 - 2x^2 + 3x = 2x^2 + 5x - 12$

47. $\left(2y^2 - 8y\right) - \left[\left(3y^2 - 3y\right) + \left(5y^2 + 3y\right)\right]$

$= \left(2y^2 - 8y\right) - \left(8y^2\right) = -6y^2 - 8y$

49. $\left[\left(9x^2 - 3x + 5\right) - \left(3x^2 + 2x - 1\right)\right]$

$\qquad - \left(x^2 - 2x - 3\right)$

$= \left(9x^2 - 3x + 5 - 3x^2 - 2x + 1\right)$

$\qquad - \left(x^2 - 2x - 3\right)$

$= \left(6x^2 - 5x + 6\right) - \left(x^2 - 2x - 3\right)$

$= 6x^2 - 5x + 6 - x^2 + 2x + 3 = 5x^2 - 3x + 9$

51.

$$\begin{array}{r} 4w^2 \qquad + 11 \\ 7w \quad -9 \\ \underline{2w^2 - 9w \qquad} \\ 6w^2 - 2w \quad +2 \end{array}$$

53.

$$\begin{array}{r} 3x^2 + 3x - 4 \\ 4x^2 - 3x - 3 \\ \underline{2x^2 - \ x + 7} \\ 9x^2 - x \end{array}$$

55.

$$\begin{array}{r} 9a^2 - 4a \\ \underline{-\left(7a^2 - 9a\right)} \\ \\ \end{array} \qquad \begin{array}{r} 9a^2 - 4a \\ \underline{-7a^2 + 9a} \\ 2a^2 + 5a \end{array}$$

57.

$$\begin{array}{r} 8x^2 - 5x + 7 \\ \underline{-\left(5x^2 - 6x + 7\right)} \\ \\ \end{array} \qquad \begin{array}{r} 8x^2 - 5x + 7 \\ \underline{-5x^2 + 6x - 7} \\ 3x^2 + x \end{array}$$

59.
$$\begin{array}{r} 8x^2 - 9 \\ -\left(5x^2 - 3x\right) \\ \hline \end{array}$$

$$\begin{array}{r} 8x^2 \qquad -9 \\ -5x^2 + 3x \\ \hline 3x^2 + 3x - 9 \end{array}$$

61. $P = 2L + 2W = 2(8x + 9) + 2(6x - 7)$

$= 16x + 18 + 12x - 14 = 28x + 4$
The perimeter is given by $28x + 4$.

63. Profit $(P) =$ Revenue $(R) -$ Cost (C)

$= \left(90x - x^2\right) - \left(150 + 25x\right)$

$= 90x - x^2 - 150 - 25x$

$= -x^2 + 65x - 150$
The profit can be represented by
$-x^2 + 65x - 150$.

65. sometimes

67. The shear polynomial S_v for a polymer after vulcanization is

$S_v = \left(0.4x^2 - 144x + 318\right)$

$+ \left(0.2x^2 - 14x + 144\right)$

$= 0.6x^2 - 158x + 452$

69. If $\alpha = 3.9 \times 10^{-3}$ and $R_0 = 1.72 \times 10^{-8}$, then

$R_t = R_0 \left(1 + \alpha t\right)$

$= \left(1.72 \times 10^{-8}\right)\left(1 + 3.9 \times 10^{-3} t\right)$

$= \left(1.72 \times 10^{-8}\right)(1) + \left(1.72 \times 10^{-8}\right)\left(1 + 3.9 \times 10^{-3} t\right)$

$= 1.72 \times 10^{-8} + 6.708 \times 10^{-11} t$
The resistance R_t of the copper piece is
$R_t = 1.72 \times 10^{-8} + 6.708 \times 10^{-11} t$.

71. $3ax^4 - 5x^3 + x^2 - cx + 2$

$= 9x^4 - bx^3 + x^2 - 2d$

$3a = 9 \quad -5 = -b \quad -c = 0 \quad 2 = -2d$

$a = 3 \quad b = 5 \quad c = 0 \quad d = -1$

Exercises 10.5

< Objectives 1 and 2 >

1. $\left(5x^2\right)\left(3x^3\right) = (5 \bullet 3)\left(x^2 x^3\right) = 15x^5$

3. $\left(-2b^2\right)\left(14b^8\right) = (-2 \bullet 14)\left(b^2 b^8\right) = -28b^{10}$

5. $\left(-5p^7\right)\left(-8p^6\right) = \left[(-5)(-8)\right]\left(p^7 p^6\right)$

$= 40p^{13}$

7. $\left(4m^5\right)(-3m) = \left[4 \bullet (-3)\right]\left(m^5 m\right) = -12m^6$

9. $\left(4x^3 y^2\right)\left(8x^2 y\right) = (4 \bullet 8)\left(x^3 x^2\right)\left(y^2 y\right)$

$= 32x^5 y^3$

11. $\left(-3m^5 n^2\right)\left(2m^4 n\right) = (-3 \bullet 2)\left(m^5 m^4\right)\left(n^2 n\right)$

$= -6m^9 n^3$

13. $10(x + 3) = 10 \bullet x + 10 \bullet 3 = 10x + 30$

15. $3a(4a + 5) = 3a \bullet 4a + 3a \bullet 5 = 12a^2 + 15a$

17. $3s^2\left(4s^2 - 7s\right) = 3s^2 \bullet 4s^2 - 3s^2 \bullet 7s$

$= 12s^4 - 21s^3$

19. $3x\left(5x^2 - 3x - 1\right) = 3x \bullet 5x^2 - 3x \bullet 3x - 3x \bullet 1$

$= 15x^3 - 9x^2 - 3x$

21. $3xy\left(2x^2 y + xy^2 + 5xy\right)$

$= 3xy \bullet 2x^2 y + 3xy \bullet xy^2 + 3xy \bullet 5xy$

$= 6x^3 y^2 + 3x^2 y^3 + 15x^2 y^2$

23. $6m^2 n\left(3m^2 n - 2mn + mn^2\right)$

$= 6m^2 n \bullet 3m^2 n - 6m^2 n \bullet 2mn + 6m^2 n \bullet mn^2$

$= 18m^4 n^2 - 12m^3 n^2 + 6m^3 n^3$

25. $(x + 3)(x + 2) = x \bullet x + x \bullet 2 + 3 \bullet x + 3 \bullet 2$

$= x^2 + 5x + 6$

27. $(m-5)(m-9)=m^2-9m-5m+45$
$$=m^2-14m+45$$

29. $(p-8)(p+7)=p^2+7p-8p-56$
$$=p^2-p-56$$

31. $(w-7)(w+6)=w^2+6w-7w-42$
$$=w^2-w-42$$

33. $(3x-5)(x-8)=3x^2-24x-5x+40$
$$=3x^2-29x+40$$

35. $(2x-3)(3x+4)=6x^2+8x-9x-12$
$$=6x^2-x-12$$

37. $(3a-b)(4a-9b)$
$$=12a^2-27ab-4ab+9b^2$$
$$=12a^2-31ab+9b^2$$

39. $(3p-4q)(7p+5q)$
$$=21p^2+15pq-28pq-20q^2$$
$$=21p^2-13pq-20q^2$$

41.
$$\begin{array}{r} 2x^2\ -x\ +8 \\ x^2+4x\ +6 \\ \hline 12x^2-6x+48 \\ 8x^3\ -4x^2+32x \\ 2x^4\ -x^3\ +8x^2 \\ \hline 2x^4+7x^3+16x^2+26x+48 \end{array}$$

< Objective 3 >

43. $(x+3)^2=(x)^2+2(3)(x)+(3)^2$
$$=x^2+6x+9$$

45. $(w-6)^2=w^2-2\bullet 6\bullet w+(-6)^2$
$$=w^2-12w+36$$

47. $(z+12)^2=z^2+2\bullet 12\bullet z+144$
$$=z^2+24z+144$$

49. $(2a-1)^2=(2a)^2-2\bullet 1\bullet 2a+1$
$$=4a^2-4a+1$$

51. $(6m+1)^2=(6m)^2+2\bullet 1\bullet 6m+1$
$$=36m^2+12m+1$$

53. $(3x-y)^2=(3x)^2-2(3x)(y)+(-y)^2$
$$=9x^2-6xy+y^2$$

55. $(2r+5s)^2=4r^2+2(2r)(5s)+25s^2$
$$=4r^2+20rs+25s^2$$

57. $(6a-5b)^2=36a^2-2(6a)(5b)+25b^2$
$$=36a^2-60ab+25b^2$$

59. $\left(x+\dfrac{1}{2}\right)^2=x^2+2\bullet\dfrac{1}{2}\bullet x+\left(\dfrac{1}{2}\right)^2$
$$=x^2+x+\dfrac{1}{4}$$

< Objective 4 >

61. $(x-6)(x+6)=(x)^2-(6)^2=x^2-36$

63. $(m+7)(m-7)=m^2-49$

65. $\left(x-\dfrac{1}{2}\right)\left(x+\dfrac{1}{2}\right)=x^2-\left(\dfrac{1}{2}\right)^2=x^2-\dfrac{1}{4}$

67. $(p-0.4)(p+0.4)=p^2-(0.4)^2$
$$=p^2-0.16$$

69. $(a-3b)(a+3b)=a^2-(3b)^2=a^2-9b^2$

71. $(3x+2y)(3x-2y)=(3x)^2-(2y)^2$
$$=9x^2-4y^2$$

73. $(8w+5z)(8w-5z)=64w^2-25z^2$

75. $(5x-9y)(5x+9y)=25x^2-81y^2$

77. $2x(3x-2)(4x+1) = 2x(12x^2-5x-2)$
$$= 24x^3-10x^2-4x$$

79. $5a(4a-3)(4a+3) = 5a(16a^2-9)$
$$= 80a^3-45a$$

81. $3s(5s-2)(4s-1) = 3s(20s^2-13s+2)$
$$= 60s^3-39s^2+6s$$

83. $(x-2)(x+1)(x-3)$
$$= (x-2)(x^2-2x-3)$$
$$= x^3-2x^2-3x-2x^2+4x+6$$
$$= x^3-4x^2+x+6$$

85. $(a-1)^3 = (a-1)(a-1)^2$
$$= (a-1)(a^2-2a+1)$$
$$= a^3-2a^2+a-a^2+2a-1$$
$$= a^3-3a^2+3a-1$$

87. $\left(\dfrac{x}{2}+\dfrac{2}{3}\right)\left(\dfrac{2x}{3}-\dfrac{2}{5}\right)$
$$= \frac{x}{2}\cdot\frac{2x}{3}-\frac{x}{2}\cdot\frac{2}{5}+\frac{2}{3}\cdot\frac{2x}{3}-\frac{2}{3}\cdot\frac{2}{5}$$
$$= \frac{x^2}{3}-\frac{x}{5}+\frac{4x}{9}-\frac{4}{15} = \frac{x^2}{3}+\frac{11x}{45}-\frac{4}{15}$$

89. $[x+(y-2)][x-(y-2)]$
$$= x^2-(y-2)^2 = x^2-(y^2-4y+4)$$
$$= x^2-y^2+4y-4$$

91. $A = LW = (3x+5)(2x-7) = 6x^2-11x-35$
The area of the rectangle is
$(6x^2-11x-35)$ cm.

93. R represents the revenue generated from the sales of x vases.
$$R = xp = x(12-0.05x) = 12x-0.05x^2$$

95. $(5x-4)^2 = 25x^2-40x+16$
There are $25x^2-40x+16$ trees in the orchard.

97. Let x be the length of the rectangle. Then $x+2$ represents its width.
$$A = LW = x(x+2) = x^2+2x.$$

99. Let x be the number.
$$(x+6)(x-6) = x^2-36$$

101. Let x be the number.
$$(x-4)^2 = x^2-8x+16$$

103. $(x+y)^2 = x^2+2xy+y^2 \neq x^2+y^2$
False

105. $(x+y)^2 = x^2+2xy+y^2$
True

107. $M = \dfrac{wc^2}{8}\cdot\dfrac{c^2}{2} = \dfrac{wc^4}{16}$
The new bending moment with a load on the cantilever is $M = \dfrac{wc^4}{16}$.

109. The allowable stress of a material after allowing is
$$\left(86.2x-0.6x^2+258\right)\left(\frac{p}{8.6}\right)$$
$$= (86.2x)\left(\frac{p}{8.6}\right)-0.6x^2\left(\frac{p}{8.6}\right)+258\left(\frac{p}{8.6}\right)$$
$$\approx 10.02xp-0.07x^2p+30p$$
(rounded to two decimal places).

111. Above and Beyond

113. Above and Beyond

115. Above and Beyond

117. $(49)(51) = (50-1)(50+1) = 2,500-1$
$$= 2,499$$

119. $(34)(26) = (30 + 4)(30 - 4) = 900 - 16$
$$= 884$$

121. $(55)(65) = 60^2 - 5^2 = 3,600 - 25 = 3,575$

Exercises 10.6

< Objective 1 >

1. $10 = 2 \cdot 5$
$12 = 2 \cdot 2 \cdot 3$
The GCF is 2.

3. $16 = 2 \cdot 2 \cdot 2 \cdot 2$
$32 = 2 \cdot 2 \cdot 2 \cdot 2 \cdot 2$
$88 = 2 \cdot 2 \cdot 2 \cdot 11$
The GCF is 8.

5. $x^2 = x \cdot x$
$x^5 = x \cdot x \cdot x \cdot x \cdot x$
The GCF is x^2.

7. $a^3 = a^3$
$a^6 = a^3 \cdot a^3$
$a^9 = a^3 \cdot a^6$
The GCF is a^3.

9. $5x^4 = 5x^4$
$10x^5 = 2x \cdot 5x^4$
The GCF is $5x^4$.

11. $8a^4 = 2 \cdot 2 \cdot 2a^4$
$6a^6 = 3a^2 \cdot 2a^4$
$10a^{10} = 5a^6 \cdot 2a^4$
The GCF is $2a^4$.

13. $9x^2y = 3x \cdot 3xy$
$12xy^2 = 4y \cdot 3xy$
$15x^2y^2 = 5xy \cdot 3xy$
The GCF is $3xy$.

15. $15ab^3 = 3ab^2 \cdot 5b$
$10a^2bc = 2a^2c \cdot 5b$
$25b^2c^3 = 5bc^3 \cdot 5b$
The GCF is $5b$.

17. $15a^2bc^2 = 5a \cdot 3abc^2$
$9ab^2c^2 = 3b \cdot 3abc^2$
$6a^2b^2c^2 = 2ab \cdot 3abc^2$
The GCF is $3abc^2$.

19. $5 = 1 \cdot 5$
$12xy = 1 \cdot 12xy$
$4ab = 1 \cdot 4ab$
The GCF is 1.

< Objective 2 >

21. $8a + 4 = 4 \cdot 2a + 4 \cdot 1 = 4(2a + 1)$

23. $24m - 32n = 8 \cdot 3m + 8(-4n) = 8(3m - 4n)$

25. $12m + 8 = 4 \cdot 3m + 4 \cdot 2 = 4(3m + 2)$

27. $10s^2 + 5s = 5s \cdot 2s + 5s \cdot 1 = 5s(2s + 1)$

29. $12x^2 + 12x = 12x \cdot x + 12x \cdot 1 = 12x(x + 1)$

31. $15a^3 - 25a^2 = 5a^2 \cdot 3a - 5a^2 \cdot 5$
$$= 5a^2(3a - 5)$$

33. $6pq + 18p^2q = 6pq \cdot 1 + 6pq \cdot 3p$
$$= 6pq(1 + 3p)$$

35. $6x^2 - 18x + 30 = 6x^2 - 6 \cdot 3x + 6 \cdot 5$
$$= 6(x^2 - 3x + 5)$$

37. $3a^3 + 6a^2 - 12a = 3a \cdot a^2 + 3a \cdot 2a - 3a \cdot 4$
$$= 3a(a^2 + 2a - 4)$$

39. $6m + 9mn - 12mn^2$

$= 3m \cdot 2 + 3m \cdot 3n - 3m \cdot 4n^2$

$= 3m\left(2 + 3n - 4n^2\right)$

41. $10r^3s^2 + 25r^2s^2 - 15r^2s^3$

$= 5r^2s^2 \cdot 2r + 5r^2s^2 \cdot 5 - 5r^2s^2 \cdot 3s$

$= 5r^2s^2\left(2r + 5 - 3s\right)$

43. $9a^5 - 15a^4 + 21a^3 - 27a$

$= 3a \cdot 3a^4 - 3a \cdot 5a^3 + 3a \cdot 7a^2 - 3a \cdot 9$

$= 3a\left(3a^4 - 5a^3 + 7a^2 - 9\right)$

45. The expression $m^2 + 1$ is a prime polynomial since it cannot be factored.

47. The expression $4x - 7$ is a prime polynomial since it cannot be factored.

< Objective 3 >

49. $-x^2 + 6x + 10 = -\left(x^2 - 6x - 10\right)$

51. $-3x^2 + 4x - 9 = -\left(3x^2 - 4x + 5\right)$

53. $-4x^2 - 16x + 4 = -4\left(x^2 + 4x - 1\right)$

55. $-4m^2n^3 - 6mn^3 - 10n^2$

$= -2n^2\left(2m^2n + 3mn + 5\right)$

57. Correct

59. Incorrect

61. Correct

63. Incorrect

65. sometimes

67. always

69. $v_0 t - 4.9t^2 = t\left(v_0 - 4.9t\right)$

71. $(2x - 6)(5x + 10) = 2(x - 3) \cdot 5(x + 2)$

$= 10(x - 3)(x + 2)$

The GCF is $2 \cdot 5 = 10$.

73. $\left(2x^3 - 4x\right)(3x + 6) = 2x\left(x^2 - 2\right) \cdot 3(x + 2)$

$= 6x\left(x^2 - 2\right)(x + 2)$

The GCF is $2x \cdot 3 = 6x$.

75. The GCF of $2a + 8$ is 2.
The GCF of $3a - 6$ is 3.
The GCF for the product is $2 \cdot 3 = 6$.

77. The GCF of $2x^2 + 5x$ is x.
The GCF of $7x - 14$ is 7.
The GCF for the product is $x \cdot 7 = 7x$.

79. $33t - t^2 = t(33 - t)$

The length of the rectangle is $33 - t$.

81. Above and Beyond

83. Above and Beyond

Summary Exercises

1. $r^4 r^9 = r^{4+9} = r^{13}$

3. $(2w)^{-3} = \dfrac{1}{(2w)^3} = \dfrac{1}{2^3 w^3} = \dfrac{1}{8w^3}$

5. $y^{-5} y^2 = y^{-5+2} = y^{-3} = \dfrac{1}{y^3}$

7. $\dfrac{x^{12}}{x^{15}} = x^{12-15} = x^{-3} = \dfrac{1}{x^3}$

9. $\left(5a^2b^3\right)\left(2a^{-2}b^{-6}\right) = 5 \cdot 2a^{2+(-2)}b^{3+(-6)}$

$= 10a^0b^{-3} = \dfrac{10}{b^3}$

11. $\left(\dfrac{m^{-3}n^{-3}}{m^{-4}n^4}\right)^3 = \left(m^{-3-(-4)}n^{-3-4}\right)^3 = \left(mn^{-7}\right)^3$

$= m^{1\cdot3}n^{-7\cdot3} = m^3n^{-21} = \dfrac{m^3}{n^{21}}$

13. $\left(\dfrac{x^3y^4}{x^6y^2}\right)^3 = \left(\dfrac{y^2}{x^3}\right)^3 = \dfrac{y^6}{x^9}$

15. $\left(2a^3\right)^0\left(-3a^4\right)^2 = 1\cdot(-3)^2\left(a^4\right)^2 = 9a^8$

17. $0.0000425 = 4.25\times10^{-5}$

19. Let $P = \$2,000$, $r = 0.027$, $t = 5$ years.

$A = P\left(1+r\right)^t = 2,000\left(1+0.027\right)^5 = 2,000\cdot1.027^5$

$= 2,284.98$

Thus, the value of the investment is $2,284.98.

21. Writing in scientific notation, 3,794,101 mi^2, rounding to one decimal place is 3.8×10^6 mi^2.

23. The percentage of U.S. surface area not covered by water $= \left(100 - 6.76\right) = 93.24\%$.

$\dfrac{93.24}{100}\times3,794,101 = 3,537,619.77$

$\approx 3,500,000$

$= 3.5\times10^6$ mi^2

25. $6x^4 - 3x$ is a binomial.

27. $4x^5 - 8x^3 + 5$ is a trinomial.

29. $-7a^4 - 9a^3$ is a binomial.

31. $5x^5 + 3x^2$ has degree 5.

33. $6x^2 + 4x^4 + 6 = 4x^4 + 6x^2 + 6$ has degree 4.

35. -8 has degree 0.

37. $\left(7a^2 + 3a\right) + \left(14a^2 - 5a\right)$

$= 7a^2 + 3a + 14a^2 - 5a = 21a^2 - 2a$

39. $\left(5y^3 - 3y^2\right) + \left(4y + 3y^2\right)$

$= 5y^3 - 3y^2 + 4y + 3y^2 = 5y^3 + 4y$

41. $\left(7x^2 - 2x + 3\right) - \left(2x^2 - 5x - 7\right)$

$= 7x^2 - 2x + 3 - 2x^2 + 5x + 7$

$= 5x^2 + 3x + 10$

43. $\left[\left(9x + 2\right) + \left(-3x - 7\right)\right] - \left(5x - 3\right)$

$= 9x + 2 - 3x - 7 - 5x + 3 = x - 2$

45. $\left(7w^2 - 5w + 2\right) - \left[\left(16w^2 - 3w\right) + \left(8w + 2\right)\right]$

$= \left(7w^2 - 5w + 2\right) - \left(16w^2 + 5w + 2\right)$

$= 7w^2 - 5w + 2 - 16w^2 - 5w - 2$

$= -9w^2 - 10w$

47. $\begin{array}{r} 9b^2 \qquad -7 \\ \underline{8b + 5} \\ 9b^2 + 8b - 2 \end{array}$

49. $\begin{array}{r} 7x^2 - 5x - 7 \\ \underline{-\left(5x^2 - 3x + 2\right)} \\ \end{array}$ $\begin{array}{r} 7x^2 - 5x - 7 \\ \underline{-5x^2 + 3x - 2} \\ 2x^2 - 2x - 9 \end{array}$

51. Let x be the number of copies of the book sold by Castor Books. Therefore, the total cost of books is $\$18.45x$. Hence, the equation to represent the weekly cost (C) for Castor Books to sell this book is $C = 18.45x + 525$.

53. $P = 21.5\left(20\right) - 525 = -\$95$ (loss)

Thus, there is a loss of $95 if they sell only 20 copies in one week.

55. $\left(3a^4\right)\left(2a^3\right) = 3\cdot2\cdot a^4\cdot a^3 = 6a^{4+3} = 6a^7$

57. $\left(-9p^3\right)\left(-6p^2\right) = 54p^5$

59. $5\left(3x - 8\right) = 5\cdot3x - 5\cdot8 = 15x - 40$

61. $(-5rs)(2r^2s - 5rs)$

$= -5rs \cdot 2r^2s - (-5rs)(5rs)$

$= -10r^3s^2 + 25r^2s^2$

63. $(x + 5)(x + 4) = x \cdot x + x \cdot 4 + 5 \cdot x + 20$

$= x^2 + 9x + 20$

65. $(a - 9b)(a + 9b) = (a)^2 - (9b)^2 = a^2 - 81b^2$

67. $(a + 4b)(a + 3b) = a^2 + 3ab + 4ab + 12b^2$

$= a^2 + 7ab + 12b^2$

69. $(3x - 5y)(2x - 3y)$

$= 6x^2 - 9xy - 10xy + 15y^2$

$= 6x^2 - 19xy + 15y^2$

71. $(y + 2)(y^2 - 2y + 3)$

$= y \cdot y^2 - y \cdot 2y + y \cdot 3 + 2 \cdot y^2 - 2 \cdot 2y$

$\quad\quad + 2 \cdot 3$

$= y^3 - 2y^2 + 3y + 2y^2 - 4y + 6 = y^3 - y + 6$

73. $(x - 2)(x^2 + 2x + 4)$

$= x^3 + 2x^2 + 4x - 2x^2 - 4x - 8 = x^3 - 8$

75. $2x(x + 5)(x - 6) = 2x(x^2 - x - 30)$

$= 2x^3 - 2x^2 - 60x$

77. $(x + 7)^2 = x^2 + 2 \cdot 7 \cdot x + 7^2 = x^2 + 14x + 49$

79. $(2w - 5)^2 = (2w)^2 - 2 \cdot 5 \cdot 2w + 5^2$

$= 4w^2 - 20w + 25$

81. $(a + 7b)^2 = a^2 + 2 \cdot a \cdot 7 \cdot b + (7b)^2$

$= a^2 + 14ab + 49b^2$

83. $(x - 5)(x + 5) = x^2 - 5^2 = x^2 - 25$

85. $(2m + 3)(2m - 3) = (2m)^2 - 3^2 = 4m^2 - 9$

87. $(5r - 2s)(5r + 2s) = (5r)^2 - (2s)^2$

$= 25r^2 - 4s^2$

89. $3x(x - 4)^2 = 3x(x^2 - 8x + 16)$

$= 3x^3 - 24x^2 + 48x$

91. $(y - 4)(y + 5)(y + 4)$

$= (y + 5)(y - 4)(y + 4) = (y + 5)(y^2 - 16)$

$= y^3 + 5y^2 - 16y - 80$

93. The area of the rectangle as a polynomial is

$A = LW = [4(x + 1)](2x + 3)$

$= (4x + 4)(2x + 3) = 8x^2 + 12x + 8x + 12$

$= (8x^2 + 20x + 12) \text{ in}^2.$

95. $x^2 - 8x + 15 = (x - 3)(x - 5)$

97. $m^2 + 8m + 12 = (m + 2)(m + 6)$

99. $p^2 - 8p - 20 = (p + 2)(p - 10)$

101. $x^2 - 16x + 64 = (x - 8)(x - 8)$

103. $x^2 - 7xy + 10y^2 = (x - 2y)(x - 5y)$

Chapter Test 10

1. $(3x^2y)(-2xy^3) = -6x^3y^4$

3. $(x^4y^5)^2 = x^8y^{10}$

5. $(3x^2y)^3(-2xy^2)^2 = (27x^6y^3)(4x^2y^4)$

$= 108x^8y^7$

7. $\dfrac{3x^0}{(2y)^0} = \dfrac{3 \cdot 1}{1} = 3$

9. $5x^2 + 8x - 8$ is a trinomial.

11. $\left(3x^2 - 7x + 2\right) + \left(7x^2 - 5x - 9\right)$

$= 3x^2 - 7x + 2 + 7x^2 - 5x - 9$

$= 10x^2 - 12x - 7$

13. $\left(8x^2 + 9x - 7\right) - \left(5x^2 - 2x + 5\right)$

$= 8x^2 + 9x - 7 - 5x^2 + 2x - 5$

$= 3x^2 + 11x - 12$

15.

$$
\begin{array}{r}
x^2 \quad\ \ + 3 \\
5x - 9 \\
+3x^2 \qquad\qquad \\
\hline
4x^2 + 5x - 6
\end{array}
$$

17. $5ab\left(3a^2 b - 2ab + 4ab^2\right)$

$= 15a^3 b^2 - 10a^2 b^2 + 20a^2 b^3$

19. $\left(3m + 2n\right) = 9m^2 + 12mn + 4n^2$

21. $-10x^3 + 5x^2 - 25x$

$= -5x \bullet x^2 - (-5x) \bullet x + (-5x) \bullet 5$

$= -5x\left(x^2 - x + 5\right)$

23. Canada's surface area in water is

$\dfrac{8.92}{100} \times 9{,}984{,}670 \approx 890{,}000 \ \text{km}^2$; in

scientific notation is $8.9 \times 10^5 \ \text{km}^2$.

25. **(a)** $C = 6.27x + 285$

  (b) $P = 8.68x - 285$

  (c) $P = 8.68(75) - 285 = 366$

  The profit that Funghi Books earns on selling 75 copies one week is \$366.

Final Examination

1. 806,015 is eight hundred six thousand, fifteen.

3.

$$
\begin{array}{r}
\overset{3}{\cancel{4}}\ \overset{11}{\cancel{2}}\ \overset{1}{5} \\
-\ \ \ 6\ 7 \\
\hline
3\ 5\ 8
\end{array}
$$

5.

$$
\begin{array}{r}
19 \\
13\overline{)247} \\
\underline{13}\ \ \ \\
117 \\
\underline{117} \\
0
\end{array}
$$

7. The amount Minh receives each month is \$30,240 divided by 12.

$$
\begin{array}{r}
2520 \\
12\overline{)30{,}240} \\
\underline{24}\qquad \\
62 \\
\underline{60}\ \ \\
24 \\
\underline{24} \\
0
\end{array}
$$

Min receives \$2520 each month.

9. $9 - 2^3 = 9 - 8 = 1$

11. $-7 - (-4) = -7 + 4 = -3$

13. $(-6) \bullet (-9) = 6 \bullet 9 = 54$

15. $-24 \div (8 - 12) = -24 \div (-4) = \dfrac{\overset{6}{-\cancel{24}}}{\underset{1}{-\cancel{4}}} = 6$

17. $23 - (-8) = 23 + 8 = 31$

19. $(-2)(-4)(-7) = -2 \bullet 4 \bullet 7 = -56$

21. $9 - 5 \bullet 3^2 = 9 - 5 \bullet 9 = 9 - 45 = -36$

23. $-2x^2 - 3x + 5 = -2(-3)^2 - 3(-3) + 5$

$= -2(9) + 9 + 5 = -18 + 14$

$= -4$

25. $1{,}260 = 2 \bullet 2 \bullet 3 \bullet 3 \bullet 5 \bullet 7$

27. $60 = 2^2 \cdot 3 \cdot 5$; $132 = 2^2 \cdot 3 \cdot 11$

$$\frac{60}{132} = \frac{2^2 \cdot 3 \cdot 5}{2^2 \cdot 3 \cdot 11} = \frac{\cancel{2^2} \cdot \cancel{3} \cdot 5}{\cancel{2^2} \cdot \cancel{3} \cdot 11} = \frac{5}{11}$$

29. $\dfrac{3}{7} \cdot \dfrac{5}{12} = \dfrac{3}{7} \cdot \dfrac{5}{2^2 \cdot 3} = \dfrac{\cancel{3} \cdot 5}{7 \cdot 2^2 \cdot \cancel{3}} = \dfrac{5}{28}$

31. $\dfrac{2}{3} \div \dfrac{4}{9} = \dfrac{2}{3} \cdot \dfrac{9}{4} = \dfrac{3}{2}$

33. $\dfrac{3}{4} + \dfrac{5}{8} = \dfrac{6}{8} + \dfrac{5}{8} = \dfrac{11}{8} = 1\dfrac{3}{8}$

35. $8\dfrac{1}{4} - 2\dfrac{7}{8} = \dfrac{66}{8} - \dfrac{23}{8} = \dfrac{43}{8} = 5\dfrac{3}{8}$

37. $6\dfrac{1}{2} - 1\dfrac{4}{5} = \dfrac{65}{10} - \dfrac{18}{10} = \dfrac{47}{10} = 4\dfrac{7}{10}$

$4\dfrac{7}{10}$ of the material is left.

39. $-6x = 42$

$$\frac{-6x}{-6} = \frac{42}{-6}$$

$$x = -7$$

41.
$$\frac{x}{4} - 9 = -3$$

$$4\left(\frac{x}{4} - 9\right) = 4(-3)$$

$$x - 36 = -12$$

$$x - 36 + 36 = -12 + 36$$

$$x = 24$$

43. $4x - 2(x - 5) = 16$

$$4x - 2x + 10 = 16$$

$$2x + 10 = 16$$

$$2x + 10 - 10 = 16 - 10$$

$$2x = 6$$

$$x = 3$$

45.
$$\overset{1\ 1\ \ 1}{17}.289$$
$$\underline{+\ \ 4.930}$$
$$22.219$$

47. $0.08)\overline{2.40}$

$$8)\overline{240}$$
$$\,30$$
$$\underline{24}$$
$$0$$

49. $113.8)\overline{2537.74}$

$$1138)\overline{25377.4}$$
$$\ 22.3$$
$$\underline{2276}$$
$$2617$$
$$\underline{2276}$$
$$3414$$
$$\underline{3414}$$
$$\ \ \ \ \ 0$$

51.
$$\overset{7\ 18\ \ 1}{189}.75$$
$$\underline{-\ \ 39.95}$$
$$149.80$$

Toni had \$149.80 left in her account.

53. The 8 is in the ten-thousandths place. The next digit to the right, (43) is less than 5, so leave the ten-thousandths digit as it is. Discard the remaining digits to the right. 0.426839 is rounded to 0.4268.

55. $\sqrt{81} = 9$

57.
$$\frac{6}{n} = \frac{9}{57}$$

$$57n\left(\frac{6}{n}\right) = 57n\left(\frac{9}{57}\right)$$

$$57 \cdot 6 = 9n$$

$$\frac{57 \cdot 6}{9} = \frac{9n}{9}$$

$$n = 38$$

59. $185\% = 185\left(\dfrac{1}{100}\right) = \dfrac{185}{100} = 1.85$

61. $1\dfrac{1}{5} = \dfrac{6}{5} = 1.2 = 120\%$

63. $\dfrac{A}{80} = \dfrac{16}{100}$

$100A = 1,280$

$A = \dfrac{1,280}{4} = 12.8$

16% of 80 is 12.8.

65. $\dfrac{A}{42} = \dfrac{7}{100}$

$100A = 294$

$A = \dfrac{294}{100} = 2.94$

The sales tax on the purchase is $2.94.

67. The amount deducted from the regular price $= \dfrac{25}{100} \times 90 = \$22.5$. The sales price is $\$90 - \$22.5 = \$67.50$.

69. $630 \text{ cm} = 630(0.01 \text{ m}) = 6.3 \text{ m}$

71. The area of a triangle A is given by the formula $A = \dfrac{1}{2}b \cdot h$, where b is the length of the base and h is the height of the triangle. From the figure, $b = 38.2 \text{ cm}$ and $h = 12 \text{ cm}$.

$A = \dfrac{1}{2}(38.2)(12) = 229.2$

So, the area is 229.2 cm^2.

73. Sum the number of words Joseph learned in weeks 3, 4, and 5. As shown in the graph, Joseph learned 8, 6, and 2 words, respectively, during these weeks. Since $8 + 6 + 2 = 16$, Joseph learned 16 words after the second week.

75. The mean of the test scores is $\dfrac{82 + 95 + 77 + 86}{4} = \dfrac{340}{4} = 85$

77.

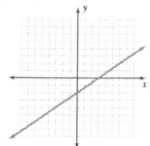

79. $\left(3x^2 - 5x - 9\right) - \left(4x^2 + 2x - 13\right)$

$= 3x^2 - 5x - 9 - 4x^2 - 2x + 13$

$= 3x^2 - 4x^2 - 5x - 2x - 9 + 13$

$= (3 - 4)x^2 + (-5 - 2)x + 4 = -x^2 - 7x + 4$

81. $(7x - 2y)(7x + 2y)$

$= 7x(7x) + 7x(2y) - 2y(7x) - (2y)^2$

$= 49x^2 + 14xy - 14xy - 4y^2$

$= 49x^2 + (14 - 14)xy - 4y^2$

$= 49x^2 + 0 \cdot xy - 4y^2 = 49x^2 - 4y^2$

83. $2x(x + 3y) - 4y(2x - y)$

$= 2x(x) + (2x)3y - 4y(2x) + (-4y)(-y)$

$= 2x^2 + 6xy - 8xy + 4y^2$

$= 2x^2 - 2xy + 4y^2$

NOTES

NOTES

NOTES

NOTES

NOTES

NOTES

NOTES

NOTES

NOTES